The International Measurement Confederation (IMEKO) is an international federation of 30 national member organizations individually concerned with the advancement of measurement technology. Its fundamental objectives are the promotion of international interchange of scientific and technical information in the field of measurement, and the enhancement of international co-operation among scientists and engineers from research and industry.

IMEKO Technical Committee events series No. 18

IMEKO TC series include publications by the 17 Technical Committees of the International Measurement Confederation:
Higher Education (TC1)
Photon-Detectors (TC2)
Measurement of Force and Mass (TC3)
Measurement of Electrical Quantities (TC4)
Hardness Measurement (TC5)
Vocabulary Committee (TC6)
Measurement Theory (TC7)
Metrology (TC8)
Flow Measurement (TC9)
Technical Diagnostics (TC10)
Metrological Requirements for Developing Countries (TC11)
Temperature and Thermal Measurement (TC12)
Measurements in Biology and Medicine (TC13)
Measurement of Geometrical Quantities (TC14)
Experimental Mechanics (TC15)
Pressure Measurement (TC16)
Robotics (TC17)

Latest editions:

IMEKO TC Series No. 7
Studies on Metrology – (TC8)
Publisher: OMIKK–Technoinform, 1428 Budapest, P. O. Box 12,
        Hungary (in two volumes)
650 pages, US $ 55.—

IMEKO TC Series No. 8
Proceedings of the 11th TC3 Symposium
on Mechanical Problems in Measuring
Force and Mass
held in Amsterdam (Netherlands), 1986
Publisher: Martinus Nijhoff, P. O. Box 163, 3300 AD Dordrecht,
        The Netherlands
333 pages, US $ 77.—

IMEKO TC Series No. 9
Proceedings of the 2nd TC12 Workshop
on Heat Flux Measurement
held in Budapest (Hungary), 1986
Publisher: OMIKK–Technoinform
325 pages, US $ 36.—

IMEKO TC Series No. 11
Proceedings of the 12th TC2 Symposium on Photon Detectors
held in Varna (Bulgaria), 1986
Publisher: OMIKK–Technoinform
615 pgaes, US $ 42.—

IMEKO TC Series No. 12
Proceedings of the 1st TC4 Symposium on Noise
in Electrical Measurements
held in Como (Italy), 1986
Publisher: OMIKK–Technoinform
221 pages, US $ 29.—

IMEKO TC Series No. 13
Proceedings of the 3rd TC8 Symposium on Theoretical Metrology
held in Berlin (GDR), 1986
Publisher: OMIKK–Technoinform
410 pages, US $ 28.—

The above prices do not include postage.

IMEKO TC Series No. 10
Proceedings of the 5th TC7 Symposium on Intelligent Measurement
held in Jena (GDR) within the series „Measurement Theory and
its Application to Practice" in 1986
520 pages

IMEKO TC Series No. 14
Proceedings of the 1st Symposium on Laser Applications
in Precision Measurement
held in Budapest (Hungary), 1986
267 pages

IMEKO TC Series No. 15
Proceedings of the 4th TC13 Conferences on Advances
in Biomedical Measurement
held in Bratislava (Czechoslovakia), 1987
Publisher: Plenum Publishing Company Limited,
          88/90 Middlesex Street, London E1 7EZ, England

IMEKO TC Series No. 16
Proceedings of the 6th TC7 Symposium on Signal Processing
in Measurement
held in Budapest (Hungary), 1987
274 pages

IMEKO TC Series No. 17
Proceedings of the 13 th TC2 Symposium
on Photonic Measurements (Photon Detectors)
held in Braunschweig (FRG), 1987
Publisher: OMIKK–Technoinform
368 pages

IMEKO TC Series No. 18
Proceedings of the 1st TC15 Conference on the Measurement
of Static and Dynamic Parameters of Structures and Materials
held in Plzen (Czechoslovakia), 1987
404 pages

IMEKO TC Series No. 19
Proceedings of the 2nd TC4 Symposium on Industrial
of Electrical and Electronic Components and Equipment
held in Warsaw (Poland), 1987

IMEKO TC Series No. 20
Collected Papers
on Technical Diagnostics

INTERNATIONAL MEASUREMENT CONFEDERATION

1st Conference of the Technical Committee (TC15)
on

MEASUREMENT OF
STATIC AND DYNAMIC PARAMETERS
OF STRUCTURES AND MATERIALS

Plzen, Czechoslovakia
May 26—28, 1987

IMEKO TC No. 18

PROCEEDINGS

Series editor: Dr. Tamás Kemény
Editor: Karolina Havrilla

NOVA SCIENCE PUBLISHERS
COMMACK

NOVA SCIENCE PUBLISHERS, INC.
283 Commack Road
Suite 300
Commack, New York 11725

STATIC AND DYNAMIC PARAMETERS OF
STRUCTURES AND MATERIALS
Proceedings of the 1st TC15 Conference
organized by K. H. Learmann in Plzen, Czechoslovakia
26–28 May, 1987.

Inquiries:
IMEKO Secretariat
H—1371 Budapest, P. O. Box 457 — Hungary

Library reference data: Measurement of static and dynamic parameters of
structures and materials — IMEKO TC15 Conference proceedings,
organized by K. H. Laermann

SERIES EDITOR: Dr. Tamás Kemény

Editor: Karolina Havrilla

Cataloging in Publication Data Available Upon Request

ISBN   0-941743-40-3

CONTENTS

In memoriam Jan Javornicky

TRANSMISSION PHOTOELASTICITY

Integrated Photoelasticity of the General Three-Dimensional Stress
State
   H. Aben and S. Idnurm   3
Investigation of Three-Dimensional Axis-Symmetrical
   Problems by the Photoelastic Method
   M. Achmetzyanov and V. Tichomirov   11
Elliptic Photoelasticity and Its Applications
   I. Bugakov, N. Drichko et al.   17
Determination of Strain Tensor Components by the
   Scattered Light Method
   W. Karmowski   23
Multiplication of Partial Isochromatic Lines
   by Special Photoprocessing
   S. Mazurkiewicz and J. Legendziewicz   29
The Analysis of Stresses in the Model of a Fractured
   Bone With the Application of Photoelastic Methods
   S. Mazurkiewicz and J. Legendziewicz   33
Photoelastic Model Analysis of Sandwich Beams at
   Different Ratio of Rigidity of Face to Core
   W. Walczak and M. Koteko   39

PHOTOELASTIC COATING

Application of Photoelastic Coating at Connecting Points
   of Bus Undercarriage
   L. Borbás and F. Thamm   47
Investigation of Local Stresses in the Mast Frame Up-
   rights of Fork Lift Trucks
   J. Videnova, K. Kostov and V. Vasilev   53

v

Stress Measurement of Sheet Constructions with
Optically Active Coating
Á. Zsáry   5 9

PHOTOPLASTICITY

Application of Photoplastic Methods in the Field of Forming
P. Macura   6 7
Measurement of the Influence of Viscoelastic Response
of Materials on Plates by Optical Methods
K.-H. Laermann   7 3

HOLOGRAPHIC INTERFEROMETRY, LASER METROLOGY

Holographic Examination of Cracking in Concrete
G.L. Dalakishvili and M.D. Nizharadze   8 5
Some Aspects of Vibration Analysis by Laser Metrology
V. Grosser, D. Vogel et al.   9 1
Holographic Interferometry Method for Assessment of Stone
Surface Recession and Roughening Caused by Weathering
and Acid Rain
C.A. Sciammarella, M.A. Ahmadshai and C.A. Youngdahl   9 7
Investigation of Elastoplastic Problems With the Help
of Methods Based on Holographic Recording and
Information
V.A. Zhilkin   1 0 7

MOIRÉ AND OPTOELECTRONIC METHODS

Optical Methods of Strain Measurements Application to Study
Biaxial Tension Specimens
F. Brémand and A. Lagarde   1 1 5
Application of Moiré with Coherent and Non-Coherent
Light to Dynamic Measurements of Deformation
C. Albertini, M. Montagnani et al.   1 2 9

Optoelectronic Measurements Using Diffraction of
Light in Dynamic Experiments
M. Drzik   135
The Development and Application of a Thermoelastic
Technique for Stress Analysis
P. Stanley   141
Theoretical and Experimental Study of Bond in
Reinforced Concrete
P. Stroeven and J. Bien   147

APPLICATION OF STRAIN GAUGES AND OTHER ELECTROMECHANICAL TRANSDUCERS

The Drift and Creep Behavior of Encapsulated Welded
Strain Gauges
C. Amberg   155
Dynamic Forces and Impedance Measurement in
Statically Loaded Structures
M. Genkin, V. Tikchonov, and V. Yablonskij   161
Evaluation of Surface Roughness Using a New Simple
Handy Device
H.A.M. Hanna   167
Vibration Amplitude Testing by Electroacoustic Means
A. Boleslav, and K. Antropius   173
FEM Optimization of the Shape of Profiled Membranes
Utilized at Strain Gauge Instrumented Transducers
D.-M. Stefanescu   179
New Information on Application of Semiconductor
Strain Gauge Sensors
J. Lukas   185

HYBRID TECHNIQUES OF STRESS ANALYSIS

Hybrid Techniques for Analyzing Nonlinear Problems in
Solid Mechanics
K.-H. Laermann   193

Experimental Stress and Strain Analysis of Machine
  Parts by Photoelasticity, Moiré Method and Speckle
  Photography
  J. Heymann, R. Meyer et al.   209
Calculated Experimental Methods for Residual Stress
  Estimation
  M.N. Dveres, A.V. Fomin et al.   215
Experimental Investigation Into the Plastification of
  Thin Webs
  M. Drdácky, P. Jaros and O. Weinberg   221
Stress Relaxation Behavior in Polymers for
  Combined Bending and Torsion
  G. Milewski and B. Targosz   227
Photoelastic and Strain Gauge Analysis of Thick-
  Walled Shell of Revolution
  Z. Goja, M. Husnjak and S. Jecic   233

COMPUTER-AIDED EVALUATION OF MEASUREMENTS

Computer-Aided Reconstruction of Displacement Fields in
  Holographic and Speckle Interferometry
  W. Osten and R. Höfling   241
Application of the Band Selectable Fourier Transform
  for Structural Dynamic Testing
  F. Wahl   251
On the Digital Processing of Fringe Patterns
  A. Pietrzyk   263
Automatic Recording of Transient Processes in Dies
  K. Vrba, M. Forejt and R. Vrba   269

FRACTURE MECHANICS

JTJ Controlled Crack Growth--Improved J-Resistance Testing
  P. Will, B. Michel and U. Zerbst   277
Dynamic Measurements of Initiation Toughness at High
  Loading Rates
  J.W. Dally and D.B. Barker   283

Test Methods in the Mechanics of the Brittle Cement
    Based Composites
    A.M. Brandt   301
Failure and Fracture Criteria in Composites
    P.S. Theocaris   307
Identification Experiment for Tracing the Cause of the
    Failure of a Stone Crasher
    F. Thamm   319
Some Remarks on the Determination of Stress Intensity
    Factors
    S. Jecic and D. Semenski   325

SPECIAL APPLICATIONS OF STRAIN ANALYSIS IN THE MECHANICAL AND CIVIL ENGINEERING PRACTICE

Experimental Determination of the Arbitrary Constants in the
    Solution for Over-Determinate Shallow Spherical Shells
    at Early Stage
    S.S. Issa   333
Long-Term Strain Observation of Concrete Bridges
    T. Jávor   339
Stress and Strain Investigation in the Umbrella Arch
    Method on a Tunnel
    F. Nazari   345
Methods of Measurement of Normal Stress in Soil
    Excited by Blasting and Their Comparison
    J. Olmer   351
Dynamic Measurements on the Earthmoving Machines--
    The Foundation of Prototype Testing
    B. Pacas   357
Model Analysis in Mechanical Engineering
    A. Lingener   363
Experimental-Numerical Method of Stress Analysis
    of Random Loads in Metallic Constructions
    M. Kopecky   369

Automated Investigation of Dynamic Parameters
of Bridges
  R. Kyska and A. Sokolik   375
Growth Energy of Ring-Shaped Fatigue Crack
  A. Solecki   381
Vibrational Test of Nuclear Power Plant Failure
  Lock Control System for Earthquake Proofness
  G. Heruvimov, N. Georgiev and P. Ivanov   387
On-Line Testing and Long-Term Investigation of
  Trends, Especially Dynamic Parameters of Machine
  Foundations in Power Engineering
  V. Bohdanecky and P. Novák   393

Author Index   399

Subject Index   401

In memoriam Jan Javornický

We have to bewail the decease of Jan Javornický, a highly esteemed
scientist in the field of solid mechanics, especially of experimental mechanics. The scientific community has lost a colleague
of international reputation. Suffering from a malicious disease
since years, he passed away too early for his family and for his
friends all over the world.

Dr. Jan Javornický has been active since decades to promote the
principles of experimental mechanics; he played a most active role
very successfully in improving international cooperation, too.
Therefore he has been strongly interested in promoting the activities of the IMEKO-Technical Committee No. 15 "Measurements in Experimental Mechanics".

Although mainly engaged in basic scientific research, Jan Javornický
never lost his links to practice, always looking how to solve real
problems. He never hovered about the clouds of mere academic thinking,
but he always picked up practical problems to include them into research, and vice versa, he permanently has been interested in transferring scientific results to practice in order to improve the economical as well as the social situation.

There is a quotation of the German philosopher Schopenhauer:
"Mankind will not make any essential progress as long as people are
listening more to those who speak very loudly instead to those who
speak most intelligently". At no time Jan Javornický has either
spoken very much or very loudly. But whenever he spoke and whatever
he said always demonstrated his outstanding intellect as well as his
ability of stringent analytical thinking. These are the reasons why
discussions with him were always highly inspiring, no matter what the
subject was.

The scientific community of the ČSSR as well as all his friends all
over the world - and to emphasize - the members of IMEKO-Technical
Committee No. 15 are indebted to Jan Javornický for his activities
and performance. Therefore we have to take care of his legacy to
strive hard for mutual understanding among human beings across all
frontiers - and to promote principles and methods in Experimental
Mechanics. All of us have to remind the message of his life and his
work, this should be our obligation in memoriam Jan Javornický.

Transmission photoelasticity

INTEGRATED PHOTOELASTICITY OF THE GENERAL
THREE-DIMENSIONAL STRESS STATE

Dr.Sc. Hillar Aben - Siim Idnurm, Cand.Sc.
Institute of Cybernetics, Estonian Academy of Sciences
21, Akadeemia tee, 200108 Tallinn, Estonia, USSR

It is shown that integrated photoelasticity permits to determine the general three-dimensional stress state. From six unknown stress components four can be expressed through the remaining two by the aid of equilibrium and compatibility equations. The latter two are determined on the basis of experimental data.

Keywords: photoelasticity, tomography

1. Introduction

In integrated photoelasticity [1] stresses in three-dimensional
bodies are determined on the basis of optical data, obtained when polarized light passes through the whole body in a usual transmission polariscope. Up to now integrated photoelasticity has been used to determine
only comparatively simple stress states (e.g., in shells and bodies of
revolution).

In comparison with the frozen-stress and scattered light method
integrated photoelasticity has several advantages: experiment can be
carried out at room temperature, the method is nondestructive, optical
measurements are comparatively simple, in principle it is possible to
investigate also non-elastic deformations. Therefore it is of practical
interest to investigate how wide is the class of stress states which
can be determined by integrated photoelasticity. Since experimental information one obtains with integrated optical measurements is limited,
investigators have been discouraged to tackle problems more complicated
than the axisymmetric stress state.

In this paper it is shown that in principle integrated photoelasticity can be used to determine arbitrary three-dimensional state of
stress. At that we consider only the case when optical measurements are

IMEKO 1st TC15 Conference, Plzeň, Czechoslovakia, May 25-28, 1987 made in sections parallel to each other, i.e., the model is to be rotated around only one axis. Although passing of light through the model in arbitrary directions may give more experimental information [2,3], it makes the experimental setup very complicated.

In this respect our approach resembles tomography [4] in which internal structure of three-dimensional objects is also determined by sections. However, while classical tomography considers determination of scalar fields, integrated photoelasticity is to be considered as optical tomography of a tensor field [5].

2. Approximation of the Stress Components

Let us assume that the section $z=z_0$ of an arbitrary three-dimensional body is located in a circle with radius R. In cylindrical coordinates r, θ, z we may express the stress components σ_{rr}, $\sigma_{\theta\theta}$, σ_{zz}, $\sigma_{r\theta}$, $\sigma_{\theta z}$, and σ_{zr} in the following way:

$$\sigma_{ij} = \sum_{mn}\sum (a_{mn}^{ij}r^n\cos m\theta + b_{mn}^{ij}r^n\sin m\theta), \quad i,j=r,\theta,z \quad (2.1)$$

where a_{mn}^{ij} are coefficients we have to determine. Stress components in the auxiliary section situated at a distance Δz from the main section will be marked by a hyphen: $\bar{\sigma}_{ij}$, $\bar{a}_{mn}^{ij}$, $\bar{b}_{mn}^{ij}$.

Boundary conditions as well as equilibrium conditions impose on the coefficients a_{mn}^{ij} and b_{mn}^{ij} certain conditions which permit to eliminate part of the coefficients.

3. Equations of the Theory of Elasticity

Equations of equilibrium in cylindrical coordinates are as follows:

$$\frac{\partial \sigma_{rr}}{\partial r} + \frac{1}{r}\frac{\partial \sigma_{r\theta}}{\partial \theta} + \frac{\sigma_{rr}-\sigma_{\theta\theta}}{r} + \frac{\partial \sigma_{zr}}{\partial z} = 0, \quad (3.1)$$

$$\frac{1}{r}\frac{\partial \sigma_{\theta\theta}}{\partial \theta} + 2\frac{\partial \sigma_{r\theta}}{r} + \frac{\partial \sigma_{r\theta}}{\partial r} + \frac{\partial \sigma_{\theta z}}{\partial z} = 0, \quad (3.2)$$

$$\frac{\partial \sigma_{zz}}{\partial z} + \frac{1}{r}\frac{\partial \sigma_{\theta z}}{\partial \theta} + \frac{\sigma_{zr}}{r} + \frac{\partial \sigma_{zr}}{\partial r} = 0. \quad (3.3)$$

Compatibility equation for the components of deformation in the plane of the section, expressed through stresses, is the following

$$\frac{1}{r^2}\frac{\partial^2 \sigma_{rr}}{\partial \theta^2} - \frac{1}{r}\frac{\partial \sigma_{rr}}{\partial r} + \frac{2}{r}\frac{\partial \sigma_{\theta\theta}}{\partial r} + \frac{\partial^2 \sigma_{\theta\theta}}{\partial r^2} - \frac{2}{r^2}\frac{\partial \sigma_{r\theta}}{\partial \theta} -$$

$$- \frac{2}{r}\frac{\partial^2 \sigma_{r\theta}}{\partial r\partial \theta} = \frac{\mu}{1+\mu}\left(\frac{1}{r}\frac{\partial \sigma}{\partial r} + \frac{\partial^2 \sigma}{\partial r^2} + \frac{1}{r^2}\frac{\partial^2 \sigma}{\partial \theta^2}\right), \quad (3.4)$$

4

where μ is the Poisson ratio, and

$$\sigma = \sigma_{rr} + \sigma_{\theta\theta} + \sigma_{zz}. \tag{3.5}$$

4. Partial Elimination of the Stress Components

Introducing expressions (2.1) into Eq. (3.2) we obtain

$$\sum_m \sum_n \{ [mb_{mn}^{\theta\theta} r^{n-1} + 2a_{mn}^{r\theta} r^{n-1} + na_{mn}^{r\theta} r^{n-1} +$$

$$+ \frac{1}{\Delta z}(\bar{a}_{mn}^{\theta z} - a_{mn}^{\theta z}) r^n]\cos m\theta + [-ma_{mn}^{\theta\theta} r^{n-1} + 2b_{mn}^{r\theta} r^{n-1} +$$

$$+ nb_{mn}^{r\theta} r^{n-1} + \frac{1}{\Delta z} (\bar{b}_{mn}^{\theta z} - b_{mn}^{\theta z}) r^n]\sin m\theta \} = 0. \tag{4.1}$$

The latter relationship must be satisfied for arbitrary r and θ. Equating to zero expressions which are multiplied to $r^{n-1}\cos m\theta$ and $r^{n-1}\sin m\theta$ in Eq. (4.1), we have

$$a_{mn}^{\theta\theta} = \frac{2+n}{m} b_{mn}^{r\theta} + \frac{1}{m\Delta z} (\bar{b}_{m,n-1}^{\theta z} - b_{m,n-1}^{\theta z}),$$

$$b_{mn}^{\theta\theta} = - \frac{2+n}{m} a_{mn}^{r\theta} - \frac{1}{m\Delta z} (\bar{a}_{m,n-1}^{\theta z} - a_{m,n-1}^{\theta z}), \quad m \neq 0. \tag{4.2}$$

Similarly, Eq. (3.1) gives, taking into account also Eqs. (4.2)

$$a_{mn}^{rr} = \frac{2-m^2+n}{m(n+1)} b_{mn}^{r\theta} + \frac{1}{m(n+1)\Delta z} (\bar{b}_{m,n-1}^{\theta z} - b_{m,n-1}^{\theta z}) -$$

$$- \frac{1}{(n+1)\Delta z} (\bar{a}_{m,n-1}^{\theta z} - a_{m,n-1}^{zr}),$$

$$b_{mn}^{rr} = - \frac{2-m^2+n}{m(n+1)} a_{mn}^{r\theta} - \frac{1}{m(n+1)\Delta z} (\bar{a}_{m,n-1}^{\theta z} - a_{m,n-1}^{\theta z}) - \tag{4.3}$$

$$- \frac{1}{(n+1)\Delta z} (\bar{b}_{m,n-1}^{zr} - b_{m,n-1}^{zr}),$$

Besides, the following relationships are also valid:

$$a_{00}^{rr} = a_{00}^{\theta\theta}, \quad a_{m0}^{rr} = a_{m0}^{\theta\theta} - mb_{m0}^{r\theta}, \quad b_{m0}^{rr} = b_{m0}^{\theta\theta} + ma_{m0}^{r\theta}, \tag{4.4}$$

Introducing Eqs. (2.1) into the compatibility equation (3.4) we obtain

$$a_{mn}^{r\theta} = f_1 \cdot (\bar{a}_{m,n-1}^{\theta z} - a_{m,n-1}^{\theta z}) + f_2 \cdot (\bar{b}_{m,n-1}^{zr} - b_{m,n-1}^{zr}) + f_3 \cdot b_{mn}^{zz},$$

$$b_{mn}^{r\theta} = f_4 \cdot (\bar{b}_{m,n-1}^{\theta z} - b_{m,n-1}^{\theta z}) + f_5 \cdot (\bar{a}_{m,n-1}^{zr} - a_{m,n-1}^{zr}) + f_6 \cdot a_{mn}^{zz}, \tag{4.5}$$

where $f_i = f_i(m,n,\mu)$.

Similarly, Eq. (3.3) relieves

$$\bar{a}_{mn}^{zz} = a_{mn}^{zz} - \frac{\Delta z}{2}\ [m(b_{m,n+1}^{\theta z}+\bar{b}_{m,n+1}^{\theta z}) + (1+n)(a_{m,n+1}^{zr}+\bar{a}_{m,n+1}^{zr})],$$

$$\bar{b}_{mn}^{zz} = b_{mn}^{zz} - \frac{\Delta z}{2}\ [m(a_{m,n+1}^{\theta z}+\bar{a}_{m,n+1}^{\theta z}) + (1+n)(b_{m,n+1}^{zr}+\bar{b}_{m,n+1}^{zr})].$$

$$(4.6)$$

Similar expressions hold for the auxiliary section, one has only to change a_{mn}^{ij} for $\bar{a}_{mn}^{ij}$, b_{mn}^{ij} for $\bar{b}_{mn}^{ij}$, and vice versa, and z for (-) z.

Thus, the number of sets of unknown coefficients in both sections is reduced from six to two.

5. Determination of the Stress Components

Let us assume that polarized light is passed through a section of the body along s (Fig. 1). At point A the state of polarization is influenced by stress components which lie in the plane perpendicular to s, i.e.,

$$\sigma_{x'x'} = \sigma_{rr}\cos^2\theta' + \sigma_{\theta\theta}\sin^2\theta' - 2\sigma_{r\theta}\sin\theta'\cos\theta',$$

$$\sigma_{zz} = \sigma_{zz},$$

$$(5.1)$$

$$\sigma_{zx'} = \sigma_{\theta z}\sin\theta' + \sigma_{zr}\cos\theta'.$$

On every ray we can experimentally measure three characteristic parameters [1]: the angles α_0^e and α_*^e which determine the primary and secondary characteristic directions, and characteristic phase retardation Δ_*^e. If perpendicular to s the body has a plane of symmetry, we have $\alpha_0^e = \alpha_*^e$. Thus, the number of experimental data we have on each ray equals the number of sets of coefficients we have to determine.

Scanning the section of the body under various angles β we can measure α_0^e, α_*^e and Δ_*^e on a great number of rays. Unfortunately, the relationship be-

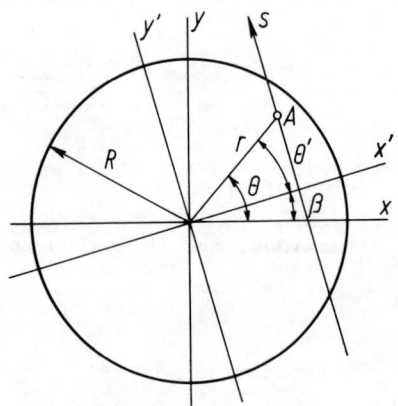

Fig. 1

Passing light through a section of the body

tween the coefficients which determine the stress distribution, and the experimentally measured α_0^e, α_*^e and Δ_*^e is very complicated and in the general case cannot be expressed in a closed form. However, giving to the coefficients a_{mn}^{ij} and b_{mn}^{ij} numerical values we can calculate "theoretical" values of α_0^t, α_*^t and Δ_*^t. Now determination of the coefficients a_{mn}^{ij} and b_{mn}^{ij} leads to a problem of non-linear minimization

$$F(\alpha_0^e - \alpha_0^t, \ \alpha_*^e - \alpha_*^t, \ \Delta_*^e - \Delta_*^t) \to \min \qquad (5.2)$$

In comparatively simple particular cases such an approach has been used previously [6,7].

Determination of stresses starts with a section where σ_{zz} is known. Coefficients $a_{mn}^{r\theta}$, $b_{mn}^{r\theta}$, $a_{mn}^{\theta z}$ and $b_{mn}^{\theta z}$ for both sections are determined on the basis of experimental data. Other coefficients are calculated from the equations derived above. After that the auxiliary section is considered as basic and determination of stress goes on recursively until the stresses are determined in all sections one is interested in.

It is obvious that solution of the problem formulated here may present considerable mathematical difficulties.

6. Case of Weak Optical Anisotropy

If optical anisotropy of the model is weak, the following relationships are valid:

$$\Delta_* \cos 2\varphi = CR \int (\sigma_{x'x'} - \sigma_{zz}) \, dy',$$

$$\Delta_* \sin 2\varphi = 2CR \int \sigma_{zx'} \, dy'. \qquad (6.1)$$

Here C is optical constant, and φ is parameter of the isoclinic in a plane polariscope.

Putting Eqs. (2.1) into Eq. (6.1) and taking into account Eqs. (4.2)-(4.6) we obtain a system of linear equations from which sets of coefficients which determine distribution of shear stresses $\sigma_{\theta z}$ and σ_{zr} can be determined. Thus, in this case the problem is much simpler. Eqs. (6.1) have been applied in [8] to develop a method for determination of the axisymmetric stress state.

In the case of low optical anisotropy interferometric measurements permit to determine distribution of the average refractive index n [9]:

$$n = n_0 + C_1 (\sigma_{rr} + \sigma_{\theta\theta} + \sigma_{zz}), \qquad (6.2)$$

where n_0 is the refractive index of the unloaded material and C_1 is an

7

optical constant.

If distribution of n is also determined experimentally, stresses in an arbitrary section (together with the auxiliary section) can be determined independently.

7. Numerical Experiment

Numerical and physical experiments to check practical applicability of the proposed method are under way and will be reported at the Conference. However, partially applicability of the method can be checked in the following way.

Let us consider determination of stresses in a circular disc loaded along a diameter by concentrated forces (the Hertz disc). This problem has analytical solution and is a classical example in two-dimensional photoelasticity. Our aim is to determine stresses in the disc on the basis of integrated optical data obtained when the polarized light passes through the disc in an unconventional way parallel to its midplane (Fig. 2). From the point of view of integrated photoelasticity we have a particular case of a three-dimensional stress state with $\sigma_{zz} = \sigma_{\theta z} = \sigma_{zr} = 0$.

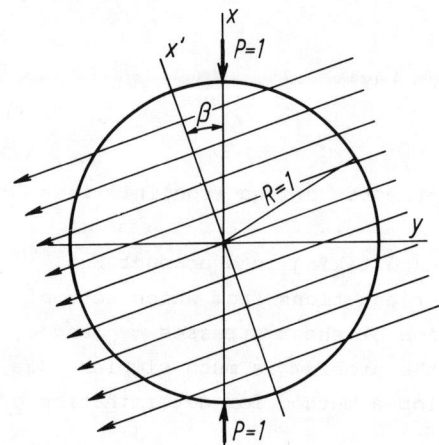

Fig. 2
Polarized light is passed through the Hertz disc parallel to its midplane

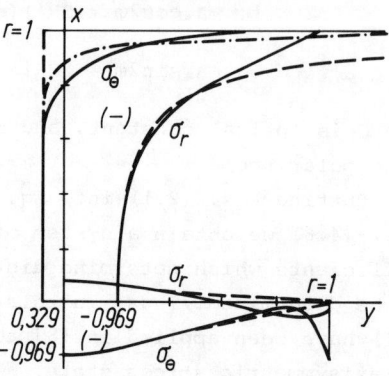

Fig. 3
Stress distribution with m_1 = 18:
—— integrated photoelasticity,
--- analytical,
-·- paper [11]

Theoretical experimental data for 10 values of β and for 20 values of x' were calculated for P=1. In this case experimental data are expressed through Eqs. (6.1) with $\varphi=0$. It is to be mentioned that for each value of β the value of Δ_* is a constant, proportional to $P\cos\beta$, if x'/cosβ< R, and equals zero if this condition is not fulfilled.

In this case it was reasonable to express stress components σ_{rr} and $\sigma_{r\theta}$ through $\sigma_{\theta\theta}$. The latter was chosen in the following form [10]:

$$\sigma_{\theta\theta} = a_{00}^{\theta\theta} + \sum_{m=2}^{m_1} (a_{m,m-2}^{\theta\theta} r^{m-2} + a_{mm} r^m)\cos m\theta. \qquad (7.1)$$

This expression was put into the right side of the first Eq. (6.1), using Eqs. (5.1). The overdetermined system of linear equations was solved by the method of least squares. In Fig. 3 the ditribution of σ_{rr} and $\sigma_{\theta\theta}$ along the diameters for m_1=18 is given. Analytical stress distributions are shown by dashed curves. In the middle part of the disc the results coincide excellently. On the x-axis near the load σ_θ differs considerably from the analytical solution but is very close to the experimental results given in [11]. On the y-axis near the boundary stress distribution deviates considerably from the analytical one. That implies the need to keep more terms in the expansion (7.1) to describe adequately a concentrated load.

8. Acknowledgement

The authors wish to express their sincere gratitude to Jüri Josepson for fruitful discussions.

9. References

1 Aben,H.: Integrated Photoelasticity. McGraw-Hill, New York, 1979.

2 Davin,M.: Sur l'exploitation de l'information donnée par un "effet résultant" recueilli sur chaque sécante traversant une éprouvette et appartenant à un ensemble donnée de sécantes. Application à la photo-élasticité. C.r.Acad.sci., 269A, No.13, pp. 543-545, 1969.

3 Kubo,H.-Nagata,R.: Determination of Dielectric Tensor Fields in Weakly Inhomogeneous Anisotropic Media. J.Opt.Soc.Am., No.4, pp. 604-610, 1979.

4 Herman,G.T.: Image Reconstruction from Projections. Academic Press, New York, 1980.

5 Aben,H.-Kell,K.-J.: Integrated Photoelasticity as Tensor Field Tomography. ZAMM, 66, No.4, pp. T118-T119, 1986.

6 Бросман,Э.,Полль,В.,Тяхт,Р.: Уточненный расчет квазиглавных направле-

ний в интегральной фотоупругости кубических монокристаллов. Изв. АН ЭССР, физика, математика, 32, № 2, 179-183, 1983.

7 Фомин,А.: К решению обратной задачи интегральной фотоупругости. Изв. АН ЭССР, физика, математика, 33, № 3, 277-284,1984.

8 Doyle,J.F.-Danyluk,H.T.: Integrated Photoelasticity for Axisymmetric Problems. Exp.Mech., 18, No.6, pp. 215-220, 1978.

9 Vest,C.M.-Ural,E.A.: The Role of Interferometry and Tomography in Stress Analysis of Transparent Media. Proc. SESA Spring Meeting, Dearborn, pp. 242-247, 1981.

10 Timoshenko,S.P.-Goodier,J.N.: Theory of Elasticity. McGraw-Hill, New York, 1970.

11 Pindera,J.T.: New Development in Photoelastic Studies: Isodyne and Gradient Photoelasticity. Opt.Eng., 21, No.4, pp. 672-678, 1982

INVESTIGATION OF THREEDIMENSIONAL AXISSYMMETRICAL PROBLEMS BY PHOTOELASTIC METHOD

Prof.Dr. Mahrat Achmetzyanov - Victor Tichomirov
Constructing Department of Novosibirsk Railway Transport
Engineering Institute
D.Kovalchuk I9I, 630023 Novosibirsk, USSR

An algorithm of finding of stresses in an axissymmetrical
loaded model is presented in the work /I/. It is based on
data of integral photoelastic method. The present work shows
that if the data of integral photoelastic method are added
to data received by a method of scattering light then the
algorithm of finding of stresses significantly simplifies
and the components of stresses in each ring-layer may be
defined independently.

Keywords: integral photoelasticity, scattering light, axis-
symmetrical

A through translucence of the model with measuring characteristic
values is done, as above /I/, along the direction y (Fig. I). The
difference in the run of light rays and primary characteristic directions are measured successively for the rays I,2,3 etc.

The run difference δ_{11} and the parameter of isocline α_{11} for the
first ring the outer radius of which coinsides with the radius of
body surface in corresponding cross-srction may be directly connected
with the difference of normal stresses $\sigma_r - \sigma_z$ and tangential stresses τ_{rz} , which are reffered to point A_1 of the first ring.

$$(\sigma_r - \sigma_z)^{(1)} = \frac{\delta_{11}}{2 C_\sigma \, y_{11}} \cos 2\alpha_{11} \ ,$$

$$\tau_{rz}^{(1)} = \frac{\delta_{11}}{4 C_\sigma \, y_{11}} \sin 2\alpha_{11} \ ,$$

(I)

11

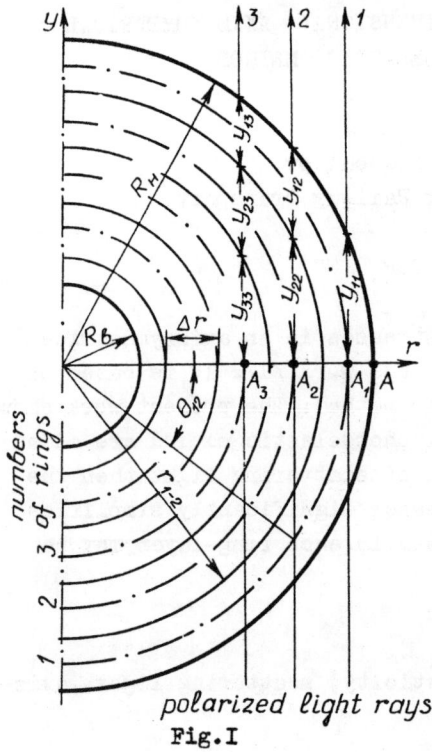

polarized light rays
Fig.I

where C_σ - optical constant of the model material, $2y_{11}$ - optical light ray path (I) in the first ring.

The primary ray of light is directed along the radius of the corresponding disc and the optical difference of ray run is measured in scattered light which in this case will be connected with stresses with the help of Wertgame integral law. It is being done for getting lack data necessary for defining of the components of tensor stresses.

$$\Delta_r = C_\sigma \int_{R_4}^{r} (\sigma_\theta - \sigma_z) d\rho . \qquad (2)$$

Optical differences of ray run $\Delta_1, \Delta_2, \Delta_3$ are measured on borders of corresponding rings. The run difference accumulated for the first ring is defined as

$$(\sigma_\theta - \sigma_z)_1 = \frac{\Delta_1}{C_\sigma \cdot \Delta r} . \qquad (3)$$

The same is done for the second disc-ring

$$(\sigma_\theta - \sigma_z)_2 = \frac{\Delta_2 - \Delta_1}{C_\sigma \cdot \Delta r} .$$

The values of $(\sigma_\theta - \sigma_r)$ for all the discs of the model ring under consideration are defined by the same way.

Boundary conditions are used to find stresses in the point of free contour (A) and the equation (4) is taken into account

$$(\sigma_1 - \sigma_2)_A = \lim_{r \to R_H} \frac{\delta}{2C_\sigma \sqrt{R^2 - r^2}} , \qquad (4)$$

where δ - optical run difference under translucence through the first ring; $\sigma_1 - \sigma_2$ - the difference of principal stresses on the model contour.

But if the contour is free of loads then

$$\sigma_{z,A} = (\sigma_1 - \sigma_2)_A \cos^2\beta \ ,$$

$$\sigma_{r,A} = (\sigma_1 - \sigma_2)_A \sin^2\beta \ ,$$

(5)

where β - the angle between a normal to the arc of contour and r .

To find $(\sigma_1 - \sigma_2)_A$ it is necessary to do several translucences in the zone of the first layer under different meanings of r and by means of extropolation to determine the needed value.

The stresses σ_z and σ_r for point A determined with the help of equation (4) and (5) in correspondence with the supposition that the stressed condition of each ring is homogeneous may be simultaneously considered as stresses for point A_1 . Therefore all the components of the tensor of stresses for point A_2 are defined.

After this the stresses are determined for point A_2 . For this purpose the results of model translucence along ray 2 are considered. This ray crosses the centre of the second ring in meridianal plane perpendicular to the direction of translucence. In this case and in all the following cases it is impossible to neglect the changing in stressed condition on the path of translucence because the ray crosses three rings, the first ring - two times and the second ring - one time. But this does not make any influence on the experiment technique because the photoelastic medium is symmetrical.

Photoelastic medium matrix for ray 2 is given as

$$U = \tilde{U}_{12} \cdot U_{22} \cdot U_{12} \ ,$$

(6)

where U_{12} - the first layer matrix corresponding to translucence of the first ring at the second ray run from the outer surface through the model, $\tilde{U}_{12}$ - transposed matrix corresponding to the first layer at the inverse ray run (the third layer), U_{22} - the second layer matrix describing passing the second ray through the second ring. Muller's matrix as well as Johns' ones may be used for this calculation.

Johns' matrix for homogeneous photoelastic medium may be represented as

$$U_{ij} = \begin{vmatrix} \cos T_{ij} + i \sin T_{ij} \cos 2\alpha_{ij} & i \sin T_{ij} \sin 2\alpha_{ij} \\ i \sin T_{ij} \cdot \sin 2\alpha_{ij} & \cos T_{ij} - i \sin T_{ij} \cos 2\alpha_{ij} \end{vmatrix}$$

(7)

13

where $\tau_{ij} = -\frac{\pi \delta_{ij}}{\Lambda}$, Λ - light wave-length.

The value of δ_{12} and α_{12} for filling matrix U_{12} are calculated with the help of expressions

$$\delta_{ij} = C_\sigma y_{ij} \sqrt{(\sigma_z^{(i)} - \sigma_r^{(i)} \cos^2 \theta_{ij} - \sigma_\theta^{(i)} \sin^2 \theta_{ij}) + (2\tau_{rz}^{(i)} \cos \theta_{ij})^2}$$

$$tg\, 2\alpha_{ij} = \frac{2\tau_{rz}^{(i)} \cos \theta_{ij}}{\sigma_z^{(i)} - \sigma_r^{(i)} \cos^2 \theta_{ij} - \sigma_\theta^{(i)} \sin^2 \theta_{ij}} \quad , \tag{8}$$

where θ_{ij} - polar coordinate of element i under j translucence. In this case

As matrix U has been obtained experimentally from the measurement data of the second ray, components of the matrix U_{22} may be defined from equation (6). For example

$$x_{11} = C_{11} \cdot a_{22}^2 - a_{22} \cdot a_{21}(C_{12} + C_{21}) + a_{21}^2 C_{22} \quad , \tag{9}$$

where the components of the needed matrix U_{22}, matrix of the whole photoelastic medium U and matrix U_{12} are denoted as x_{ij}, C_{ij} , a_{ij} respectively.

The difference of phases are obtained using x_{11}

$$\cos \tau_{22} = Re(x_{11}) \tag{I0}$$

and azimuth (isocline parameter) is obtained from

$$\cos 2\, \alpha_{22} = Im \left(\frac{x_{11}}{\sin \tau_{22}} \right) \tag{II}$$

which correspond to the second layer after the second translucence.

Using the data given above the values of $\sigma_r - \sigma_z$ and τ_{rz} at point A_2 of the model are defined with the help of expression (I) and the like. Thus, adding to these data the data of scattered light method we can find the values of τ_{rz} , $\sigma_r - \sigma_z$, $\sigma_\theta - \sigma_z$ and by this way the value of $\sigma_r - \sigma_\theta$ (as the difference of the two last expressions). The equation of strain co-locality

$$\frac{\partial \mathcal{E}_\theta}{\partial r} - \frac{\mathcal{E}_r - \mathcal{E}_\theta}{r} = 0 \tag{I2}$$

14

at terminal differences for the point which lies on the boundary of the first and the second layer are used for separate definition of normal stresses at this point A_2. After this we have

$$\mathcal{E}_{\theta, A_2} = \mathcal{E}_{\theta, A_1} - \frac{(\mathcal{E}_r - \mathcal{E}_\theta)_{1-2}}{r_{1-2}} \quad , \qquad (13)$$

where

$$(\mathcal{E}_r - \mathcal{E}_\theta)_{1-2} = \frac{(\mathcal{E}_r - \mathcal{E}_\theta)_{A_2} + (\mathcal{E}_r - \mathcal{E}_\theta)_{A_1}}{2} \ , \qquad r_{1-2} = \frac{r_{A_1} + r_{A_2}}{2}$$

According to Hoock's law we shall have

$$\mathcal{E}_r - \mathcal{E}_\theta = \frac{E}{1+\mu}(\sigma_r - \sigma_\theta) \quad , \qquad (14)$$

$$\mathcal{E}_\theta = \frac{1}{E}\left[\sigma_\theta - \mu(\sigma_r + \sigma_z)\right] . \qquad (15)$$

Substituting the strain in the right part of the equation (13) by already known meaning of stresses we shall find $\mathcal{E}_{\theta, A_2}$, and then using the equation (15) we obtain

$$\sigma_z = \frac{E \mathcal{E}_\theta - (\sigma_\theta - \sigma_z) + \mu(\sigma_r - \sigma_z)}{1 - 2\mu} \qquad (16)$$

Then all the separate meanings of stresses at point A_2 are easily defined.

Now we define the stresses at point A_3 of the meridianal cross-section. Ray 3 passes through the first and the second rings twice and through the ring 3. This system is modeled by the equation

$$U = \widetilde{U_{13}} U_{23} \ U_{33} \ U_{23} \ U_{13} \qquad (17)$$

Parameters of matrix U_{13} and U_{23} are calculated with the help of expressions (7) and (8) and then the separate products of these matrix are found

$$U_{123} = U_{13} \cdot U_{23} \qquad (18)$$

Then the expression (17) may be represented as

$$U = \widetilde{U}_{123} U_{33} \cdot U_{123} . \qquad (19)$$

15

The components of matrix U_{33} are found with the help of expression (9) and by analogy the stresses at point A_3 are obtained.

The algorithm given allows to calculate stresses at all the other points of the model layer under consideration, and then if needed to investigate the stressed condition through the whole volume of the model.

Reference

I Ахметзянов М.Х., Соловьев С.Ю. Способ исследования пространственного осесимметричного напряженного состояния методом интегральной фотоупругости. В кн. "Механика деформируемого тела и расчет транспортных сооружений". Межвузовский сборник научных трудов (НИИЖТ), Новосибирск, 1984, с. 56-60

ELLIPTIC PHOTOELASTICITY AND ITS APPLICATIONS

Dr. Illya Bugakov, Dr. Nadezda Drichko,[x] Dr. Sofia Golubeva,[x]
Dr. Elena Urchenko
Leningrad State University
Bibliotechnaya pl. 2, 198904, Petrodvorez, Leningrad, USSR
[x]State Optical Institute, USSR

The theoretical and experimental fundamentals of the method
of elliptic photoelasticity are presented. The peculiarities of birefringence in a loaded quartz disk are studied.
Likewise the characteristics of birefringence are studied
and the stresses determined in a loaded quartz disk with
an initial circular birefringence.

Keywords: photoelasticity, gyrotropy, quartz

1. Gyrotropic materials

In connection with advances in the physics of solids and experimental technics, the use of gyrotropic materials has become widespre-
ad in scientific reseach and technical applications. Gyrotropic materials are generally characterized by elliptic birefringence (EBR):
with elementary electromagnetic vibrations throughout the medium polarized elliptically. In particular cases circular birefringence (CBR)
and linear birefringence (LBR) are observed. Materials with CBR are
called optically active.

In [1] a formulation typical for the methtod of photoelasticity,
Wood shows the experimental and theoretical study of the characteristics of EBR in loaded cuts of quartz with initial CBR (direct problem). A plane reflection polariscope was used. For a pure bending
beam experimental results correlated well with calculations.

2. Elliptic photoelasticity

The classic method of photoelasticity is applied only to objects
with LBR. In [2] the theoretical possibility of determination of the
stresses in plane mechanical isotropic and anisotropic gyrotropic

IMEKO 1st Conference, Plzen, Czechoslovakia, May 25-28, 1987 objects is shown. The method of photoelasticity as it applies to materials with EBR is called elliptic photoelasticity. Its theory, based on the results of [1-3], is summarized in [4]. It was applied to the solution of direct problem for a quartz cut with initial CBR. In this work the direct problem for a quartz cut with initial EBR is solved, as well as the direct and reverse problem for a quartz cut with CBR.

Aben's primary and secondary characteristic directions with azimuths α_0 , α_* and characteristic phase retardation 2γ [3] are used as integral parameters of optical anisotropic spatial inhomogenious medium. The angle $\psi = \alpha_* - \alpha_0$ is called the characteristic angle or effective rotation.

It is known that with EBR the phase retardation Δ may be represented as $\Delta^2 = \Delta_1^2 + \Delta_2^2$, where Δ_1 is the phase retardation of linear birefringence and $\Delta_2 = 2\chi$ is the phase retardation of circular birefringence with the rotation of plane of polarization at angle χ . The phase retardation is expressed through characteristic parameters as follows:

$$\Delta = 2 \, arc \, tg(a/\cos\psi) \qquad \left(a = \sqrt{tg^2\gamma + \sin^2\psi}\right),$$

$$\Delta_1 = \Delta \cdot a^{-1} tg\gamma, \qquad\qquad \Delta_2 = \Delta \cdot a^{-1} \sin\psi. \qquad (1)$$

Following [1], we will describe the ellipticity of elementary vibrations in terms of the parameter $\tau = \Delta_1/\Delta_2$. Thus, according to the Eqs. (1) , we get :

$$\tau = tg\gamma / \sin\psi. \qquad (2)$$

The azimuth of the secondary direction in relation to an arbitrary system of coordinates we will call y . By analogy with the plane problem of classical photoelasticity we will call y the parameter of the isokline. Aben [3] shows that the secondary directions coincide with the bisector of angle ψ , so in an arbitrary system of coordinates connected with the cut, we have

$$y = \alpha_0 + \psi/2. \qquad (3)$$

Optical values may be expressed by using τ and χ :

$$\Delta = 2\chi p \qquad\qquad \left(p = \sqrt{1 + \tau^2}\right),$$

$$\sin\gamma = \tau p^{-1} \sin(\Delta/2), \qquad\qquad tg\psi = p^{-1} tg(\Delta/2). \qquad (4)$$

The most simple case occurs when mechanically isotropic material has an initial CBR and induced LBR. Then in the initial unstrained state $\Delta_1 = \Delta_{10} = 0$, $\Delta_2 = \Delta_{20} = 2\chi$, $\tau = \tau_0 = 0$, $y = y_0 = 0$; in the strained state $y = y_\sigma$ or $y = y_\sigma + 90°$, where y_σ is the azimuth of the greatest principal normal stress σ_1 , $\tau = C(\sigma_1 - \sigma_2)$, (5) where C is the reduced optical coefficient of the stresses, and $\sigma_1 - \sigma_2$ is the difference of the principal stresses.

3. Measuring apparatus

A variety of means are known for determining the characteristic parameters of optical inhomogeneous and anisotropic medial [3-7]. For our study measurements of characteristic parameters were made using modified scattered light phase polarimeter with a wave length 633 nm [7]. Just as in the study of spatial bodies by means of the scattered light method, at each point of the plane optical inhomogeneous and anisotropic gyrotropic object, six values of phase characteristics of electrical signals f_{ij} are measured (i =1,2,j =1,2,3). According to the values f_{ij}, the characteristic parameters may be determined as shown in [6,7]. We note that with Wood's apparatus it is possible to measure only $\angle_0$, $\angle_*$, but not γ .

4. The object of study

Quartz (class 32) was studied,which is the most thoroughly reseached gyrotropic uniaxial chrystal. In unstrained quartz in the direction of the optical Z-axis CBR is observed; in the direction 56^0 to the optical axis LBR, and in all remaining cases EBR is observed.

Study was conducted on two plane round disks, cut from synthetic right rotating quartz: an inclined-cut and a Z-cut. The disks were loaded along the diameter. The inclined-cut was cut out at an angle of 86^0 to the optical axis, its diameter was 40.0 mm, thickness 5.1mm and the line of the compression was set in the plane containing the normal to the cut and the optical axis. The diameter of the Z-cut was 45.1 mm, thickness 12.8 mm; the line of the action of compression was arbitrarily oriented in relation to the chrystallographic axes X,Y. The disks were lit along the normal to their plane. The Descartes system of coordinates x and y was used,situated in the plane of the disk with the origin in the center. The y-axis was directed along the line of the action of load.

5. Study of the inclined-cut

On this disk, with initial EBR, the peculiarities of induced birefringence (IBR) were studied. The values f_{ij} were measured in the center of the disk at different loads and in the transverse central section y=0 at a load of F= 1.40 kN. Then the values of $\angle_0$,2γ, ψ were determined,after which,according to the Eqs. (1) and (3),Δ_1, Δ_2,

ι and Ψ_σ were found (Fig.1,2). According to the symmetry, $\Psi_\sigma = 0$. Since $\Delta_{10} \neq 0$, $\iota \neq 0$ when $F = 0$. It was observed that Δ_2 and Ψ are weakly dependent on F (Fig. 1), which allows us to suppose a certain degree of ellipticity in IBR. Note, that IBR is subtracted from the initial effect. When $F = 0.9$ kN, CBR was observed. There is a correlation between the graphs in Fig. 1 and 2, since the stresses on the x-axis increase from the perifery towards the center.

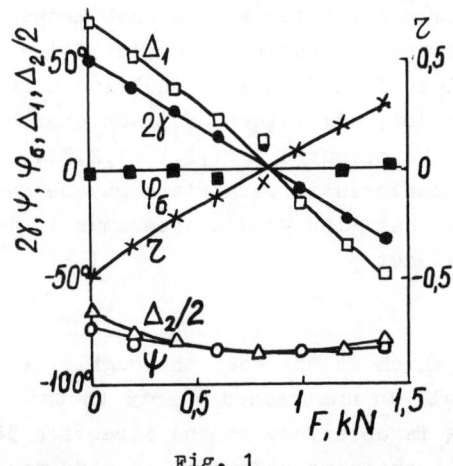 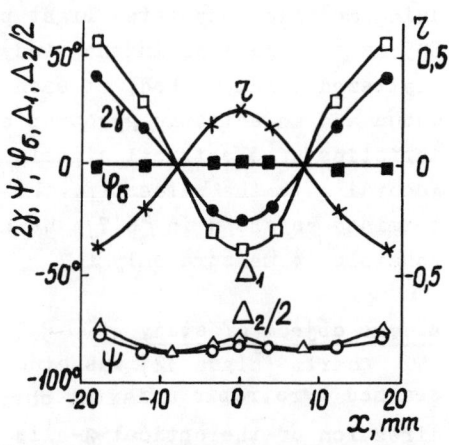

Fig. 1

Dependence of optical values on load in the center of an inclined-cut

Fig. 2

Distribution of optical values in an inclined-cut along line $y = 0$

6. Study of the Z-cut

On this disk with an initial CBR ($\chi = \chi_0 = -240°$) the peculiarities of IBR were studied as well as stresses. Upon measurement of the values f_{ij} in the center of the disk at different levels of F optical values were found. In this problem $\Delta_2 = $ const, but Δ_1 and ι are lineary dependent on F (Fig. 3); IBR is linear. The theoretic stress distribution for the given problem is well known. According to the data given in Fig. 3 and Eqs.(5), we arrive at a value of $C = 4.2 \times 10^{-2}$ MPa^{-1}.

For the solution of direct and reverse problems we got the connection of 2γ and $\Psi - \chi_0$ with ι (Fig. 4). It can be deduced from the graphs that, in order not to lose accuracy when determining ι , it is preferred to apply the dependence 2γ (ι , χ_0) when ι is small, and in the extremum of this function, the dependence Ψ (ι , χ_0).

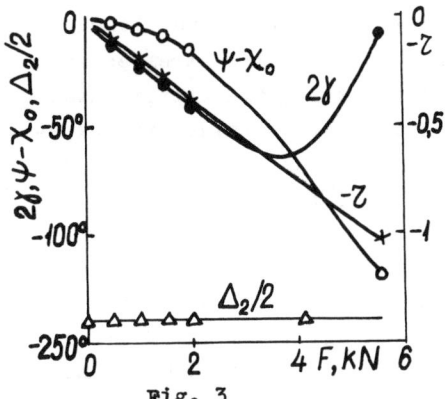

Fig. 3

Dependence of optical values
on load in the center of Z-cut

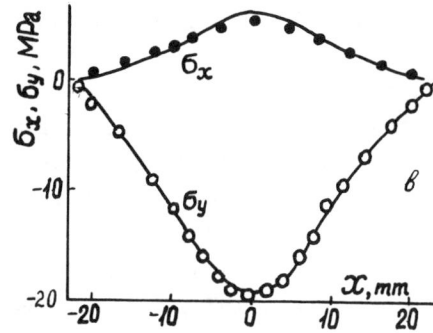

Fig.4

Dependence of 2γ and $\psi - \chi_o$ on
τ for initial CBR, induced
LBR when $\chi_o = -240^\circ$

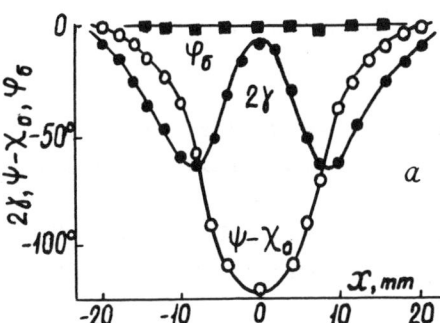

Fig. 5

Distribution of optical values (a) and stresses (b) in a Z-cut
along line y = 0.

Direct and reverse problems of photoelasticity were solved for
line of the disk y=0 and y= -11,7 mm, where F = 5.58 kN. A greater
load was chosen than for the inclined disk, because of the greater
size of the Z-cut. The results for the line y=0 are represented in
Fig.5. The points show experimental results and the lines express
theoretical results. The theoretical curves were arrived at as follows.   From the known apriori stresses with the help of (5) we
converted to τ. Then with the help of the graphs in Fig.3 we conver-
ted to 2γ and $\gamma - \chi_o$. According to the symmetry of the problem $\psi_\sigma = 0$.
Thus, as can be seen in the results shown in Fig.5a, there is close
agreement between the experimental and theoretical solution of the
direct problem.

When   solving the reverse problem, with the help of Fig. 3, we converted from 2γ and $\psi - \chi_o$ to γ. Then with (5) we found the distribution of the difference $\sigma_1 - \sigma_2$. Separation of the stresses was accomplished using method [8], which doesn't require experimental data for one more parallel line. In this way we obtained the experimental result indicated by points in Fig. 5b. The theoretical results are indicated by lines. The error of measurement at point x = 0 was 12.5% for σ_x and 3.5% for σ_y. The condition of equilibrium is satisfied with an error of 5%.

Note that the least exact results are obtained by the method of photoelasticity in the line where y = 0. At the point where x= 0, y =-11.7 mm the error was 7.3% for σ_x, 2.1% for σ_y. The condition of equilibrium for the line y =-11.7 mm is satisfied with an error of 0.5%.

The results obtained show the good future prospects of the method of photoelasticity as applied to plane gyrotropic objects.

7. References

1 Wood,A.F.B.: On the photoelastic Examination of Vibrating Bodies and the Photoelastic Effect in Optically Active Media. J.Mech. Phys. Solids, 8. No 1, pp. 26-38, 1960.

2 Бугаков,И.И.: Применение гиротропных моделей в фотомеханике. Материалы УШ Всесоюзн. конф. по методу фотоупругости, Таллин, СССр, 25-28 сент. 1979, т.1У, с. 34-38.

3 Абен,Х.К.: Интегральная фотоупругость. Валгус, Таллин, 1975.

4. Бугаков,И.И.-Причко,Н.М.-Юрченко,Е.Г.: Применение метода фотоупругости к гиротропным объектам. Экспериментальные методы исследования деформаций и напряжений.Сб. статей, Киев, 1983, с. 26-34.

5 Lagarde,A.: Some Aspects of Developments of Photoelastic Measurements. Proc. IUTAM Symposium on Optical Methods in Mechanics of Solids. Poitier, France, Sept.10-14, 1979,SIJTHOFF & NOORDHOFF, The Netherlands - USA, 1981, pp.1-40.

6 Александров,А.Я.-Ахметзянов,М.Х.-Плешаков,Ф.Ф.: Об исследовании пространственных задач поляризационно-оптическим методом. Тр.Новосибирск.института инж.жел.-дор.трансп.,вып.167, 1975,с.175-187.

7 Причко,Н.М.-Голубева,С.Г.-Анищенко,В.В.-Лейкин,М.В.: Поляриметр для определения параметров пространственно неоднородной оптической анизотропии.Оптико-механическая промышленность, №6, с.28-30, 1985.

8 Гузь,А.Н.-Вологжанинов,Ю.И.: Определение напряжений на основе ограниченного объема экспериментальных данных. Докл.АН СССР, 277, №3, с. 563-565, 1984

DETERMINATION OF STRAIN TENSOR COMPONENTS BY THE SCATTERED LIGHT
METHOD

Wojciech Karmowski,MSc.MEng
Stanisław Mazurkiewicz,Prof.D.Sc,MEng
Politechnika Krakowska
Inst. of Mech. and Mach.Design
ul.Warszawska 24
31-155 Kraków Poland

In the paper a new photoelastic method to measure full
strain tensor in 3D case has been presented. Three light
beams propagate through a investigated body in three non
co-planar directions. Intensity of scattered light in the
investigated point is measured by photometric device. The
experiments with beams propagating in two mutually reverse
directions and in both simultaneously enable the calculation
all components of strain tensor.

Keywords: photoelasticity,scattered light,strain

1. Introduction

The last of subsequent improvements of the scattered light method
- well known in photoelasticity since 1938 [1]- have caused it to be
informative in analyzes of components of strain tensor in transparent
models [2,3,4,5]. Substantial results have been obtained in 2D
problems, the value of one of the diagonal components of strain can be
measured with the use of isodyne technique [6], and both of the remain
components can be calculated from supplementary relationships of the
strain tensor. The authors have done it in 1981 [7] reaching the
complete description of the strain field by integrating the equations
of equillibrium and analyzing the isodyne field. The fact that the
discrete-type isodyne fringes must numerically be smeared on the
entire field being analyzed [8] is the significant disadvantage of the

23

procedure. In the method being presented now the authors propose the photometric recording of the light scatttered while three non co-planar light beams passing through the object under the test are intersecting mutually in the point being analyzed. The method utilizes the complete information available when the light beam passes through the transparent model. Since the light passes the object through in two mutually reverse directions and simultaneously in both, the calculating of the strain tensor becomes to be possible. The last mentioned potentiality is the major superiority of the method presented by the authors in 1983 [9], when hydrostatic component was not distinguishable. 9 equations with 6 unknown quantities (components of the strain tensor) can be obtained from the measurements in three non co-planar directions. The complete strain tensor is getatable by solving the overdetermined system of mentioned equations.

2. Measurements and calculation of strain

The electric field intensity in the considered point of the beam path along z axis can be expressed by the following quantities

A — amplitude $\qquad A=\sqrt{E^2}$ $\qquad\qquad\qquad\qquad$ (1),

τ — angle for which $\tan\tau = b/a$, where b and a denote axes of polarization ellipse,

$\emptyset$ — rotation angle to such position that the longer axis is parallel to "x" axis,

ψ — constant phase between E and "x" axis,

τ — normalized time $\tau = \omega\,t$.

The electric field intensity vector can be expressed as

$$\vec{E}=\sqrt{2}A\check{T}^{\omega}\check{D}\tau\check{\tau}-\psi\vec{\theta}$$

$\qquad\qquad\qquad\qquad\qquad\qquad\qquad\qquad\qquad\qquad\qquad\qquad$ (2),

where

$\check{T}^{\omega}$ — rotation tensor $\qquad \check{T}^{\omega}=\begin{pmatrix} \cos\emptyset & -\sin\emptyset \\ \sin\emptyset & \cos\emptyset \end{pmatrix}$

$\check{D}\tau$ — matrix $\qquad\qquad\quad \check{D}\tau=\begin{pmatrix} \cos\tau & 0 \\ 0 & \sin\tau \end{pmatrix}$

24

$\vec{\theta}$ - vector $\qquad\qquad \vec{\theta}=\begin{pmatrix} \cos\tau \\ \sin\tau \end{pmatrix}$

The passing of the light through the model (quasiisotropic medium) can be conceived as the propagation of two waves in the planes mutually perpendicular and simultaneously parallel to the secondary principal directions in the point considered. In general the secondary principal directions vary point to point. Considering the form of the dielectric constant of the medium as

$$\breve{\mathcal{X}} = \breve{\mathcal{X}}_0 + \mathcal{X}_1 \breve{\epsilon} + \mathcal{X}_2 \, tr\breve{\epsilon} \qquad\qquad (3)$$

man can present the wave vectors of both rays as

$$k^{(1)} = \frac{\omega}{c}\sqrt{\mathcal{X}_0} + \frac{\omega\mathcal{X}_1}{2c\sqrt{\mathcal{X}_0}}\,\epsilon^{(1)} + \frac{\omega\mathcal{X}_2}{2c\sqrt{\mathcal{X}_c}}\,tr\breve{\epsilon} \qquad\qquad (4),$$

where the integer 1 numbers the principal directions, and the secondary principal strain is denoted as $\epsilon^{(1)}$. In the isochromatic method $\epsilon^{(1)} - \epsilon^{(2)}$ is the only subject to be photoelastically measured, and in the method being proposed $\epsilon^{(1)} + \epsilon^{(2)}$ is that subject also. The direction of recording of the light scattered in the point under consideration is perpendicular to the rays and forms angle α with positive x axis and light intensity is

$$I(\alpha) = I_0 A [1 - \cos 2\tau \cdot \cos(2\theta - 2\alpha)] \qquad\qquad (5).$$

Quantity "Q" introduced by the authors in aforecited paper [9], at which place has been termed by them the complex photoelastic parameter, is created of values of $I(0), I(\pi/4), I(\pi/2)$ according to formula

$$Q = \frac{i[I(0) - I(\pi/2)] + [I(0) + I(\pi/2) - 2I(\pi/4)]}{I(0) + I(\pi/2)} \qquad\qquad (6).$$

The quantity "Q" can be also denoted as

$$Q = -i \cos2\tau \; e^{\pm i\phi} \tag{7}.$$

That quantity is connected directly with the strain tensor throughout the expression

$$\frac{\omega \chi_1}{2c\sqrt{\chi_0}} \, \widetilde{\varepsilon} = -\frac{\text{sgn}\tau}{\sqrt{1-Q*Q}} \, \frac{dQ}{dz} \tag{8}.$$

where $\widetilde{\varepsilon}$ - consistently termed by the authors [9] the complex directional strain is defined as

$$\widetilde{\varepsilon} = (\varepsilon^{(1)} - \varepsilon^{(2)}) e^{\pm i\vartheta} \tag{9}$$

(ϑ - the angle of transformation of laboratory co-odinate system to the principal one). Quantity $\widetilde{\varepsilon}$ brings two pieces of information about projection of the strain tensor on to the plane perpendicular to the rays. For the lack of the third relationship, the light intensity measurements at the angle $\alpha=0,\pi/4,\pi/2$ have to be done for two separately, reversely propagating rays (corresponding quantities are marked with indices + and - indcating conventionally positive or negative directions). The third measurement (the quantities without any mark) concerns the very same angle α values but has to be done for both rays propagating simultaneously and therefore interfering. The quantities being obtained as the result of the measurements satisfy the equation

$$A^2 Q = A_+^2 Q_+ + A_-^2 Q_-$$
$$- 2A_+ A_- \exp[i(\emptyset_+ + \emptyset_-)][i\cos(\tau_+ + \tau_-)\cos(\psi_+ - \psi_-) + \sin(\tau_+ - \tau_-)\sin(\psi_+ - \psi_-)] \tag{10}.$$

The only unknown quantity in the equation (10) is the expression
$\psi_+ - \psi_-$. For each of the rays individually the following equation is fulfilled

$$\pm \frac{k^{(1)} + k^{(2)}}{2} = \frac{d\psi}{dz} - \frac{1}{\sin2\tau} \frac{d\emptyset}{dz} \tag{11}$$

(the left-hand term is positive or negative, according to the rays directions). In that case the relation (12) connecting the sum of principal strains with photoelastic quantities is true

$$\frac{2\omega}{c}\sqrt{\chi_c} \ (1+ \frac{\chi_1+2\chi_2}{4\chi_c}\varepsilon'') =$$

$$=\frac{d}{dz}(\psi_+ -\psi_-) - \frac{1}{\sin 2\tau_+}\frac{d\emptyset_+}{dz} + \frac{1}{\sin 2\tau_-}\frac{d\emptyset_-}{dz} \cdot \qquad (12),$$

In the foregoing

$$\varepsilon'' = \varepsilon^{(1)} + \varepsilon^{(2)} + \frac{2\chi_2}{\chi_1+2\chi_2}\varepsilon_{33} \ .$$

The equation (12) is the missing one in the procedure of calculation of strain tensor components in 2D problem. The 3D problem can be resolved by carrying out the measurements in three directions as it has been stated above.

3. References

1 Weller.R A New Method for Photoelsticity in Three Dimensions. J.Appl.Phys. 10.4(1939)266

2 Aben.H..Integrated Photoelasticity.Mc Graw-Hill.New York.London.1979

3 Robert.A..The Application of Poincare s sphere to Photoelasticity, Int. J.Solids and Struct..6(1970)423

4 Laermann.K.H..Das Prinzip der integrierten Photoelastizitat.Angewandt auf die experimentalle Analyse von Platten mit nichlinearen Formannderungen.Proc.Seventh Int.Conf. on Exp. Stress Anal..Haifa, 1982.301

5 Theocaris P.S..Gdoutos E.E Matrix Theory of Photoelasticity in Springer-Verlag Berlin Heidelberg New York 1979

6 Mazurkiewicz.S.B-Pindera.J.T Integrated-plane Photoelastic Method-Application of Photoelastic Isodynes. Exp.Mech.19(1979)225

7 Karmowski.W-Mazurkiewicz.S.B The 2D stress field evaluation form isodyne pattern by use of scattered light method (in Polish), Mech.Teor.i Stos 20.1-2(1982)87

8 Karmowski.W-Orkisz.J Fitting of curves and surfaces based on interaction of physical relations and experimental data. Appl.Math.Modelling 7.2(1983)65

9 Karmowski.W-Mazurkiewicz.S.B A New Method of 3D strain field evaluating by use of scattered light method.(in Polish).Mech.Teor. i Stos. 21.2/3(1983)371

MULTIPLICATION OF PARTIAL ISOCHROMATIC LINES
BY SPECIAL PHOTOPROCESSING

Prof.Dr Stanisław Mazurkiewicz - Jacek Legendziewicz M.Sc
Technical University of Cracow
Poland

The paper presents the method of multiplication of partial
isochromatic lines. The method consists in the special kind
of photoprocessing with the use of the so called Sabattier's
effect. Owing to its simplicity, the method may find application in the photo-elastic research that demands the derivation of fractional isochromatic lines.

Keywords: photo-elasticity, isochromatic lines, photoprocessing, Sabattier's effect

Introduction

Photo-elastic research often demands the multiplication of the
measurement results precision that can be reached in the way of generation of partial isochromatic lines. The compensation methods allow
their spot measurement and they are limited to static loads. Loads
that very in time, or a greater number of spots demand the application of automatic measurement. The field method consists in optical
multiplication of isochromatics that is based on summing the photoelastic effect due to the multiple passage of light through the model.
That method requires the application of additional equipment and it
produces satisfactory results for small gradients of stresses. This
paper presents the method of generation of partial isochromatic lines in the way of special kind of photoprocessing.

Concept of method

The method is based on the so called Sabattier's effect. In the
development process combined with an additional exposure takes place
tone inversion and there appear bright contour lines, termed "Sabattier's lines" on the fringes of fields that differ in optical density. So far, that effect has not been completely explained. It finds
application in artistic and technical photography to achieve the so
called pseudopolarization effect. In this paper its application produced partial isochromatic lines. The order of a partial isochromatic

depends on the time of the first exposure. Hence, different values of isochromatic lines can be reached by changing the time of exposure. Further concentration of partial isochromatic lines can be achieved by contact printing of the thus obtained negative at different exposure times, producing the so called second order partial isochromatic lines. Isochromatic lines of the third and further orders may be generated by the succesive repetition of this procedure.

Calibration

The family of partial isochromatic lines that has been produced by this method contains only the information on the succesion of their occurrance. They can be labeled with the isochromatic line order tag only following the calibration. Here may serve e.g. the calibration with a disk made of the same photo-elastic material, loaded with two concentrated forces, subjected to the identical photo-processing as the tested model.

Still another way is the derivation of information on the isochromatic lines orders by the compensation method.

Calibration can be also made by the elaboration of the model characteristics of the Sabattier's effect as the function of one changing parameter as e.g. the time of secondary exposure.

As is well known, the change of light intensity by circular polarization and the dark field of view are expressed with the formula:

$$J = J_1 + J_2 \sin^2 \pi m$$

where: J_1 and J_2 are the background intensity of the polariscope and the maximum intensity of light fringes N, respectively.

For the linear interval of the light sensitive material characteristics the intensity of light is transformed into the density change according to the formula:

$$D = D_0 + \gamma \left(\log J \cdot t - \log E_0 \right)$$

where: D_0 and E_0 - the fog density and its corresponding light energy, t - exposure time, and γ - the slope of the linear portion of the film characteristic curve.

Experiment

Stresses in hardened glass.

Fig. 1 presents the isochromatic lines in a glass plate to hardening process and the isochromatic lines in the calibrating disk. State of polarization; circular, sodium light.

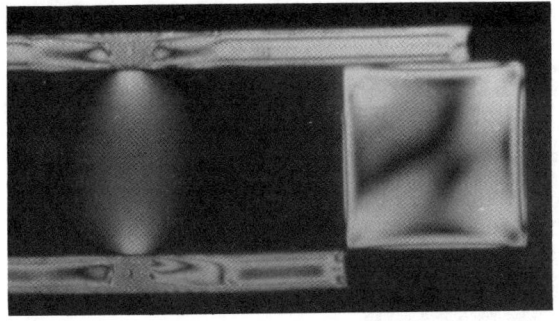

Fig. 1 Isochromatic
lines in tempered
glass plate and in
a calibration disc

Partial isochromatic lines of the 1st and 2nd order were obtained in
the way of photoprocessing. Fig.2 presents one of the many obtained
isochromatic lines of the 1st and 2nd order.

Fig. 2 Isochromatic
lines of the 1-th
and 2-nd order

A comprehesive map of these lines is shown in Fig.3. Their digital
values were derived with the use of the calibrator. The dependency
of the isochromatic lines' order along the diameter of the calibrating
disk may be expressed with the formula:

$$\frac{\sigma_1 - \sigma_2}{\sigma_0} = \frac{2}{\pi}\left(4 - \frac{4x^2}{R^2}\right) = \frac{m \cdot c}{\sigma_0} = M$$

31

where: $\tau_o = \frac{P}{b \cdot D}$; C – photo-elastic constant; D = 2R – disk diameter; b = its width; P – loading force.

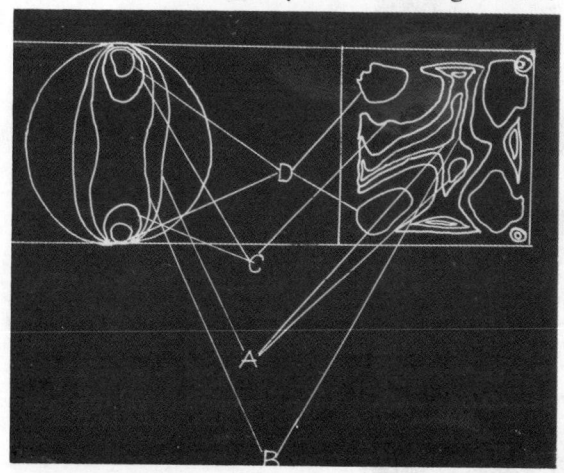

Fig.3 The map of partial isochromatic lines

The values of M for the succesive partial isochromatic lines in spots where they intersect the disk diameter may be read from M=M(R). Then those values are carried over to the respective fractional iso-chromatic lines obtained from the tested model.

Conclusions

The presented method allows to multiply the precision of the isochromatic lines reading in a simple way. Its advantage is also the fact that multiplication is made in the way of photoprocessing of isochromatic lines picture, that is following the conclusion of the experiment. That is quite significant for the elaboration of results of rapidly changing or shocking processes.
Additionally, it may also be helpful in all these cases where, for some reasons, the tests require material of low photo-elastic sensi-tivity.

THE ANALYSIS OF STRESSES IN THE MODEL OF A FRACTURED
BONE WITH THE APPLICATION OF PHOTO-ELASTIC METHODS

Prof. dr Stanisław Mazurkiewicz - Jacek Legendziewicz M.Sc.
Technical University of Cracow
Poland

The method of fixing fractured bines exerts a great influence on the process of covalescence. In particular, it is
quite essential to maintain stable hold down pressure in the
area of the fracture under various kinds of external loads.
Three kinds of fixing aids were tested and the application
of model photo-elastic methods showed their advantages and
disadvantages. It has been proved that the best available
fixing is the Mini-Fixateur-Externe.

Keywords: Bone, fracture, photo-elasticity

1. Introduction

Photo-elastic research has been finding an ever increasing application on modelling of loads and stresses to be found in bone systems
and prostheses.
The research has been very helpful in getting answers to the following
questions:
- what is the relation between the character and the value of load,
  and the kind of fractured or broken bones.
- what loads and stresses exist in the area of prostheses application,
- what are the most advantageous solutions of prostheses with regard
  to the stresses of bone tissue, hence their impact on such biological processes as e.g. bone adhesion.
The last problem shal be described in detail on a specific example.

2. The methods of fixing fractured bones

Modern traumatology employs various kinds of osteosynthetic aids
such as wires, nails, screws and splints. Their function is to make,
jointly with the fractured bone, such system that will be as mechanically strong as a good bone. On the other hand, that system should
facilitate in every possible way the process of cohesion.
Satisfaction of these conditions requires a thorough knowledge of the
forces needed during the assembly of the system, the forces and stres-

ses following the assembly, as well as the forces and stresses that the system will convey under useful loads.

These provisions impose the following tasks: the definition of the optimum directions of screwing in the screws in relation to the plane of fracture, their depth and spacing. The currently applied methods of fixing skew and torsional fractures are either by screwing in the screws in the spot of fracture or beyond it, with the use of a splint. Both methods require an operation and they do not allow a precise setting of fractured bones. When the alien matter is grown over with the bone tissue, it is difficult to remove.

Problems also arise with open fractures where infection constitutes an additional hazard. Instability of these methods of fixing makes it also necessary to apply a plaster cast.

In 1975, Jaguet invented a new system of external fixing of fractured bones, called Mini-Fixateur-Externe (MFE) (1) (2) . His method consists in screwing in the screws some distance away from the fracture and then connecting their parts protruding above the body with a bracing element. The provision of this system is that the loads are conveyed through a "bypass" that keeps most of the loads away from the fracture spot, as well as it also stabilizes the tensions generated there.

3. Experiment verification

Photo-elastic model tests were carried out to verify the provisions assumed for the MFE system. Flat models were made of epoxy resin. To maintain the similarity of conditions, the bracing elements were made of magnesium alloy; in result the moduli of elasticity of materials of the real connection (steel-bone) and the model (epoxy resin-magnesium alloy) were quire similar. The scheme of the connections made according to MFE system is presented in Fig. 1.

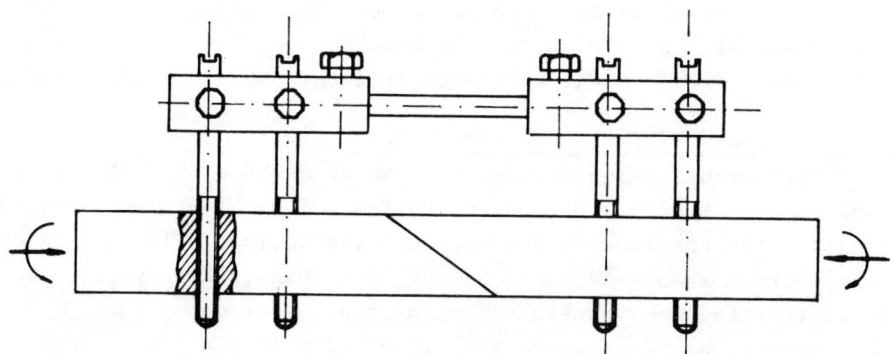

Fig.1 The scheme of connections made according to MFE system

The models were subjected to compression and bending in line with the possible loads that are generated in a bone due to the muscle work. Only the torsional forces were neglected, as the tests were conducted on a flat model. The method of multiplication isochromatic lines by special photo-processin has been used (3).

The tests indicated the change of isochromatics in the area of fixing for all three examples of bracing, caused by the change of loads. In the case of bracing with one screw only, the very screwing it in generates stresses that are unevenly distributed along the connection line (uneven hold down). Compression force and, in particular, the bending moment cause significant and unfavourable changes in the distribution of stresses (see Fig.2). That indicates that the hold down is unstable.

In particular, in the case of bending, the stresses in the spot of fracture grow on one side of the screw and they become smaller on the other side, what may lead to detachment of the two parts of the fractured bone.

The application of the second method of fixing (the use of four screws and a splint) produces a uniform hold down. Compression results in the growth of stresses in the spot that is most remote from the splint. Bending decisively changes the conditions, since the hold down significantly grows around the splin (see Fig.3). MFE fixing produces stresses from the screwed in screws in a significant distance from the fracture spot. Both parts of bone are uniformly held down. The most important thing is that the values of the holding down force remain practically constant during the changes of values and character of loads (see Fig.4). Theoretical calculations were to prove the obtained results. The scheme of the system in the form of a closed, rectangular frame loaded with the force P is presented in Fig.5. To simplify the procedure, the assumed model of the bone was a tube with the module E_k = 20.000 MPa, the diameter 18 mm and the wall 4.5 mm thick. MFE elements were given the following parameters:

E = 210000 MPa, the diameter of screws: 5 mm; the diameter of the connecting bar: 10 mm.

The dependencies listed below were obtained with the application of the energetic method:

$$M_a = -0.042 M + 0.778 P , \qquad N_a = -0.016 M + 0.973 P$$
$$N_b = 0.016 M + 0.027 P , \qquad M_b = -0.031 M - 0.827 P$$

where: $M = P \cdot e$

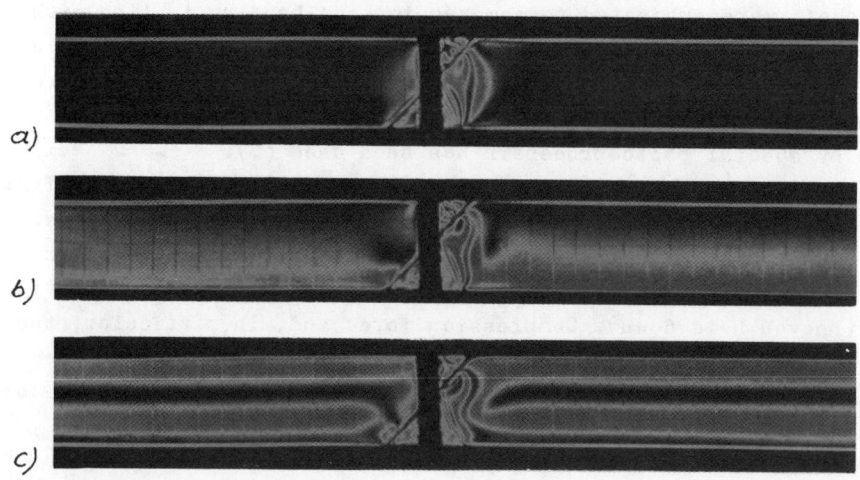

Fig.2   Isochromatic lines of one screw connection
        a) without loads, b) compression,   c) bending

Fig. 3   Isochromatic lines of 4 screws and splint connection
         a) without loads, b) compression, c) bending

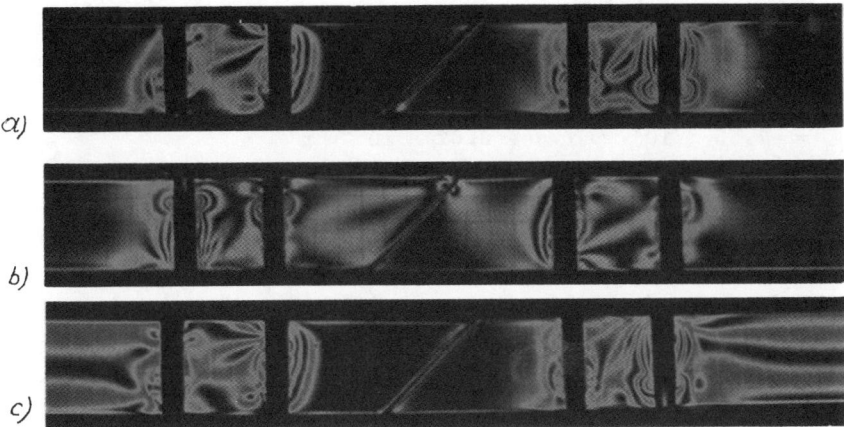

Fig.4   Isochromatic lines of MFE connection.
a) unloaded, b) compression, c) bending

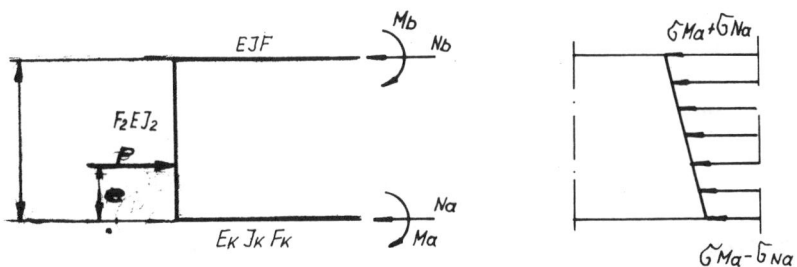

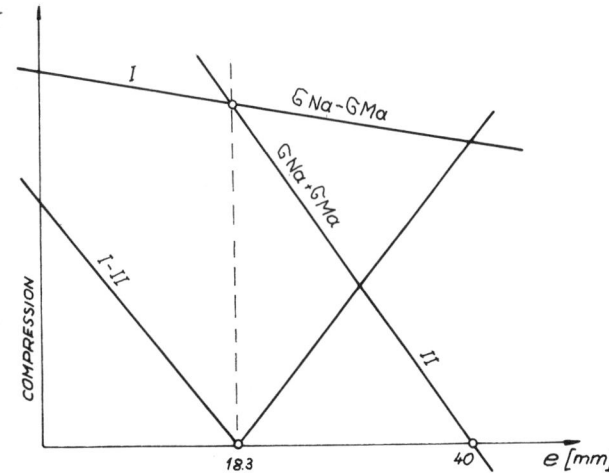

Fig.5
The scheme of loading

Fig.6
Changes of the extreme
stresses in the model
of bone in dependence
of load eccentricity
"e"

The stresses generated in the extreme fibres of the bone model section are described with the following dependencies:

$$\sigma_{Ma} - \sigma_{Na} = -7.55 \cdot 10^6 \cdot p \cdot e + 3.67 \cdot 10^{-3} \cdot p$$

$$\sigma_{Na} + \sigma_{Ma} = -1.66 \cdot 10^{-4} \cdot p \cdot e + 6.57 \cdot 10^{-3} \cdot p$$

Fig. 6 shows the changes of stresses in the extreme fibres of the bone model in dependence from the eccentricity of load "e". As it can be seen in the diagram, for eccentricity smaller than 40 mm compression stresses will be generated along the whole section of the bone. However, for the eccentricity P equal 18.3 mm, the obtained distribution of stresses is perfectly uniform along the whole section of the bone. Needless to say, that is of crucial importance for the success of the whole recuperation period.

Bending moment influence for changes of stresses in the bone is negligible what was proved in the experiment.

4. Conclusions

The photo-elastic model research promoted the explanation of the positive and negative characteristics of different types of bracing of the fracture spot. Within the three methods under analysis, the MFE system proved to be superior as it ensures a stable hold down, regardless the loads of the bone system.

The simplifications assumed for the model comprise the stresses (flat stresses); also the influence of the torsional moment has been neglected.

References

1 Ashe G., Burny F., Akt. Troumatol. 12(1982) 103-110.
2 Ashe G., Hanchirurgie 15.38-42(1983) Hippokrates Verlag GmbH.
3 Mazurkiewicz S., Legendziewicz J., proc. 1st Conference of
   IMEKO - Technical Committe 15, Czech.Plzen, May 25-28.1987

PHOTOELASTIC MODEL ANALYSIS OF SANDWICH BEAMS
AT DIFFERENT RATIO OF RIGIDITY OF FACE TO CORE

Doc. Dr Władysław Walczak - Dr Ing. Maria Kotełko
Mechanical Department, Technical University of Łódź
ul. Żwirki 36, 90-924 Łódź, Poland

The results of photoelastic investigations carried out to determine an influence of the ratio of rigidity of face to core on a stress distribution in sandwich beams are presented. Experimental results have been compared with the results of analysis based on two different theoretical models.

Keywords: sandwich structures, photo-elasticity

1. Introduction

One of the most important problems in designing the three-layered sandwich structures is selection of a proper relation between the stiffness of the faces and of the core. The ratios of elastic moduli E_f/E_c and of thickness h_f/h_c of face to core are the factors of the highest influence on the character of the collaboration of face and core and on the stress distribution in faces and core.

In the case when the stiffness of the core - in comparison with the stiffness of the faces - is negligible weak, in theoretical analysis we can assume that the bending moment is resisted by the faces alone, while the core is under shear only [1],[2]. In the case of some structural cores /f.ex. honeycomb construction/ and cores produced from advanced composite materials with high rigidity, the bending stiffness of the core can't be neglected.

The analytical solution for the three-layered beam with rigid core - under the assumption, that a part of the total bending moment is carried by the core - has been given in [3]. According to [4], the bending stiffness of the three-layered beam's core can be neglected, when $E_f h_f / E_c h_c \geqslant 5$. This criterion however hasn't been verified in the experimental way. If the faces are relatively thin in comparison with the core, the additional assumption may be considered in both above mentioned

IMEKO 1th TC15 International Conference, Plzen, Czechosłovakia, May 25-28, 1987 theoretical models. It is assumed, that the bending moment in cross-sections of faces is negligible.

Kemmochi and Uemura [5] undertook photoelastic investigations of three-layered sandwich beams for the purpose of determining the core's and faces participation in the bending moment-carrying. They investigated beams under pure, four-point bending.

The object of the paper was to determine in the experimental way an influence of the ratios of Younge's moduli and of thicknesses of face to core on a stress distribution in layers of the three-layered sandwich beam under three-point bending /fig. 1./. Photoelastic investigations have been carried out on two groups of models: with very soft, lighter core at the established value of the ratio E_f/E_c, but at different ratios h_f/h_c and – the other group – with the relatively rigid core, at different ratios E_f/E_c and h_f/h_c. The experimental results have been compared with the results of the analysis based on two three-layered beams theories under the assumption of lighter and of rigid core respectively.

2. Photoelastic investigations

Photoelastic investigations of sandwich beams on the eight models have been carried out. Three of these models – with very light cores – have been completely made from optical active materials. Faces of these models have been made from epoxy resin and cores – from epoxy resin with addition of a plasticiser. Ratio of Young's moduli E_f/E_c of these materials was constant.

Faces of the five left models have been made from thin sheet aluminium or steel sheet, and their cores – from optical active epoxy resin. Faces were glued with core using the same epoxy resin from which cores of the beam models have been made.

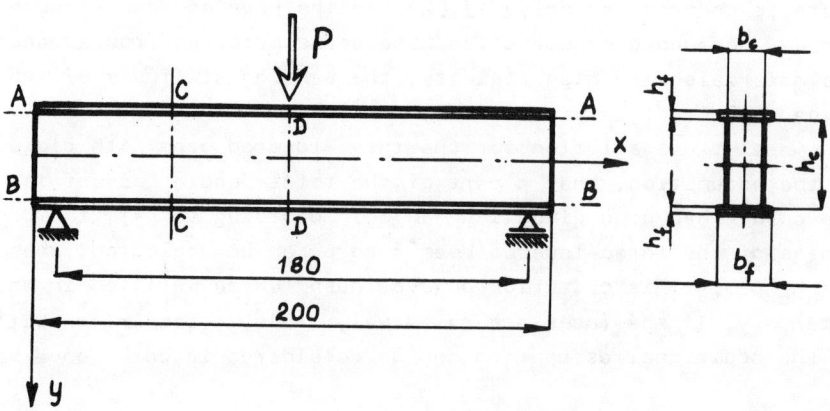

Fig. 1   Loading scheme of sandwich beam model

Dimensions of the investigated models and also properties of used materials have been put together in the table 1.

Table   1

Material properties and dimensions
of sandwich beam models

Nr of the model	Sizes				Elastic properties		Photoelastic properties	
	b_c	b_f	h_c	h_f	E_f	E_c	K_f	K_c
	mm				daN/cm^2		daN/cm^2	
1	7,5	7,5	40	10	27904	64,365	10,603	0,211
2	7,5	7,5	30	5	27904	64,365	10,603	0,211
3	7,5	7,5	24	2	27904	64,365	10,603	0,211
4	6	10	30	2,5	$2,1 \cdot 10^6$	21950	–	16,791
5	6	10	30	2	$2,1 \cdot 10^6$	21950	–	16,791
6	6	10	30	1,5	$2,1 \cdot 10^6$	21950	–	16,791
7	3	8	30	1	$2,1 \cdot 10^6$	27904	–	10,360
8	6	10	30	1,5	$0,7 \cdot 10^6$	21950	–	16,791

Photoelastic investigations of the sandwich beam models in the passing through light have been carried out. The isochromatic and izoclinic fringe patterns were registered. Tardy's compensation method has been applied at point measurements of the order of isochromatics.

The measurements of strains on the faces of models Nr 4 → Nr 8 have been also carried out. In these measurements the foil strain gauges have been applied.

3. Analysis of the experimental results

The components of the stress state in the model Nr 1 were determined on the basis of data, directly obtained from photoelastic investigations. However, an analysis of results of photoelastic measurements, obtained for the left investigated models caused some difficulties. These difficulties resulted from a lack of photoelastic data for faces of the models Nr 4 → Nr 8, and also from incomplete photoelastic data for faces of the models Nr 2 and Nr 3 with regard on a small thickness of these faces.

In order to determine components of a stress state the additional relations have been utilized. These relations – following from equilibrium conditions and from compatibility condition of strain on the boundary line between a face and core – have the next form for $y = \pm \frac{1}{2} h_c$:

$$\varepsilon_x^f = \varepsilon_x^c \; ; \qquad \sigma_y^f = \sigma_y^c \; ; \qquad \tau_{xy}^f = \tau_{xy}^c \; . \qquad /1/$$

Obtained results of photoelastic investigations and strain-gauge measurements – together with utilising the conditions /1/ – made it possible to separate completely the components of a stress state in the all investigated models. Examplary diagrams of distributions of the stress state components along the line of connection of faces with a core in models Nr 3 and Nr 7 in figure 2 are presented.

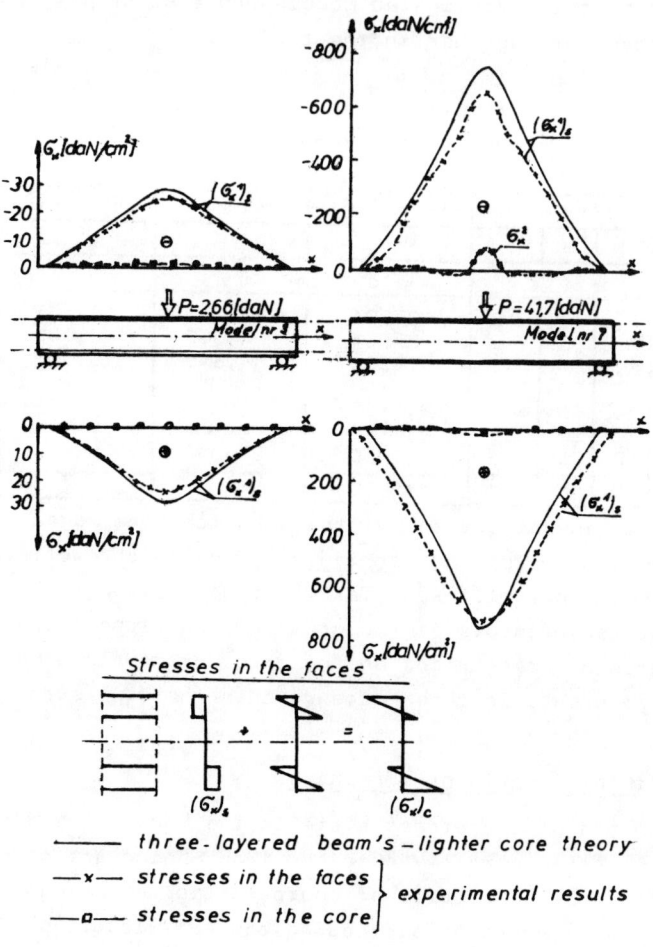

Fig. 2   Distributions of stresses σ_x
along the line A-A and B-B

Experimentally determined distributions of stresses in the cross sections
of the beam models permitted to determine values of the following moments: of the moment M_p - beeing a couple of forces be derived from resultants of normal stresses in the cross sections of faces, of own moments M_f of the faces and of a moment M_c in a core.

Mentioned above moments: M_p, M_f and M_c are the components of the
total bending moment M, transfered by the whole cross section of a beam.
Thanks to the knowledge of these moments it was possible to determine an
efficiency of a beam sandwich construction, defined by a ratio $\alpha = M_p/M$.
Examplary diagrams of the coefficient α, as the function $\alpha = \alpha(h_f/h_c)$ -
drawn up for the models Nr 1 ÷ Nr 3 - on the figure 3 are presented.

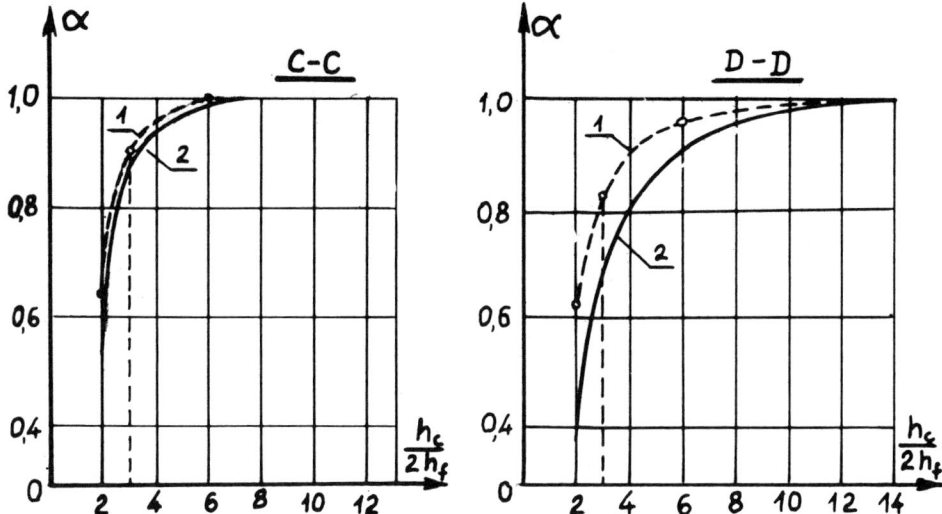

Fig. 3   Diagrams of relations $\alpha = \alpha(h_f/h_c)$ for the
models Nr 1 ÷ Nr 3: experimental curve 1 and
theoretical curve 2 – in the sections C-C
and D-D

4. Conclusions

Results of carried out investigations of sandwich beam models enable
to draw the following conclusions:
- flexural rigidity of faces of sandwich beams can be acknowledged as
  negligible small when ratio $h_c/h_f \geqslant 12$ – for sandwich beams loaded
  by a concentrated force in the way, shown in fig. 1;
- participation of a core in transfering of bending moment appeared to
  be important in the models Nr 7 and Nr 8, for which the ratio $E_f h_f/E_c h_c$
  is equal adequately, 2,5 and 1,6.

5. References

1   K. Stamm, H. Witte: Sandwichkonstruktionen Berechnung, Fertigung,
    Ausführung. Springer Verlag, 1974.
2   F.I. Plantema: Sandwich constructions. The bending and buckling of
    sandwich beams, plates and shells – John Wiley and Sons Inc., 1966.
3   S. Majewski: Zginanie wielomateriałowych ustrojów klejonych.
    Archiwum Inżynierii Lądowej – Tom XVIII Z. 1/1972.
4   Stateczność Konstrukcji Przekładkowych – Wyd. Politechnika Wrocław-
    ska, Wrocław 1972.
5   K. Kemmochi, M. Uemura: Measurements of stress distribution in sand-
    wich beams under four – point bending – Experimental Mechanics, Nr 3,
    1980

Photoelastic coating

APPLICATION OF PHOTOELASTIC COATING AT CONNECTING POINTS OF BUS UNDERCARRIAGE

Dr. Borbás, L. - Dr. Thamm, F.
Technical University, Budapest
Hungary

Bus undercarriages of three-dimensional frame design are usually designed computer-aided considering the frame consisting of 7-degree-of freedom beams. This type of computation does not take into account the actual shape of the beam connections and other transition points of external forces, thus neglecting the stress concentrations at these points. To get sufficient design data about the behaviour of such points, auxiliary investigations have to be carried out. As mounting loads affect the state of stress in mounted junction pieces considerably and to a more or less uncontrolled extent, only a measurement on actual specimens can be considered sufficiently reliable. As an example of such investigations the tests on two types of junctions are described.

1. Cast, one rod junction

The tested piece has to transmit compressive and brake forces developing on the carriage into the frame. The junction piece investigated is of cast type for the transmission of a rod force. It is mounted on the autobus frame by four locking screws. The position of the junction piece is determined by dowel pin on the autobus frame that is welded on the piece. Strain and pressure load affect the junction piece depending on push or brake operation. The tests were carried out at maximum rod forces which were determined by strain gauge measurements on the junction rod during operation.

First the stress developing from mounting the piece on the frame was determined. As the specimen is a structural element having relatively thick walls, fastening on the frame did not cause any significant strain. Then the effect of operation load was tested.

Critical points of the structure were around the dowel pin which is transferring the load – despite suitable fastening of the upper screws of the piece – on the autobus frame. Figure 1 shows the development of load along the critical cross-section.

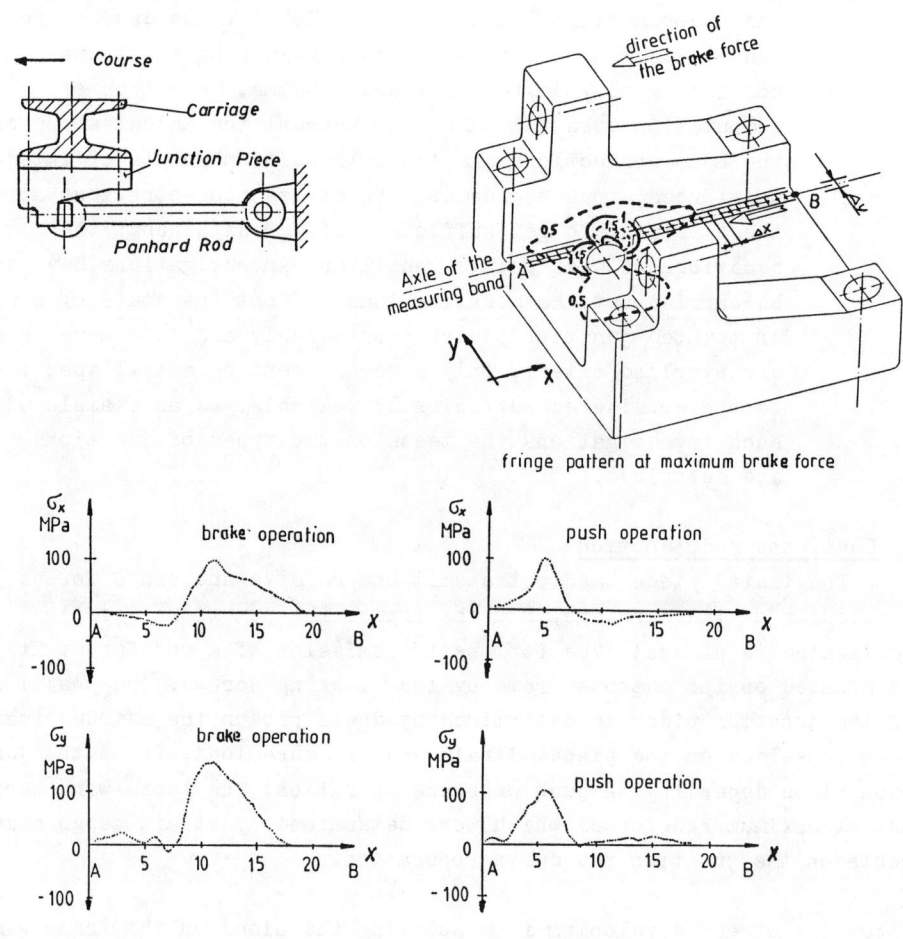

Figure 1

Fringe pattern distribution of the critical zone of the specimen at maximum brake force and the development of stress along the measuring band on the base of the piece

Measurement results have shown that significant load in other zones of the piece does not develop /except for the vicinity of rounding off radii/.

On the basis of the measurement results the construction of the structure was modified. The middle part of the base of the piece which contains the dowel pin was replaced by using fitted locking screws. Significant quantity of material was also saved by removing material from the shoulders of the piece. The measurement of the modified piece proved more homogeneous surface stress distribution.

2. Twin rod welded junction piece

The function of the tested junction piece besides the transition of compressive and brake forces developing on the carriage /longitudinal axis forces, Panhard junction/ is taking up vertical forces caused by the bend stabilizer. The junction piece is welded. Because of the forces acting in different directions its load is more complex than that of the junction piece mentioned above.

Isochromatic pattern originated by loads caused by mounting the junction piece on the frame and mounting of junction rods is shown in Figure 2. Significant loads developed in different surface zones of the piece.

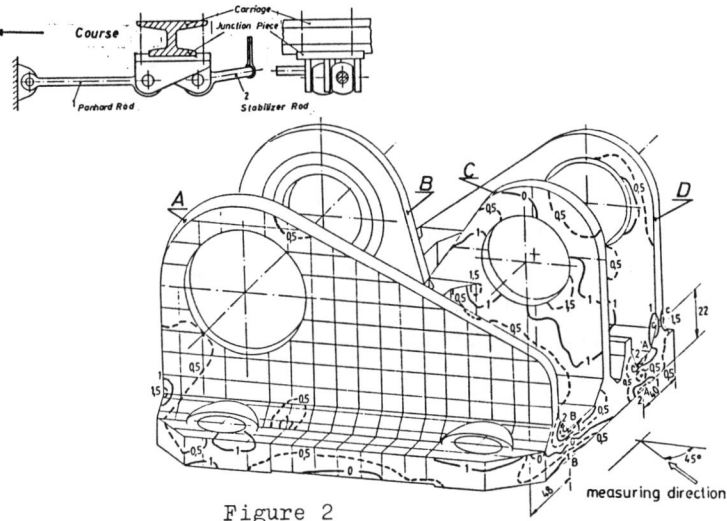

Figure 2

Fringe pattern resulting from the mounting of the junction piece and the rods in front view

Considerable load appeared on the base of the piece and in the vicinity of the welding seams of the ribs in different phases of mounting. They are shown in Figures 3 and 4.

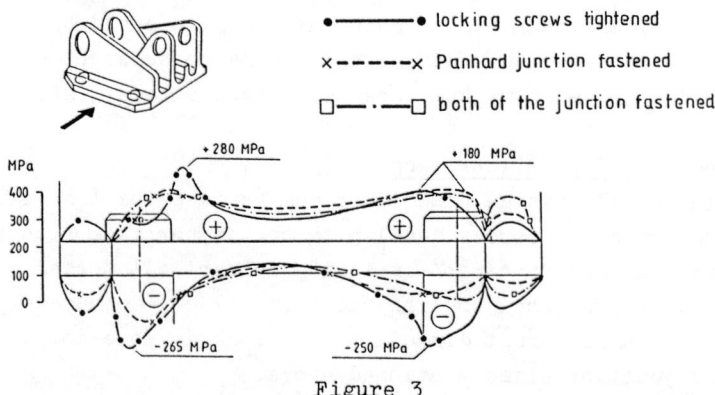

Figure 3

Load on the side of the base of the piece in different phases of mounting

Figures 2 and 3 show the majority of stress peaks at the edges of welding seam. Simplified control calculations carried out to determine stresses developing in the junction ribs during mounting the Panhard rod and stabilizer rod have shown considerable agreement with the measurement results.

It can be concluded from the results of the measurement and the checking calculation that the load of the welded piece is decisively determined by the fastening of the locking screws and the prestress appearing during connecting the rods. The stress peaks are near the yield limit. The stress distribution of the piece can be improved by the modification of the design using cast or forged elements.

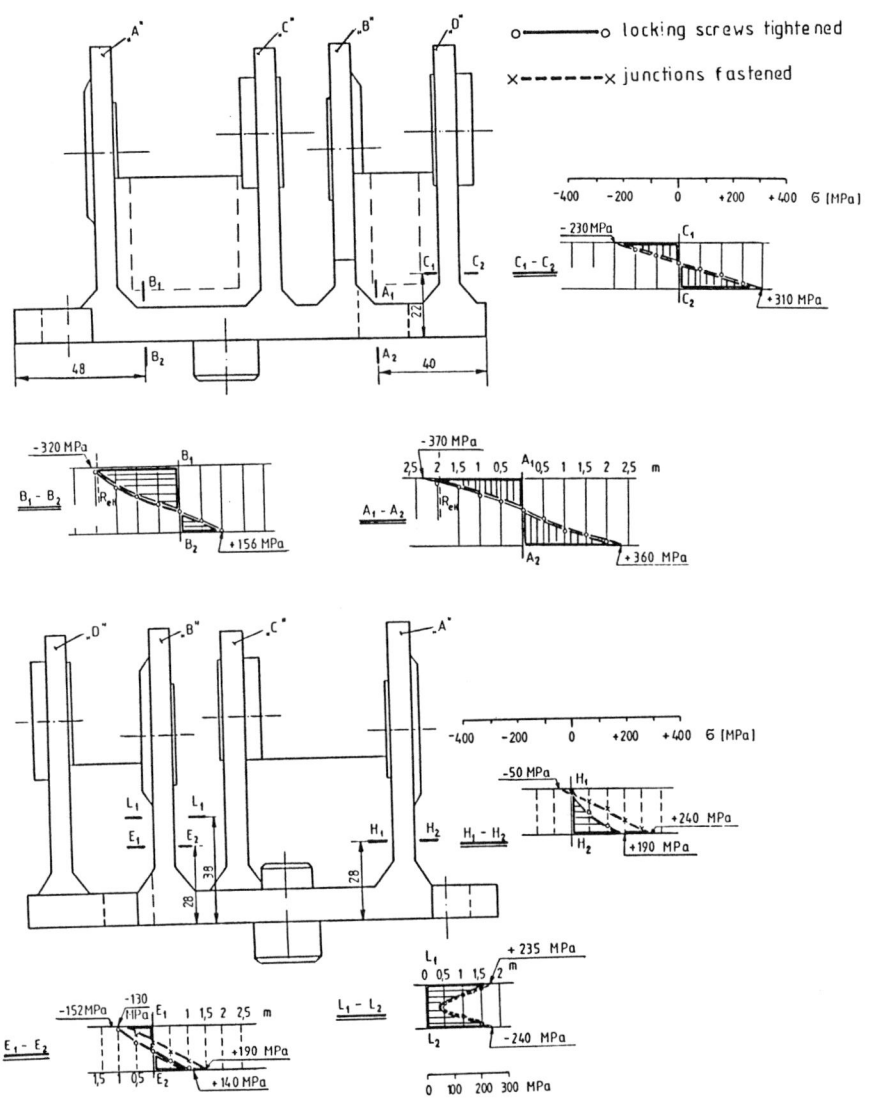

Figure 4

Load in different cross-sections of the piece in different phases of mounting

References

1   Thamm,F. - Borbás, L.: Checking the Shear Difference Method
                           in case of Photoelastic Coating
                           Proc. VIII. Int. Conf. on Experimental
                           Stress Analysis, 12-16 May, 1986
                           Amstredam

2   Borbás, L.: Photostress Analysis of Machine Parts and
                           evaluability of the measurement results
                           Gép. Sept. 1986. Budapest

3   Stress Analysis of Junction Piece
                           Measurement Report at the Department
                           of Machine Elements of Transport
                           Engineering, Technical University,
                           Budapest, July, 1986

Investigation of Local Stresses in the
Mast Frame Uprights of Fork Lift Trucks

Dipl.Eng.J.Videnova, Dr.Eng.K.Kostov,
Dr.Eng.V.Vasilev   Sofia, Bulgaria

The creation of lightier and at the same time more
reliable constructions depends on the precise knowledge of the strained and deflected state of the load bearing elements.Object of this
work is the determination of the stresses in the mast uprights
which are the heaviest elements of the lift mast. The investigation
were carred out on mast upright and lift mast samples. It is known
that by the strength evaluation of the mast uprights the following
stresses must be taken into consideration:
        - normal stresses due to bending moments Mx and My and
longitudinal forces Pz;
        - normal and tangential stresses due to primari St.Ve-
nan and secondary - buckling torsion.In relation to the torsion the
mast uprights are considered as thin walled ones with an open section;
        - tangential shearing stresses due to the transversal
forces Px and Py;
        - local Hertz stresses duo to the concentrated appli-
cation of the external forces by rollers;
        - local bending stresses in the supporting rollers'
action area;
        Most interesting are the local bending stresses, which
form the main part of the equivalent stresses in the section's most
loaded points. These stresses affect an area with a thickness equal
to the whole flange thickness and a length which is several times
larger than the width.
        The published investigation of local stresses - (4),
(7), (8) and others deal mainly with thin wall profiles loaded by point or line load. They are solving the problem by the methods of
the elasticity theory. The loaded area is considered as a thin elastic plate by suitable boundary conditions. The criterion for a thin
wall is that the relation between the flange thikness "h" and his
width "b" should not exceed 0,2. For the investigated mast lift uprights the relation " h/b " was from 0,35 to 0,54. In (6) is shown
that under these conditions it is possible to use the known equations for thin wall profils.

$$\sigma_{mx} = C_x . P / h^2 , \quad \sigma_{mz} = C_z . P / h^2 \qquad (1)$$

The coefficients Cx and Cz are determined experimen
tally. The main purpose of this work is the determination of these
coefficients depending on the wall thickness of the mast uprights -
the relation "h/b" depending on the load intensity - the parameter
"v/b".

53

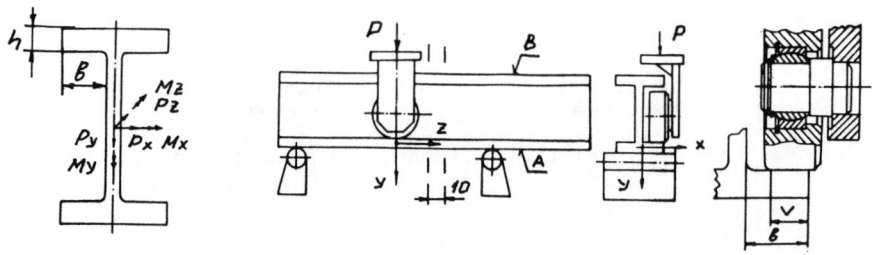

Fig.1   Fig.2

        The experimental works were carried out on test   samp-
les of mast upright profiles and lift masts. As test samples of mast
upright profiles were used 700 mm long I - profiles. They were laid
free on the roller supports of the bending test equipment of a test
mashine type " ZD 20 Pu ", Fig. 2.   The pressing column is provided
with a loading consol having a cylindrical roller,laid upon a self-
aligning bearing in order to assure a line loading. The local bending stresses obtained through these test conditions have values near
to the real ones and the part of the stresses due to bending and
mixed torsion is known and relatively small. The load width "v" was
selected by transversal shifting of the test sample whereby " v/b "
varries from 0,25 to 0,75. There were used tensil resistores type WG
12/025 - 160 with a base of 2,5 mm, which were placed upon the surface "A" in two lines at a distance of 10 mm along the axes Ox and
Oz. Through a longitudinal shifting of the test sample the load was
applied on different distances from the strain gauges. Through loading of all profil flanges it was possible to obtain a complete picture of the mast upright strained state. The relative deflections
 εx and εz due to a given applied force which is shown on
the test mashine's scale were registered by tensile amplifier type
"Hottinger KWS 3050" complete with a swich "UMK SO/D" and a printer
"KINZLE D-30". The same measuring equipment was used also for the
tensile research on the uprights of real lift masts as a part of a
more complete investigation including the determination of the section sizes in the connecting girder.
        Fig.3 shows the strain gauge arrangement on a lift
mast.

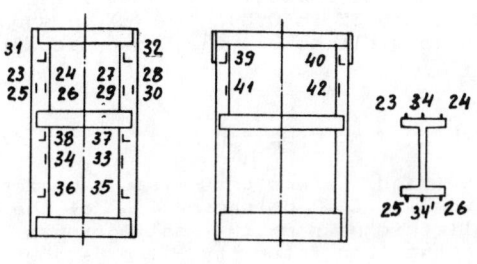

Fig.3

Points 33 - 34 and 41-42 are
designed as half bridges in
the mast uprights middle of
the two lift frames in order
to avoid local stresses due
to mixed torsion. Points 31-
32, 35-36, 37-38 and 39-40
are gauge pairs arranged in
the   rollers   action area.
Points 23-30 are single tensile resistors placed in the
mast uprights' edge area.

54

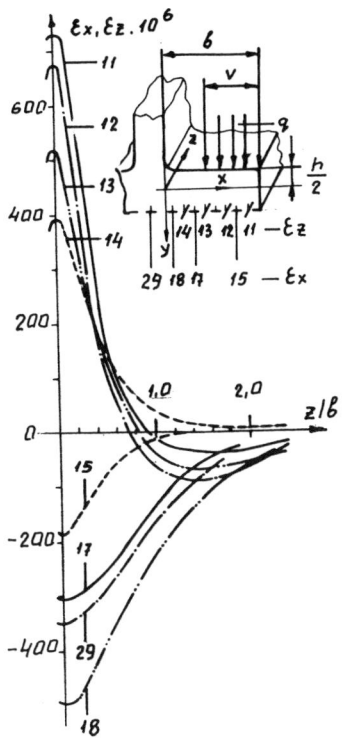

Fig.4

Simultaneously with the tensile research a photoelastic stress analysis on symetric areas of the mobil and immobil lift masts was carried out. The investigation was carried out with an equipment type " 030 Vishay" complete with a zero-compensator.

We used Bulgarian optically active coatings 3mm thick. The width was equal to the mast upright width and the length 100 to 150 mm. The total coating surface was about 0,06 m . In both investigations on real lift trucks were determined the stresses at several lifting hights, vertical and inclined lift mast and centrically placed load as well as with an eccentricity of 140 mm. The precision of the results obtained at the tensile investigations is 10 % and at the photoelastic stress analyses 5 – 10 %.

Fig.4 shows the relative deflections ε_x and ε_z in some points obtained from the laboratory testing on a mast upright sample depending on the application line of the force, (z = 0). The character of these curves remaines the same for all load widths " V "and for all test samples. In order to isolate the deflections ε_{mx} and ε_{mz} caused by the local stresses it is assumed that the deflections on the external wall of the not loaded flange (surface "B" on Fig.2) are caused by the bending and some mixed torsion for the force application point lies outside the bending centre. The local bending stresses are calculated using the known equations.

$$\sigma_{mx} = \frac{E}{1-\mu^2} (\varepsilon_{mx} + \mu \varepsilon_{mz})$$
$$\sigma_{mz} = \frac{E}{1-\mu^2} (\varepsilon_{mz} + \mu \varepsilon_{mx}) \quad (2)$$

By analogy with the thin elastic plates the stresses calculated using (2) were standardized with the value F/h. This way the values for C_x and C_z are obtained. Fig.5 shows these coefficients depending on the profil width in the load application line (z=0) for one of the test samples.

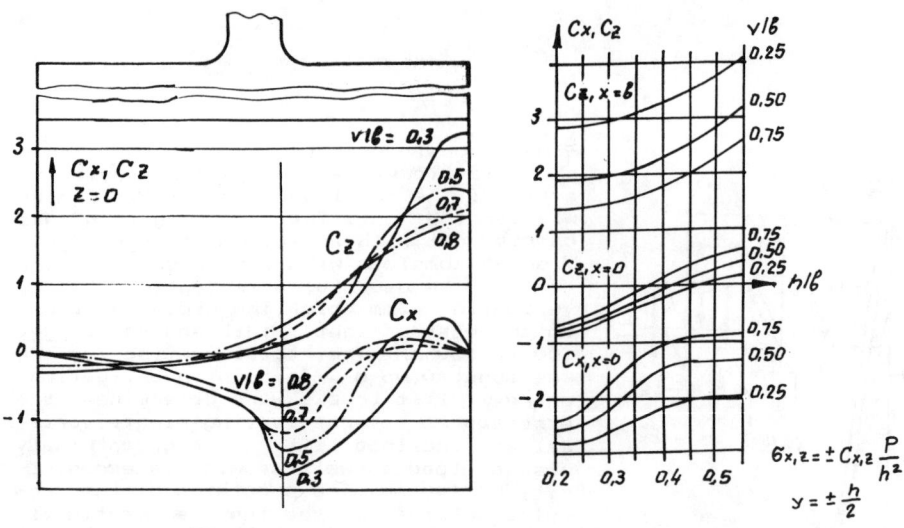

Fig.5   Fig.7

$$6_{x,z} = \pm C_{x,z} \frac{P}{h^2},$$

$$y = \pm \frac{h}{2}$$

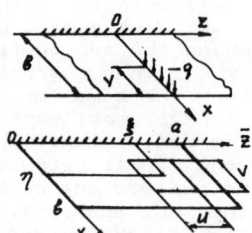

Fig.6

In order to receive a comparison with thin wall profil was drawn the solution of the diferential equation

$$\frac{\partial^4 w}{\partial x^4} + 2 \frac{\partial^4 w}{\partial x^2 \partial z^2} + \frac{\partial^4 w}{\partial z^4} = \frac{q}{D} \qquad (3)$$

in analytical form for boundary conditions, including one clamped edge (x = 0),(Fig.6), one opposite free edge (x = b) and free supporting of the rest two sides.The general solution concerns a load wich is uniformly distributed upon a square surface with the dimensions "U x V" and is arbitrary situated on the plate. The equations for the line load are obtained as a special case through a boundary transition u -> 0. For b/a -> 0, ξ = a/2 we receive the solution, corresponding to an infinitly long plate in direction Oz loaded with a line load at (z = ξ). Using the method described in (4) the calculation of the moments was reduced to the determination of the values of an infinite integral. For the clamped side (x = 0) and free edge (x = b) we received the following values of the coefficients at ($\bar{z} = \xi$) :

x	v/b	0	0,25	0,50	0,75	1,00
o	Cx	-3,06	-2,81	-2,60	-2,44	-2,30
b	Cz	–	2,866	1,913	1,395	1,057

We used these coefficients as beginnig values for h/b < 0,2 in wor-

king out the diagram on Fig.7 where the curves have horizontal tan-

gents. This figure shows the variation of C_x and C_z at ($x = 0$) and
($x = b$) depending on the wall thickness of the profile which is
characterized by " h/b " and by the intensity of the applied line
load beginning at the free edge, expressed by the parameter " v/b ".
       Using the data from Fig. 7 and the known expressions
laid down in (1) for the mixed torsion of open sections we calculated the stresses in the mast upright's characteristic points. Comparing these stresses with the measured ones by means of the tensile
method and the photoelastic stress analysis on real constructions,
taking into consideration the real distribution of the load upon the
left and right side rollers we found that the diference (the error)
is of the order of the measurment error.
       Resulting from these research the following conclu-
sions can be made:
   1. The influence of the increasing of the flange thickness of
      I-profiles in the range $0,2 < h/b < 0,54$ on the local ben-
      ding stresses σ_{mx} and σ_{mz} could be summerized as fol-
      lows:
      1.1. The following was established valid also for thin wall
           profils:
           - σ_{mx} and σ_{mz} act upon a length along Oz which is
             4 to 5 time bigger than the width "b" . The maximum
             values occur at the load application line,($z = o$);
           - In the section ($z = 0$) the maximum values of σ_{mx}
             are at the transition point from the flange to the
             stem and of σ_{mz} – at the free edge ($x = b$).
             These two points are decisive in the strength evalu-
             ation.
           - σ_{mx} and σ_{mz} in opposit points of the loaded fla-
             nge ($y = \pm h/2$) are practically equal as absolute
             values and have contrary signs;
           - σ_{mx} and σ_{mz} affect the whole flange thickness "h"
             and the exceeding of the proportoinality boundary
             could lead to plastic deflections of the flahge. For
             this reason they must be considered equaly to the
             stresses due to the general bending and mixed torsion.
      1.2. In contrast to the thin wall profils the following was
           established:
           - In the transition point from the flange to the stem
             ($x = 0$), $\sigma_{mz} \neq_M \sigma_{mx}$ and the difference is conside-
             rable.With increasing of "h/b" σ_{mz} changes its sign.
           - In this same zone C_x decreases with increasing of
             " h/b".
           - At the free edge C_z has larger values which increase
             with increasing wall thickness. Nevertheless σ_{mz}
             in this point at a given width "b" does not increase
             because " h^2 " increases more rapidly then C_z;
   2. The magnitude of the technological tolerances concerning
      both the dimentions as well as the deviation from the sur-
      face disposition affects considerably the loading symetry
      of the lift mast at a symetricaly applied external load.
         The carrying out of experimental works on profil sam-
      ples with different sizes and configuration as well as the
      researches using the end element method will give the possi-
      bility for a precizing of the now obtained resultts and for
      a generalisation of their validity.

References

1   Власов В.З.,Избранные труды том II,Издателство академи наук
    СССР, Москва, 1963

2   Виденова Ж.,Якостно пресмятане на мачтите на повдигателните
    уредби на карите с отчитане на местните напрежения в профи-
    лите,Национална научно-техническа конференция, Пловдив 1985

3   Костов К.,   Славчев Ц.,   Якостно пресмятане на профилите на
    повдигателните уредби на карите,Национална научно-техническа
    конференция, Пловдив 1982

4   Папкович П.Ф.,Труды по прочности короєля,Судпромгиз, Ленин-
    град, 1956

5   Тимошенко С.П., Войновски-Кригер С., Пластинки   и   обслочки,
    Физметгиз 1963

6   Beisteiner F., Maisch E.,Die Beanspruchung in Gabelstapler-
    hubgerusten, Fordern und Heben, 6/1981

7   Jaramillo T.J., Deflections and momenents due to a concen-
    trated load on a cantilever plate of infinite length, Jour-
    nal of appl.mechanics, III 1950

8   Mendel, Berechnung der Tragerflancs-Beaspruchung mit Hilfe
    der Plattentheorie, Fordern und Heben, 14,15 / 1972, 13 /
    1970

STRESS MEASUREMENT OF SHEET CONSTRUCTIONS WITH OPTICALLY ACTIVE COATING

Ass. Prof. Dr. Eng.   Árpád Zsáry
Department of Machine Parts, Faculty of
Transportation, University of Technology
Budapest H-1111 Hungary

One type of the stress measurement processes is
represented by optical stress analysis which is
carried out with the help of an optically active
coating put on the surface of an actual machine
part or sheet construction by using a reflection
polariscope. A special coating material was developed for the investigations and using it measurements were carried out on a part of an autobus
frame to determine their stress distribution and to
develop a more favourable layout of the construction.

Keywords: optically active coating, autobus frame

1. Introduction

The calculation, dimensioning work can never be separated
from shape determining activity in course of designing machine
parts.
In many cases the design engineer is justified to use some kind
of experimental stress analysis besides different calculation procedures /e.g. finite element methods/ by coordinating calculation
and measurement.
From the viewpoint of the design engineer, engineer, however the
use of any method of experimental strength analysis may be necessary in different phases of the design process of a given machine
part.
The design engineer, engineer however not always have to determine
the actual magnitude or stresses and strain developing under load.
Testing the developing stress distribution, stating the magnitude and
location of stress peaks, determination of the direction of stress
and clearing how homogeneous the stress state can be regarded give
enough information for them. On the basis of this information the construction can be modified and the ready modified spare part can be
tested again.

2. The measurement method

The surface strain and stress measurement method carried out by optically active coating fixed on the surface of the actual structural element belongs to to optically active coating methods of experimental strength and strain analysis. The measuring gauge of the optically active coating method is the reflection polariscope that can be used for the determination of strain in the coating developing under load and for the determination of stress distribution.

An important means of measurement is a coating having reliable parameters. A coating having suitable measurement technological parameters made of Hungarian basic materials has been developed at the Department of Machine Parts, Faculty of Transportation, University of Technology, Budapest.

Artificial resins having two components can be used as coatings. An important requirement is the possibility of the interruption of the polymerization process of the artificial resin mixture. The coating is epoxy resin binding material of 360...500 mole mass with 160...220 epoxide equivalent that are hardened with cycloaliphatic or aromatic amino, their adducts or polyamine. This layer is casted from the casting resin mixed in this way in semipolymerized state and this semi produced material can be stored for practically unlimited period at − 18...−20 $^{\circ}$C. Before usage this thin layer becomes plastic again at room temperature and can be shaped on the spare part to be coated. In full polymerization state it is hardened, can be fixed and is ready for measurement. The coating is to be calibrated by bending specimen to state the numerical values.

Young's modulus of elasticity   : E = 2850 MPa
Poisson number   : ν = 0.41
Strain optical factor   : F_{ε} = 1070...1380 mm/mm fringe

The coating method uses a relatively simple evaluation procedure until the determination of stress measured along the free edges and strain values meet the requirement. If the numerical determination of stresses in an arbitrary point is however necessary different approximation evaluation methods or the numerical integration of balance equation known from literature can be used.

3. The investigated nodal point and the loading equipment

Several methods are known in literature for the determination of welded frame structures, stress distribution of the bar stress of

autobus frames. The stress distribution of a tested nodal point can
be determined by the method of finite element but the suitable refinement of a finite element network takes long machine time. Thus the
surface coating measurement procedure seems to be very useful and
provides substantially more information within short time. Figure 1
shows the test of the critical nodal points of the door frame of an
autobus in simplified sketch. Load developing in the nodal points
was tested from load generated by dislocating one of the bars of the
structural unit welded from pressed and bent sheets of closed section.

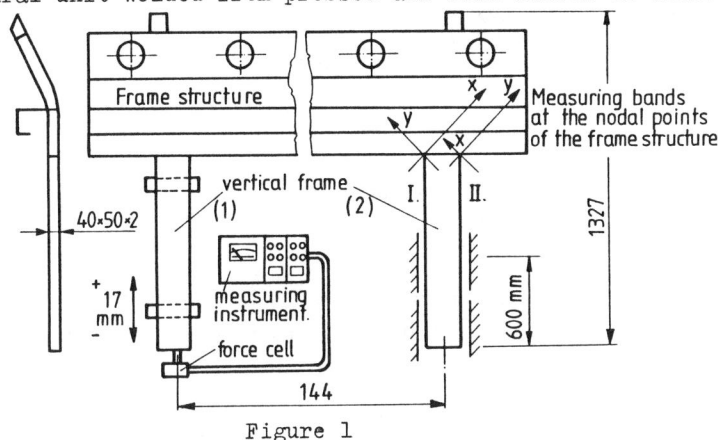

Figure 1

Load was developed by ± 1 7 mm displacement of door frame 1 in
vertical direction. Displacement was measured as the basic point of
the door frame on the transition point of external force of load cell.

Besides the load of the vertical frame in vertical axis direction it
was necessary to investigate the effect of load causing displacement
from its own plane that is the bolt stress affecting vertical frame 2.
The tested item is a cold drawn closed section of material quality
A 38, the minimum yield point is R_{eL} = 245 MPa and the maximum is
R_{eH} = 270 MPa.

4. Measurements and evaluation

The door frame was investigated horizontally. To avoid displacement from its own place because of dead load rolling support was used.
Force belonging to existing displacement was determined by load cell
and a gauge /DAFO comparator/.
Along measuring bands I. II. /Figure 2 / in case of positive displacement of the door frame /+17 mm/ and negative displacement /- 17 mm/

full evaluation was carried out i.e. stress components σ_x, σ_y and
the main stress components / σ_1, σ_2/ in certain points of the measuring bands were determined. Evaluation was carried out by numerical integration in the rectangular coordinate system of the basic equation
of the plane stress state.

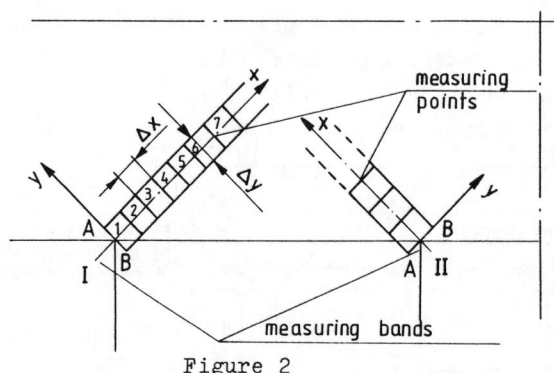

Figure 2

In course of the measurements two types of door frames were investigated. One of them was taken out of an autobus so it was exposed to load,
fatigue cracks in the nodal points were welded. The other one was a
new door frame.

5. Measurement results and conclusions

On the frontal area of the door frame along the measuring bands
I., II. stress values σ_x and σ_y were determined on the basis of
fringe number distribution. Figures 3 4 5 6 are given as examples
where the changes of stresses are shown along the measuring point in
case of fatigued and new door frames.

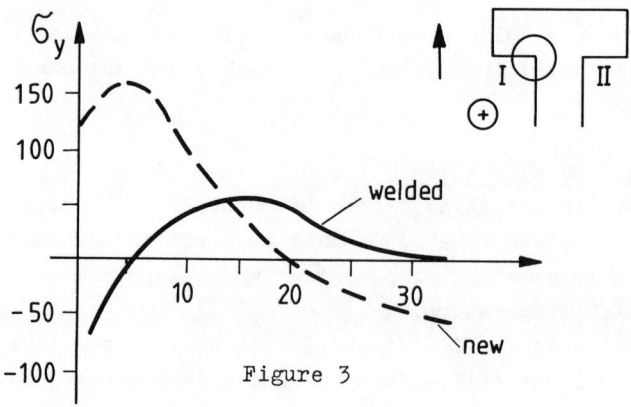

Figure 3

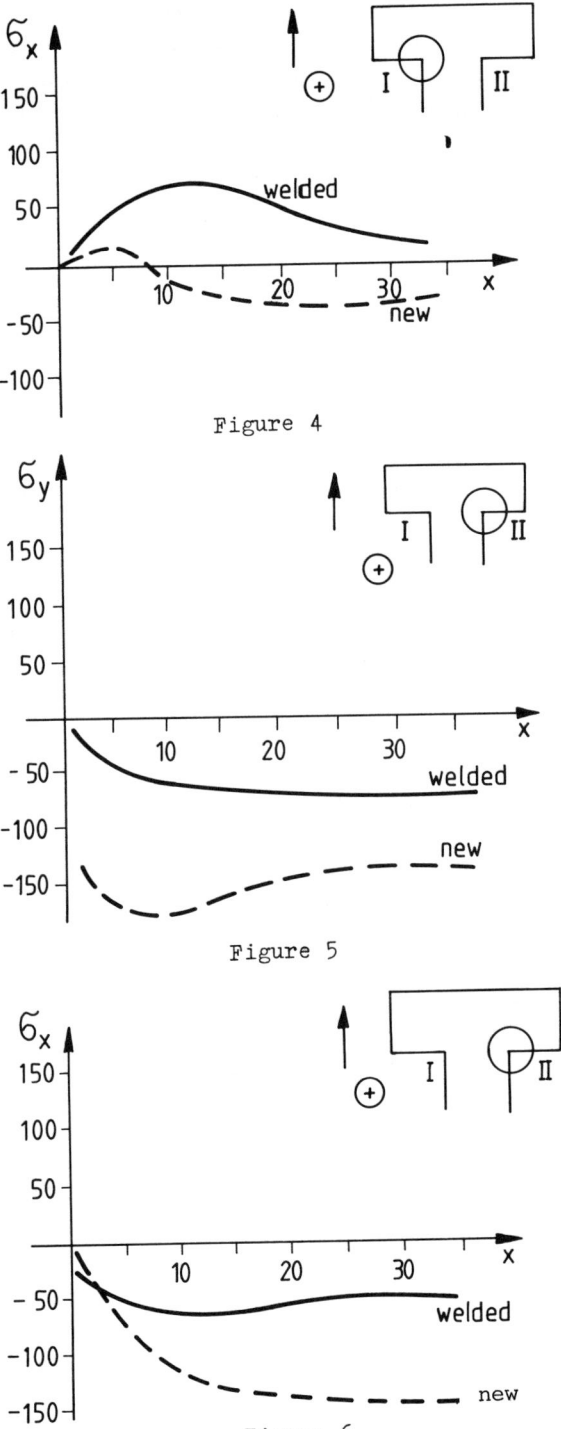

Figure 4

Figure 5

Figure 6

Considering all the measurement results the following can be concluded
- the measurement method is suitable for overall investigation of big surfaces, for evaluation of the construction by analyzing the surface stresses, for carrying out the necessary modification.
- the method is suitable for the investigation of shell body elements taking correction factor into consideration.
- using the method intact, cracked nodal points and ones with initial crack can be investigated by analyzing the isochromatic picture remaining in the coating after the disappearance of load.
- it seems to be expedient to decrease the rigidity of the structure in the vicinity of corner joints. In this way the stress peak appearing in the nodal points under load is distributed on a bigger area by increasing the permissible deformation. This modification would cause slight decrease of the rigidity of the structure but the value of stress peaks would decrease substantially thus the reliability of the structure would increase.
- the spanning, bored plate above the two door frames causes too great rigidity to the spanning structural element.If it has no other function in the frame the rigidity of the upper spanning frame could be decreased. In this way the value of the stress developing in the corner joints could also be decreased.
- besides the investigation of the frontal area of the support the side surfaces were also checked from the viewpoint of the development of stress distribution,on the basis of the conclusion from the investigation the peak stress measured on the side does not reach the value measured on the frontal area.
- the new frame was investigated for torsional stress too. On the basis of the conclusions from the measurement this load does not cause substantial measurable load in the tested nodal points.
- the stress values obtained along the measurement bands I. II. of the nodal points of the fatigued and new door frames were similar. Welding the fatigued cracks might cause the difference.

6. References

1 Borbás, L.: Basis and Experimental Possibilities of the theory of Surface Photo Stress Analysis. Department Publication No. 37. ISSN 0 139-16-15
2 Wolf, H.: Spannungsoptik 2. völligneubearbeitete Auflage. Band 1 Springer Verlag Berlin-Heidelberg-New York, 1976
3 Zandman, F., Redner, S., Dally. I.W.: Photoelastic coatings Ames Iowa, Society for Experimental Stress Analysis. Westport, Connecticut. 1977

Photoplasticity

APPLICATION OF PHOTOPLASTIC METHODS IN THE FIELD OF FORMING

Ing. Pavel Macura, CSc.
Iron and Steel Research Institute, Forming-Technology Dept
739 51 Dobrá, Czechoslovakia

The paper presents the principle of two photoplastic methods
of experimental analysis of stress and their usage for solution of the problems of stress fields in a billet when rolling by smooth rolls. The first method is a combination of
the photoplastic method of the surface foils with the classic planar photo-elasticimetry, while the second method is
based upon experimental determination of the tensor strain
field at forming a low-modulus epoxide resin and at consecutive evaluation of the stress field with the help of a theory of plasticity.

Keywords: photoplasticity, stress field, rolling

1. Introduction

The stress field in the zone of deformation at forming is one of
factors influencing substantially the formability of material at rolling. A precise analytic solution of the stress fields in a formed material is rather difficult due to nonlinear relation among the components of the strain and stress tensors in the zone of plastic deformations, due to intricated boundary conditions at the contact area of
the formed material with the forming tools and due to further problems. For this reason a more correct path was sought for making use
of the methods of experimental analysis of stress. Two experimental-
-calculating methods were proposed and utilized in which the fundamental knowledge on the changes in form and in tensor strain fields
is determined experimentally on the basis of photoplastic methods and
evaluation of the tensor stress fields is carried out analytically by
some of the theories of plasticity. Because of the complex character
and the expenditure of work at evaluation the entire solution of the
stress fields is made by computers. The principle of the two methods,
the procedure of solution, the mode of evaluation and some acquired

67

results of solution are given here after as example of rolling by smooth rolls.

2. Application of the photoplastic method of surface foils

The principle of method and of applied instrument is shown in detail in Fig.1. The lead specimens 2 are rolled between rolls 1 made of optically active material CR 39; the lead specimens have optically sensitive foil 3 stuck-on to the lateral area. The rolls refer to a laboratory rolling mill installed, instead of the common loading frames, between the polarization filters of a newly proposed transmission-reflection polariscope. In the course of the rolling process some interference phenomena encounter in the rolls and in the deformed foil and these phenomena are measured by means of optical methods of the experimental analysis of stress. The stress in rolls is measured by the principle of classic photo-elasticimetry where the light beams from a luminous source 4 are passing through the polarizer 5, the quarter-wave plates 6 and 7, the cylinders 1, the analyzer 8 and the interference phenomena in rolls are recorded by means of camera 9. For the sake of measurement of the strain fields in the stuck-on foil the instrument is equipped with the second light source 10 and the light beams are passing through the condensing lens 11, the analyzer 12, the quarter-wave plate 13, the foil 3 and are reflected from the lead-specimen surface. The reflected rays are passing through common quarter-wave plate 7 and analyzer 8 and the interference phenomena in the

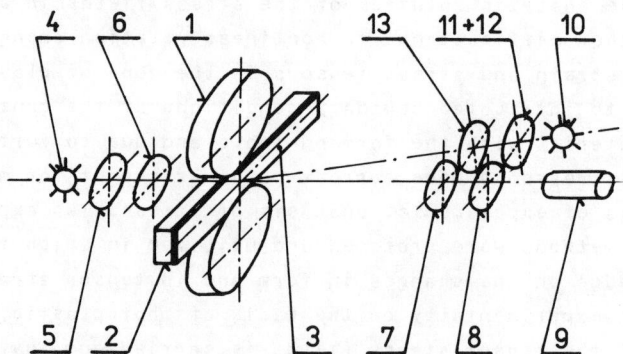

Fig.1 The principle of method and scheme of the applied instrument for measurement at rolling by smooth rolls

stuck-on foil are recorded together with the phenomena in the rolls
with the help of camera 9.

From the evaluated stress field in the rolls one has determined
the course of the contact stress between rolls and the formed material; this can be applied as the boundary conditions for solution of the
stress fields in a plastically deformed billet. The theory of elastic-
plastical deformations was applied here, and in similar manner the method and mode of evaluation of measurements are described elsewhere
[1], [2].

3. Proposal of method with application of a low-modulus optically active material

The above-mentioned method can be applied only at cold forming or
at application of spare materials such as lead used for simulation of
the hot-forming process. For analyzing the stress fields at hot forming our own experimental-calculating method was proposed, based on
forming of low-modulus optically active material. Experimentally the
tensor strain field is evaluated in formed material; then the tensor
stress field is analytically evaluated by means of one of the theories
of plasticity. This method is based upon the working hypothesis saying
the tensor strain fields to be approximately identical in the formed
low-modulus optically active material and in metallic materials subjected to hot forming. For example Fig.2 shows the shape of a billet

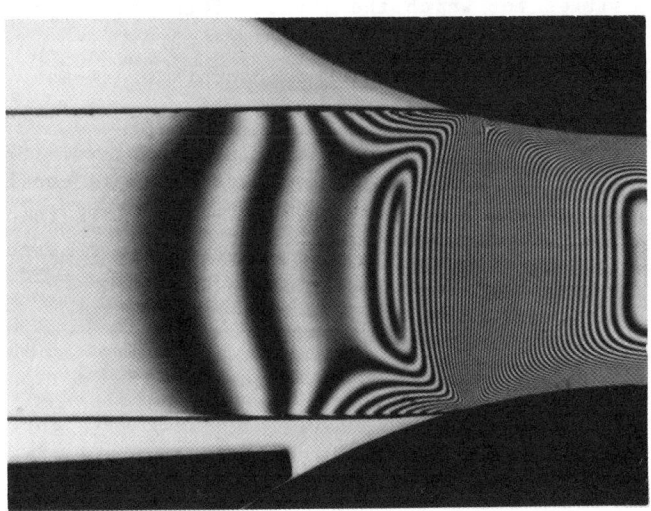

Fig.2   The isochromatic lines of semiorders in a billet

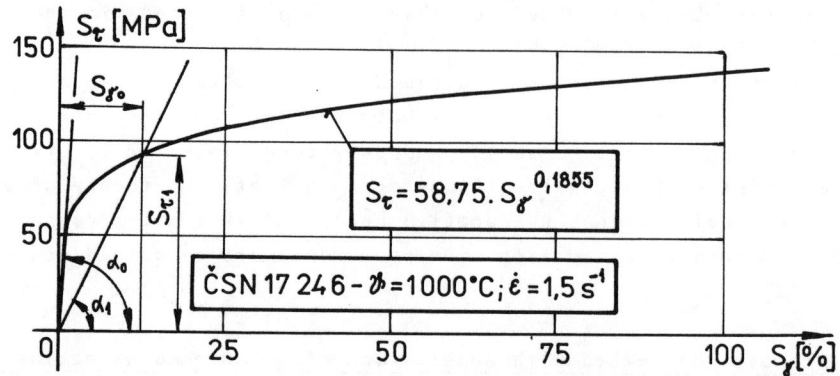

Fig.3 Curve of hardening of stainless steel to ČSN 17246-grade

and the course of isochromatic lines of semiorders at rolling a low-
-modulus material by smooth rolls. It can be seen here that the deformations originate already far before the inlet of rolling.

The tensor stress fields of various formed materials will be different and will depend on the curves of hardening of such materials.
Figure 3 shows the functional dependence of intensity of shear stresses S_τ on the intensity of shear strains S_γ , evaluated of the curve
of hardening of a stainless steel by Ref.[3]. This figure shows also the
thermodynamic conditions of forming and the analytic dependence of
such intensities for which the stress field was further evaluated for
the rolling conditions given in Fig.2.

As the relation between the components of stress tensor and the
strain tensor in the field of plastic deformations is not linear, the
theory of plasticity has to be used for evaluation of stress fields.
As relatively large zone of small deformations is available in front
of the input rolling plane as shown in Fig.2. This theory is based
among others on the principle that the deviator of stress D_6 and of
strain D_ε are similar and coaxial :

$$D_\varepsilon = \psi . D_6 \qquad\qquad /1/$$

In this respect one can suitably derive, with application of the
tensor number, the formula for calculation the difference of the principal stresses [4] :

$$(6_1 - 6_2) = \frac{2 S_\tau}{S_\gamma} (\varepsilon_1 - \varepsilon_2) \qquad\qquad /2/$$

This equation is the basic formula for calculation the stress fields according to the proposed method. The background for solution is both the dependence of the intensities of stress S_τ and strain S_γ, determined experimentally by Fig.3 and also the experimentally found difference of the principal relative strains $(\varepsilon_1 - \varepsilon_2)$ as follows :

$$(\varepsilon_1 - \varepsilon_2) = n.f \qquad\qquad /3/$$

where $\underline{n}$ is the order of an isochromate in the investigated point and $\underline{f}$ is the s.-c. value of order of applied low-carbon optically sensiti- ve material.

Stress separation is carried out by solution of the static condi- tions of equilibrium. As this is the case of a system of nonlinear partial differential equations, the solution is made numerically by the consecutive approach method. In the first approach a linear depen- dence is assumed among the components of tensors of stress and strain over the entire field of investigation the formed sample. Then, even the dependence between the intensities S_τ and S_γ is linear and is shown by a straight line with a gradient G_o , see Fig.3 :

$$\frac{S_\tau}{S_\gamma} = G_o = tg\,\alpha_o = const. \qquad\qquad /4/$$

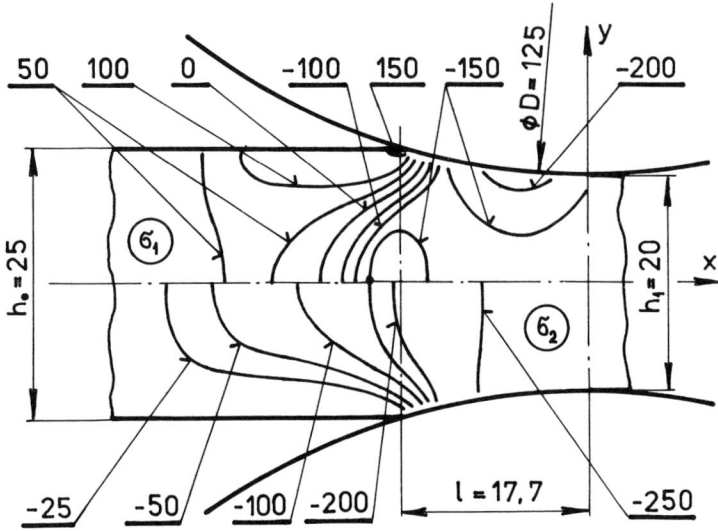

Fig.4 The equiscalar levels of the principal stresses in a billet

71

Under such preconditions and with the help of the method of shear-stress difference, the stress separation is made and the intensity of shear strain $S_{\gamma 0}$ is calculated. The relevant intensity of shear stress $S_{\tau 1}$ is determined by means of the formula given in Fig.3

$$S_{\tau 1} = a \left(S_{\gamma 0} \right)^{b}$$ /5/

and new line slope G_1 for further approach is

/6/

$$G_1 = \frac{S_{\tau 1}}{S_{\gamma 0}} = tg\alpha_1$$

The calculations are repeated with new line slope G_1 and the calculation is made so long till the desired accuracy is reached.

Figure 4 shows for example the results of evaluation of stress field for the case of rolling as illustrated in Fig.2. The stress field is shown plotted by the equiscalar levels of main stresses and it points clearly out the qualitative and quantitative distribution of the main stresses in the true zone of deformation [MPa].

4. References

1 Macura, P.: Application of the photoplastic method of the surface foils at solution of the problems of forming. Strojirenstvi, 32, Dec. 1982, No.12, PP.665-671.
2 Macura, P.: Penetration of plastic deformation at forming. Strojirenstvi, 32, 1982, No.6-7, PP. 346-351.
3 Fritsch, G. and Siegel, R.: Kalt- und Warmfliesskurven von Baustählen. ZIF 68, Karl-Marx-Stadt, 1965.
4 Macura, P.: Analysis of the stress fields at rolling of shaped billets. Transactions of the Mining and Metallurgical University Ostrava, in printing

MEASUREMENT OF THE INFLUENCE OF VISCOELASTIC RESPONSE

OF MATERIALS ON PLATES BY OPTICAL METHODS

Prof. Dr.-Ing. Karl-Hans Laermann

FB Bautechnik, BUGH Wuppertal

Pauluskirchstr. 7

D-5600 Wuppertal 2

In order to analyze plane stress-strain states considering time-depending
response of material experimentally, the method of photoviscoelasticity may
be used. Beside the mechanical properties, the optical response of material
must be measured also. How to measure and to evaluate the respective data
is described for modified epoxy resin, the time-depending response of which
can be varied according to the problem to be analyzed.

Keywords: viscoelasticity, photoviscoelasticity, material-testing

Rational design of structural components exhibiting viscoelastic response can
be done numerically as well as experimentally. If an experimental analysis is carried out by the optical method of photoviscoelasticity, not only the mechanical properties, but also the optical response of the model material must be considered. To
study the problems and effects of photoviscoelasticity, different high polymers can
be used. Pindera [1] has performed basic research in this field, but mainly to obtain informations on the reliability of photoelastic experiments. He did not look
for specially modified polymers to analyze viscoelastic phenomena experimentally.
Theocaris [2] however has looked for the essential mechanical and optical properties
of pure and plasticized cold-setting epoxy polymers. He came to the conclusion, that
such modified polymers follow to a satisfactory degree of accuracy a linear-viscoe-
lastic behavior, provided that the applied load quantities do not exceed certain
limiting values, i.e. stress and strain must be limited in a range that Boltzmann's
principle of superposition is still valid. The stress-optical relations then hold
[3], [4], [5]

$$\sigma_{12}(t) = \frac{1}{2} \left\{ C_N^*(0^+) \, g(t) - \int_{0^+}^{t} g(\tau) \frac{\partial}{\partial \tau} [C_N^*(t-\tau)] \, d\tau \right\}$$

$$\sigma_{11}(t) - \sigma_{22}(t) = C_N^*(0^+) \, \bar{g}(t) - \int_{0^+}^{t} \bar{g}(\tau) \frac{\partial}{\partial \tau} [C_N^*(t-\tau)] \, d\tau \tag{1}$$

with

where

$$g(t) = \Delta(t) \sin 2\psi_N(t), \quad \bar{g}(t) = \Delta(t) \cos 2\psi_N(t),$$

$$\Delta(t) = \lambda \frac{\delta(t)}{d_o [1 + \bar{\varepsilon}(t)]} \, .$$

IMEKO Int. Conf. MEASUREMENT OF STATIC AND DYNAMIC PARAMETERS OF STRUCTURES AND
MATERIALS, Plzeň/ČSSR, May 25-28, 1987

73

C_N^* denotes the optical relaxation function, $\delta(t)$ the isochromatic fringe order, ψ_N the angle of isoclinics, i.e. the principal axes of the refraction tensor, and $\bar{\varepsilon}(t)$ the lateral contraction.

How to determine the mechanical and optical response of the model material, especially the optical relaxation function C_N^* as well as the isochromatic fringes and the isoclinics as the loci of constant directions ψ_N, will be discussed in the following.

Usually uniaxial tensile creep tests with constant stress σ_{11} are carried out to determine the material properties. Neglecting lateral contraction, the optical creep compliance holds approximately

$$C_N(t) \approx \frac{\lambda}{d_o} \frac{1}{\sigma_{11}(o)} \delta(t) \, . \tag{2}$$

But because of laterial contraction, the cross section of the test specimen does not remain constant over time. Applying a constant load P, the uniaxial stress therefore does not remain constant also:

$$\sigma_{11}(t) = \frac{P}{b_o d_o} \left[1 - \nu(t) \cdot \varepsilon_{11}(t) \right]^{-2} \, . \tag{3}$$

However, as has been proved e.g. by Dill and Fowlkes /6/ and Weber /7/, Poisson's ratio may be assumed to be approximately constant over time $t : \nu(t) \approx \nu \approx$ const. With

$$k(t) = \left[1 - \nu \cdot \varepsilon_{11}(t) \right]^{-2} \quad \text{and} \quad f_1(t) = \lambda \frac{b_o}{P} \left[1 + \nu \cdot \varepsilon_{11}(t) \right]^{-1} \cdot \delta(t)$$

then the stress-optical relation yields /4/

$$f_1(t) = C_N(t) \cdot k(o) + \int_{o^+}^{t} C_N(t-\tau) \frac{\partial}{\partial \tau} \left[k(\tau) \right] d\tau \, . \tag{4}$$

This Volterra-integral equation can be solved in a discrete recurrent procedure for equal time intervals Δt, yielding the optical creep compliance

$$C_N(t_n) = \left[k(o^+) + \frac{1}{2} \Delta k_1 \right]^{-1} \left\{ f_1(t_n) - \frac{1}{2} C_N(o^+) \Delta k_n - \frac{1}{2} \sum_{i=1}^{n-1} C_N(t_i) \left(\Delta k_{n-i+1} + \Delta k_{n-i} \right) \right\} \tag{5}$$

$$\Delta k_j = k(t_j) - k(t_{j-1}) \, .$$

This formula demands equal time intervals Δt over the whole range of considered time. As for practical reasons the experimental data are taken at different time intervals, shorter ones at the beginning, longer ones towards the end of the measuring process, it is possible to proceed as shown in Fig. 1 for evaluation of eq. (5).

The optical creep compliance may also be determined by a shear test (Fig. 2). Then the fringe order $\delta(t)$ must be measured only

$$C_N(t) = \frac{\lambda}{d_o} \frac{1}{2\sigma_{12}(o^+)} \delta(t) \, , \tag{6}$$

74

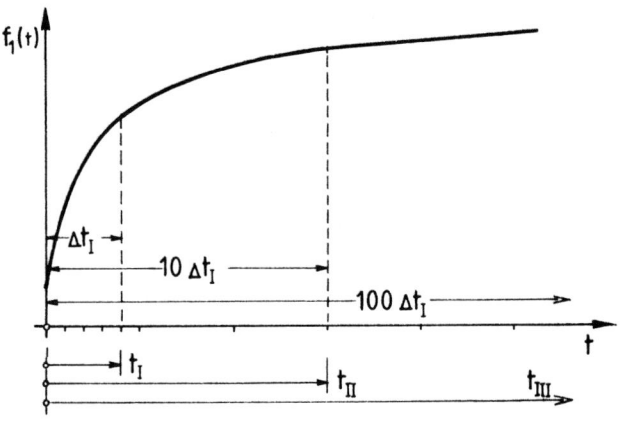

Fig. 1

Time intervals to integrate Volterra's integral eq. (4)

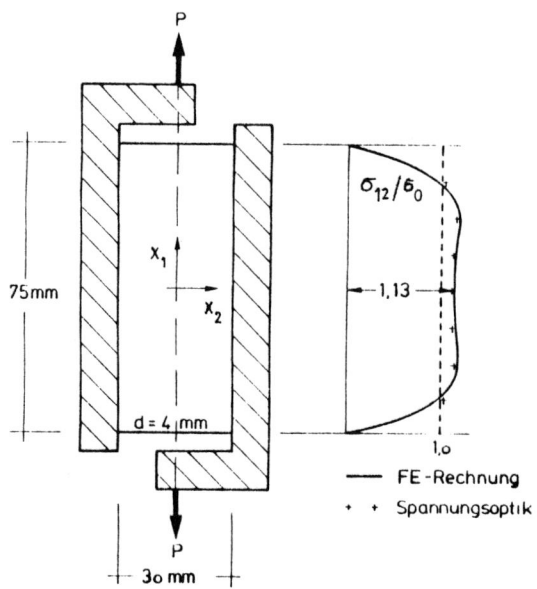

Fig. 2

Principle of the shear test to determine $C_N(t)$

as the thickness remains unchanged because of the pure shear stress state. Fig. 3 shows the results based on eq.s (2), (5), and (6) for one of the several mixture ratios investigated.

To evaluate eq.s (1), the optical relaxation function C_N^* is needed instead of the creep compliance. Between these two functions, the relation reads in the Laplace transform

$$C_N(p) \cdot C_N^*(p) = p^{-2} . \tag{7}$$

Considering the convolution to be commutative, the inverse yields a Volterra-integral equation of the second kind:

$$C_N^*(t) = \frac{1}{C_N(o^+)} \left[1 + \int_{o^+}^{t} C_N^*(\tau) \frac{\partial}{\partial \tau} \left[C_N(t - \tau) \right] d\tau \right] . \tag{8}$$

Discrete solution according to /8/ holds

$$C_N^*(t_n) = \frac{1}{C_N(o^+)} \left\{ 1 + \frac{1}{2} \sum_{i=1}^{n} \left[C_N(t_{i-1}) + C_N(t_i) \right] \left[C_N(t_n - t_i) - C_N(t_n - t_{i-1}) \right] \right\} \tag{9}$$

with the initial value of C_N^*

$$C_N^*(o^+) = 1/C_N(o^+) . \tag{10}$$

Results are given in Fig. 4.

Because of the fading memory of photoviscoelastic materials, the simple relation

$$C_N^*(t) \approx 1/C_N(t) \tag{11}$$

yields an approach of acceptable accuracy (Fig. 5), but generally not for all materials.

The measurement of material properties and the experiments have been carried out with modified epoxy resin ARALDITE F with hardener HY 951. To model viscoelastic response, two different plasticizers, ARALDITE CY 208 and THIOKOL LP3, in different mixture ratios were added. In order to determine the fringe order correctly and continuously over time in the tensile and shear test respectively, circular polarized He-Ne-laser light was used. The light intensity was measured by a photomultiplier and recorded by automatic plotting of the analogous signals. The material-testing set-up is shown in Fig. 6 and 7.

Special care must be taken during the loading process. The realization of step loading (Heaviside step function) experimentally is very difficult. To avoid dynamic effects, the load must be applied continuously from zero to the final value. This requires practically one second. Therefore it is impossible to take exact data of the different quantities at time t = 0. As an approach, the actual measurements are started in one or two seconds after having initiated the loading process; then the intial data at t = 0 are approximated by extrapolation, accepting some uncertainties

76

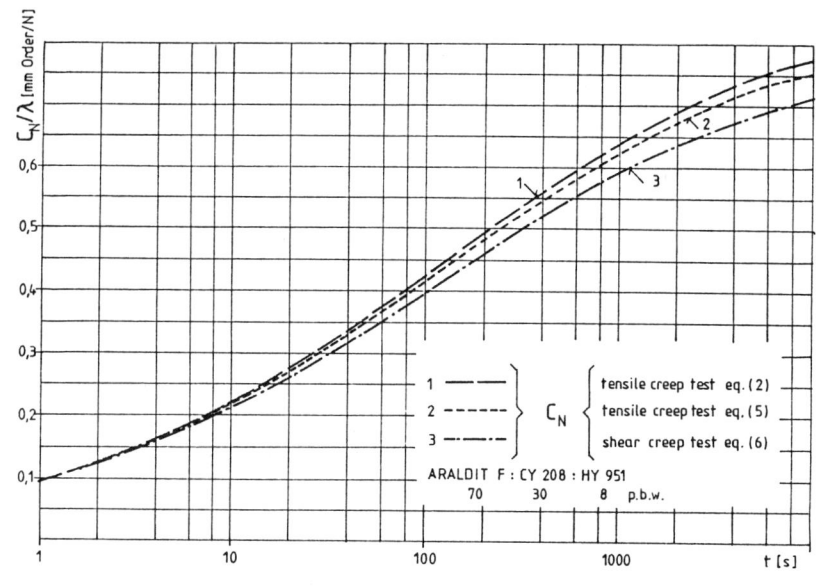

Fig. 3

Optical creep compliance $C_N(t)$ taken from different testing procedures

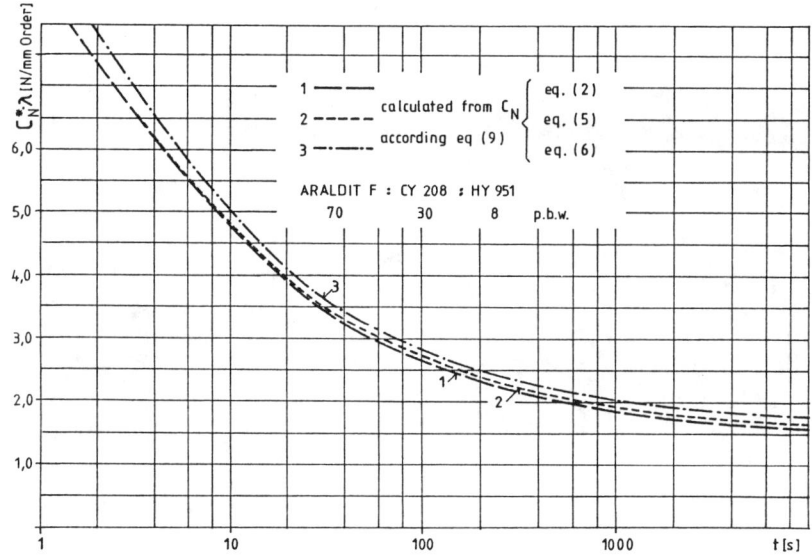

Fig. 4

Optical relaxation function $C_N^*(t)$ based on the results shown in Fig. 3

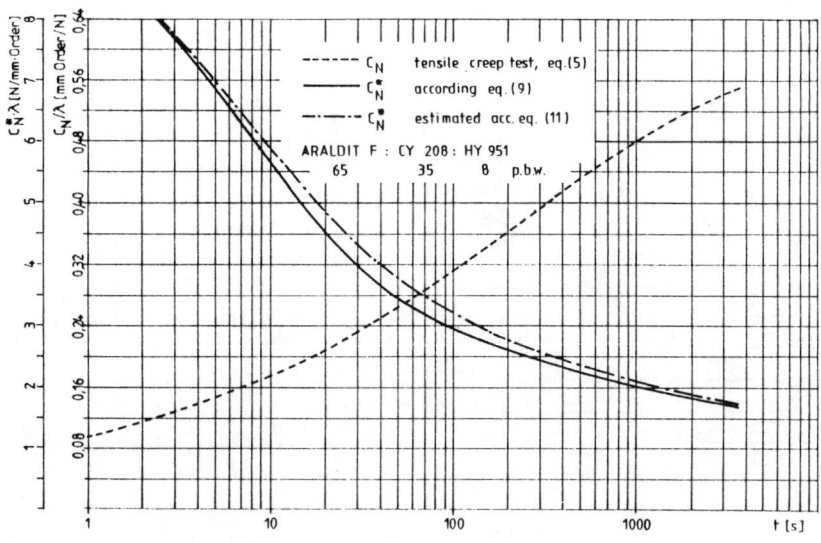

Fig. 5

Comparison of the exact optical relaxation function with estimated values

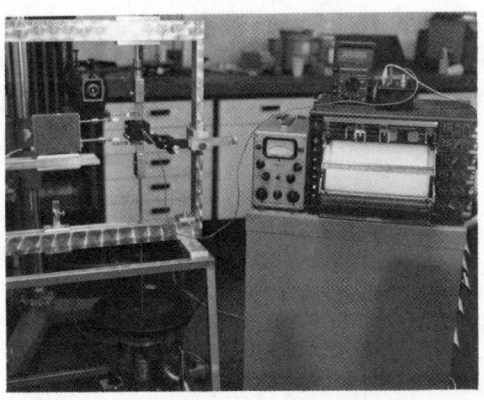

Fig. 6

Experimental set-up for the mechanical
and optical tensile creep test

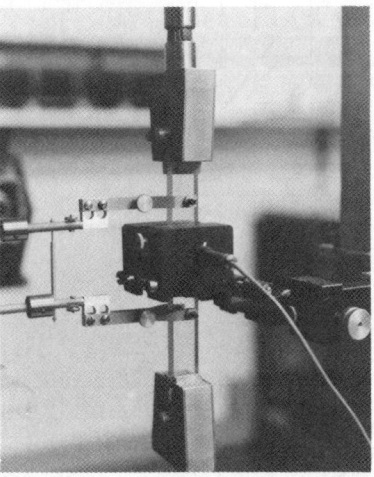

Fig. 7

Detail: Transducers for measuring
the longitudinal strain and the
optical retardation (isochromatic
fringe order)

the influence of which is decreasing in time due to the fading memory of the material.

To record the isochromatic fringe pattern at different time t_i in circular polarized light does not create any problems.

As the process of creep runs over some hours, photographic recording can be done as normal in photoelasticity. However, compensation is not possible especially shortly after load has been applied. Special care must be taken to keep the temperature constant during the whole measuring process, even avoiding the influence of thermal radiation caused by the light source.

Measuring of $\psi_N(t)$, the directions of the principal axes of the refraction tensor (isoclinics), is still a major problem. In numerous experiments it has been found out /5/ that the alteration of $\psi_N(t)$ is mostly in the range of accuracy of measurement. Only in the vicinity of isotropic points and in regions of the object with considerable changes in the strain state alterations of $\psi_N(t)$ must be observed. It has been proved that $\psi_N(t)$ can be determined by logarithmic interpolation having taken the respective data at time t_A at the beginning and at time t_E towards the end of the measuring process:

$$\psi_N(t) = \psi_N(t_A) + [\psi_N(t_E) - \psi_N(t_A)] \log \frac{t}{t_A} \left(\log \frac{t_E}{t_A}\right)^{-1}. \tag{12}$$

Fig. 8 shows the isochromatic pattern taken at different time t, and in Fig. 9 the isoclinics at time t_A and t_E are shown for a plate on an elastic subgrade. A mixed creep - relaxation problem has been considered.

Fig. 8
Isochromatic pattern from
$t_A = 5$ sec to $t_E = 1800$ sec

80

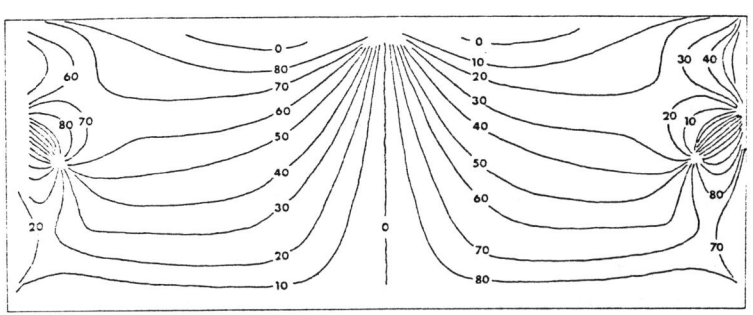

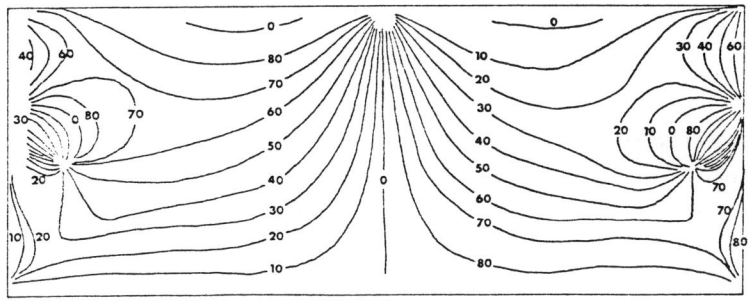

Fig. 9

Isoclinics at time t_A = 5 sec and t_E = 1800 sec

81

References

1 Pindera, J.T.: Remarks on Properties of Photoviscoelastic Model Materials. Exper. Mech. 6, 1966.

2 Theocaris, P.S.: A Review of the Rheo-optical Properties of Linear High Polymers. Exp. Mech. 5, 1965.

3 Coleman, B.D., and Dill, E.H.: Theory of Induced Birefringence in Materials with Memory. J. Mech. Phys. Solids 19, 1971.

4 Laermann, K.-H.: Problems in Application of Photoviscoelasticity in Practical Exp. Stress Analysis. Report-Labor f. exp. Spannungsanalyse u. Meßtechnik, BUGH Wuppertal, 1983.

5 Bakić, A.: Über die rechnerische und experimentelle Bestimmung des Spannungszustandes in elastisch gebetteten Scheiben unter Berücksichtigung viskoelastischen Materialverhaltens. Diss. BUGH Wuppertal, 1984.

6 Dill, E.H., and Fowlkes, C.: Photoviscoelastic Experiments. The Trends in Engineering 16, Univ. of Washington-Publication, 1963.

7 Weber, H.: Formulierung rheo-optischer Gesetze und deren Anwendung zur experimentellen Analyse ebener Spannungs- und Deformationszustände in Kunststoffbauteilen. Abschlußber. zu DFG-Forschungsvorh. WE 700/2-4, Univ. Karlsruhe (TH), 1981.

8 Lee, E.H., and Rogers, T.G.: Solution of Viscoelastic Stress Analysis Problems using measured Creep or Relaxation Functions. J. Appl. Mech. 30, 1963

Holographic interferomety,
            laser metrology

HOLOGRAPHIC EXAMINATION OF CRACKING IN CONCRETE

Dalakishvili G.L., cand.eng. - Nizharadze M.D.,
Institute of Structural Mechanics and Earthquake
Engineering, Ac.Sc. Georgian SSR,
1 Z.Rukhadze Str., Tbilisi - 380093, USSR

The effect of reinforcement distribution upon cracking in
concrete was studied using both the conventional method of
strain measurements with electric resistance sensors, and
the method of holographic interferometry. Examination of
the stress-strain state and cracking of concrete has proved
a clear advantage of the holographic interferometry technique which makes it possible to reveal cracks at earlier
stages of loading and obtain quantitative characteristics
of concrete surface in a contactless way.

Keywords: crack, stress, strain, interferometry, sensor

The strength criteria of most of reinforced structures is the
disturbance of an element integrity due to low tensile strength of
concrete, which actually specifies the work of the reinforced concrete under loading. The moment of cracking coinciding with exhaust-
ion of load-bearing capacity of concrete in the tension zone is the
principal factor determining the work of the structure.

The study of the processes of initiation of crack development,
the regularities of distribution of innternal forces under complex
conditions of work of a reinforced structure call for improved experimental techniques. It requires extensive experimentation to check
the current ideas on the behaviour of the material and to study more
thoroughly its qualitative and quantitative characteristics.

Test beams, 50 x 70 x 600 mm, were made of Portland slag cement
M 400, quartz sand and crushed aggregate of maximum particle size
10 mm. The compression strength of the concrete was 370 kg/cm^2.

One-side longitudinal reinforcement was made with class A-1
smooth reinforcement. Some of the test beams were reinforced with
1 rebar of d 12 mm, others with 2 rebars of d 8 mm, and still
others with 4 rebars of d 6 mm. The reinforcement comprised about

3% of all the test-beams. The protective layer was 1.5 cm in all the cases. To ensure destruction due to bending and not shearing, the support sections in the thirds of the span were reinforced with wire, d 4 mm. The beams were tested on two supports and were loaded with two concentrated loads in the thirds of the span (Fig.1). A special device was made for these tests to examine the butt-end opposite to that which accepted the load.

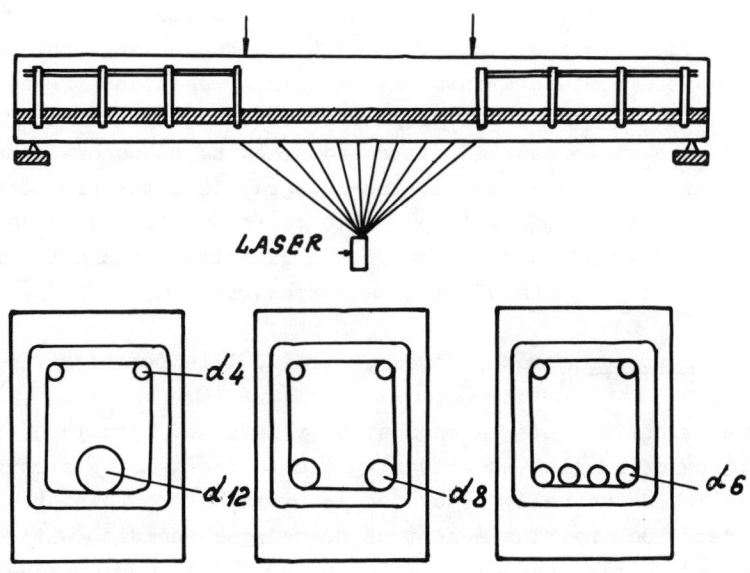

<u>Fig. 1</u> Reinforcement and holographic pattern of the test-beam

Besides the conventional method of strain measurements with electric resistance sensors, the method of holographic interferometry was used. The test-beams were examined using a He-Ne laser to get the Leight-Upatnieks hologram. The strains were measured on the 4 mm base in the upper, middle and lower parts of the beam height. The crack initiation zone part was examined and compared with its symmetrical part at the opposite end of the beam (Fig.2). In this case the crack initiation was studied in relation to the kind of strain, considering the number of rebars.

Fig.2 Examined symmet-
rical sections
of the beams

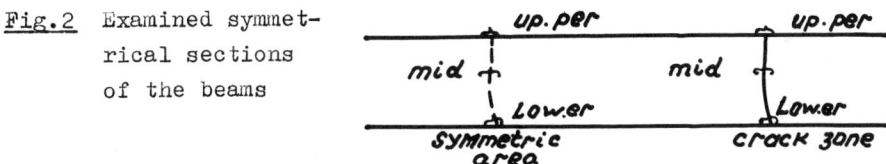

In the course of the experiments using the a.m. technique, the
load was added by 0.17 MPa increments (a preselected load to record
cracks under given forces).The experimental results lead to the following conclusions: the method of holographic interferometry revealed
cracks in the test-beam at earlier stages of loading. In beams with 1
and 2 rebars a crack was observed in the beam upper plane under 5.64
MPa load; in a beam with 4 rebars cracks formed under 5.13 MPa in the
same plane (unlike the other beams, cracking in a beam with 4 rebars
evolved along the whole height under greater loads). Fig. 3a shows
strains in the crack initiation zone before actual cracking, and strains in their symmetrical parts (Fig. 3b).

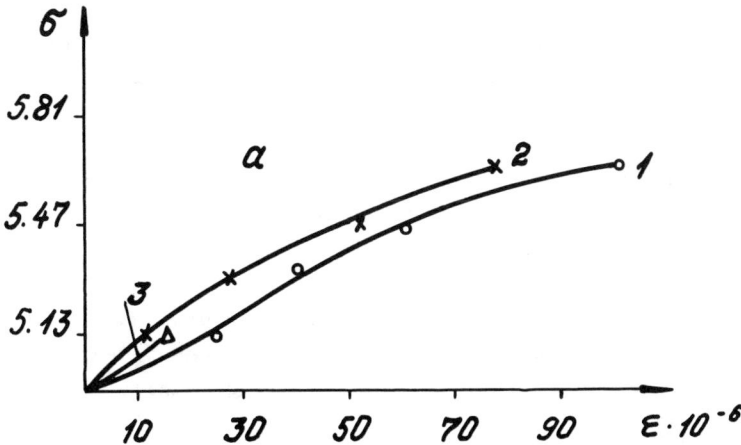

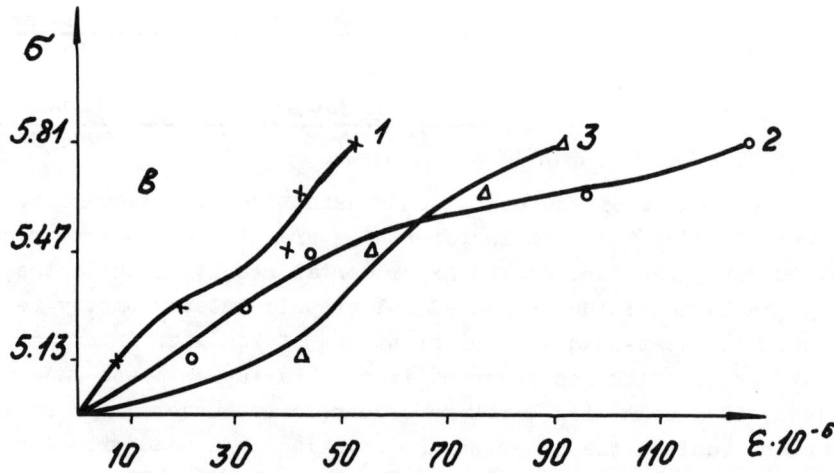

Fig. 3 Strains in the symmetrical parts of a beam
a) crack zone   b) symmetrical zone
1. beam with 1 rebar; 2. beam with 2 rebars;
3. beam with 4 rebars

When electric resistance sensors (20 mm base) were used to measure strains in the test-beams, the moment of crack formation was determined by the break in curve $\sigma - \varepsilon$ and recorded under higher loads than in the former case: in the beams with 1 and 2 rebars cracking was observed under 9.5 MPa loads, in the beams with 4 rebars, under 8.3 MPa loads.

Holographic interferometry was used to evaluate crack opening under constant load (for 1 hr) for all the reinforcement types, and provided the following results: one load value, 8.36 MPa, was preselected and maintained constant for 1 hr; i.e., the first exposure was taken after this load was applied; then, after 1 hr of loading the second exposure was taken, and thus an interferogram was obtained for a particular period of time under constant loading. Maximum opening of cracks was recorded in a beam with 1 rebar, while minimum one in a beam with 4 rebars (Fig.4). Crack opening is more intensive than under small loading steps. The a.m. load of 8.36 MPa is close to the load at which occurs the break in curve $\sigma - \varepsilon$ plotted with the help of electric resistance sensors.

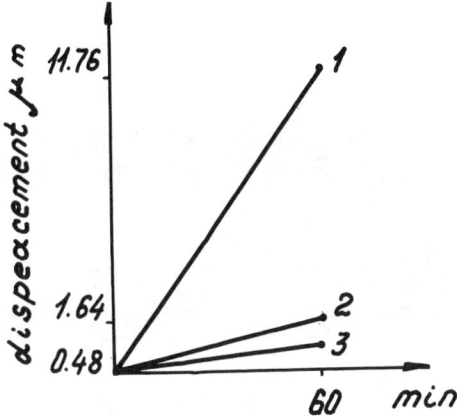

Fig. 4 Opening of cracks under constant load of 8.36 MPa
applied during 60 minutes
1. beam with 1 rebar
2. beam with 2 rebars
3. beam with 4 rebars

Fig.5 shows interferograms of beams under 14 MPa load. The examined part of the beams reveals clearly a certain number of cracks
which destray the interference band pattern, according to which one
can determine the number of units into which the examined section is
divided. Separate units seem to deform independently in the interferograms. It should be noted that of the cracks recorded in the a.m. interferograms, one occurred under low loads.

a) with 1 rebar

b) with 2 rebars

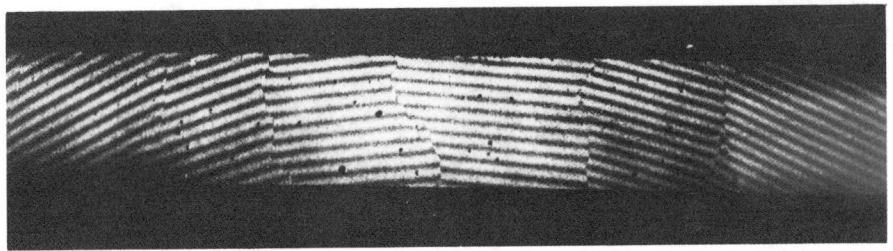

c) with 4 rebars

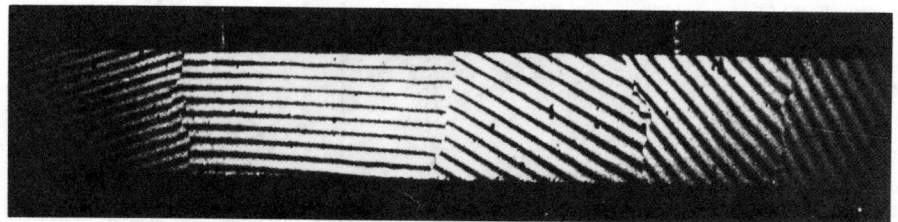

Fig. 5 Photographs of interferograms of beams cracked into
units under 17.2 MPa load

Using the method of holography in a property arrange experiment
and with a clear statement of the problem, one can follow the growth
of deformation in any point of a test piece under study, record a
crack at the moment it forms, prior to its further development, avoiding contact with the test-piece. The holographic interferometry
method provides better and deeper information on what is going on in
the concrete structure, than the method of strain measurements with
electric resistance sensors.

References
1 Collier R.J., Burchardt C.B., Lin H.L.: Optical Holography. Bell
  Telephone Labs., Murray Hill, New Jersey, 1971.
2 Tsilosani Z., Dalakishvili G., Kakichashvili Sh.: The Effect of
  Cement Composition on the Cracking Resistance. Proc. 7th International Congress on the Chemistry of Cement, Paris, v.III, 1980.
3 Tsilosani Z., Dalakishvili G., Kakichashvili Sh.: Shrinkage of
  Concrete at Early Stages of Hardening. Proc. Rilem International
  Conference on concrete of early ages, Paris, Apr.6-8, 1982, v.I,
  pp.71-75

SOME ASPECTS OF VIBRATION ANALYSIS BY LASER METROLOGY

Grosser,V.;Vogel,D.;Höfling,R.;Chmielewski,R.*;Meisel.U.*

Institute of Mechanics, Academy of Sciences,Karl-Marx-Stadt,GDR,9010
*Wilhelm-Pieck-University,Rostock,GDR,2500

Vibration analysis is a major field of application of laser metrology. This paper presents some experimental results and industrial applications, which are obtained by holographic time-
average methods, by time-average speckle metrology and by double-pulse speckle pattern photography.

Keywords: vibration analysis, laser metrology

1. Introduction

Laser metrology is a good tool for mapping and measurement of vibrations of industrial objekts. In literature can be found many different
methods, which were applied. We describe a holographic time-average
method for determination of the 3-D-amplitudes vector and two methods
for in-plane vibration measurements.

2. Vibration analysis by holographic interferometry

Holographic interferometry is a type of nondestructive testing. This
method is applicable in order to determine the natural frequencies and
the vector field of the vibration amplitudes from original structures
or components in mechanical engineering. In our case a blade of a
radial turbo supercharger has been inspected. In a first step the
natural frequencies of a blade were determined by the holographic
real-time method. In Fig.1 the results are shown.
The vector field of the vibration amplitudes can be found by holographic time average method. The used optical arrangement is schematically drawn in Fig.2. The experiments were done with a Ar-Ion-Laser.
The optical system has four directions of illumination and only one
direction of observation. Such a arrangement is well-suited for
computer aided evaluation of the interferograms, because of leypassing
perspective problems. The computer aided quantitative evaluation was
done by a method published by /1/. Fig.3 shows the results of the
quantitative evaluation for the most critical natural fequency.

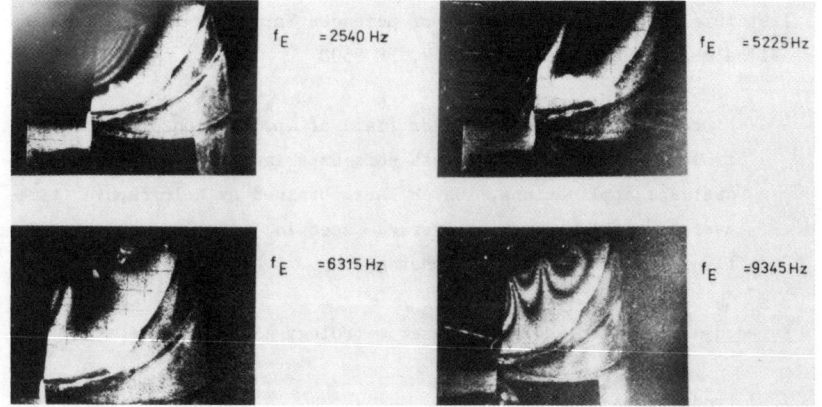

Fig.1 Natural frequencies of the blade

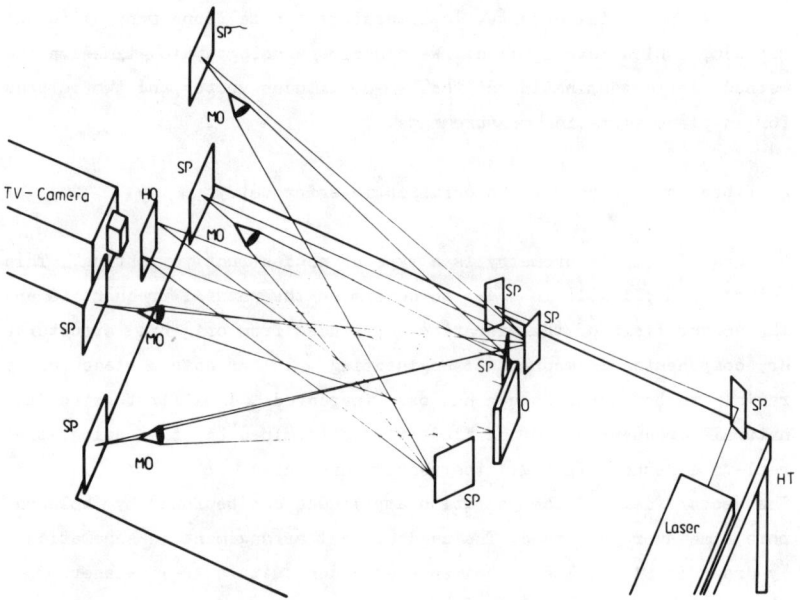

Fig.2 Scheme of the optical system
SP-mirror, O-sample, MO-lenses,
HO- holographic plate holder, HT- table

If the vector field of the vibration amplitudes is known, a computer

92

simulation is possible in order to determine the strain and the stress
in the turbocharger blade.

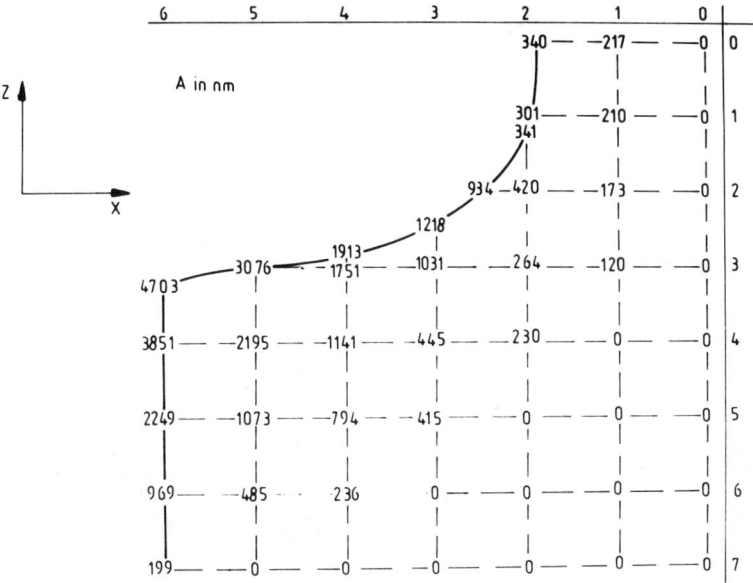

Fig.3 Vibration amplitudes of the blade at 2540 Hz

3. In-plane vibration mapping by time-average speckle metrology

Holographic interferometry is commonly used for mapping out-of-plane
vibration amplitude fields. The sensitivity vector is close the surface normal of the objekt under investigation and therefor the interference fringes correspond to lines of equal amplitude vector components. Recently, in connection with new non-metallic materials, vibration states with main direction along the surfaces of a vibrating body
became of high interest. Holographic interferometry is not well suited
for mapping those in-plane vibration modes. Speckle metrology may
close this map /2/. In order to get some experiments, we applied time-
average speckle pattern photography as well as interferometry to a
periodically vibrating tuning fork, elektromagnetically excited with
frequency ν . Looking perpendiculary to the vibration direction, the
tuning-fork undergoes a "in-plane" vibration whereas looking from a
side the same motion is "out-of-plane". We recorded time-average
specklegrams and holograms simultaneously be illuminating with two

93

lasers (Ar-Kr(500mW) and He-Ne(30mW)) and observing from different directions .

In our investigations, two methods of speckle metrology have been exploited: speckle pattern photography (SPP) and double aperture-speckle pattern interferometry (DASPI) /3/. SPP has the advantage of short exposure times (about 5 s) and the possibility of point-by-point evaluation. DASPI needs longer exposure (about 3 min) but the fringe contrast after fourier processing is superior to that of SPP.

The results of scanning a SPP-specklegram by a thin laser beam are shown in Fig.4. · The in-plane component of the amplitude $a=(a_1, a_2, a_3)$ may be obtained from fringe spacing and direction. We found, however, that the fringe spacing not always satisfies a Bessel function as it should if the vibration is harmonious.

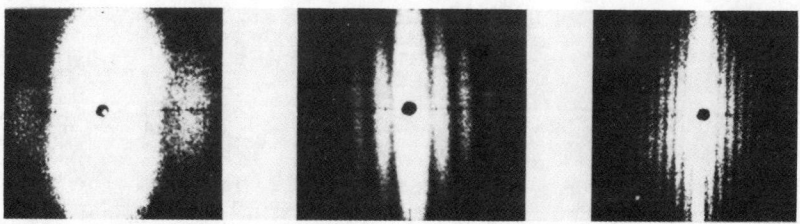

Fig.4 Some results of point-by-point analysis in time average SPP
A: $a_1 = 9,5$ µm , B: $a_1 = 30,3$ µm , C: $a_1 = 93,0$ µm

It seens interesting, that this is an additional information that is not available in time-average interferometry.

Fig. 5 gives some fourier-filtering results of specklegrams in comparison to a holographic interferogram. It becomes obvious that the sensitivity of SPP is much lower that of holographic interferometry (Fig.5,A and B). The fringe contrast is improved by DASPI (Fig.5- E).

4. Investigation of in-plane vibration by double pulse speckle pattern photography

Double pulse speckle pattern photography is an other useful tool for the study of the technical important case of in-plane vibrations, especially taking into account the problems of access by other measurement methods Specklegrams received by a double laser shot from dynamic objekts allow to evaluate in-plane displacements of objekt surface araes for the time interval between laser pulses. It is possible to obtain vibration modes of motions within the objekt surface

94

by this displacement measurement. Furthermore, with known phase location for the laser trigger or by the help of statistics of the process vibration amplitudes and vibration phases of different object areas may be determined. Special advantages of dynamic SPP are the application to aperiodic vibrations and the comparatively low sensitivity to unwanted object motions.

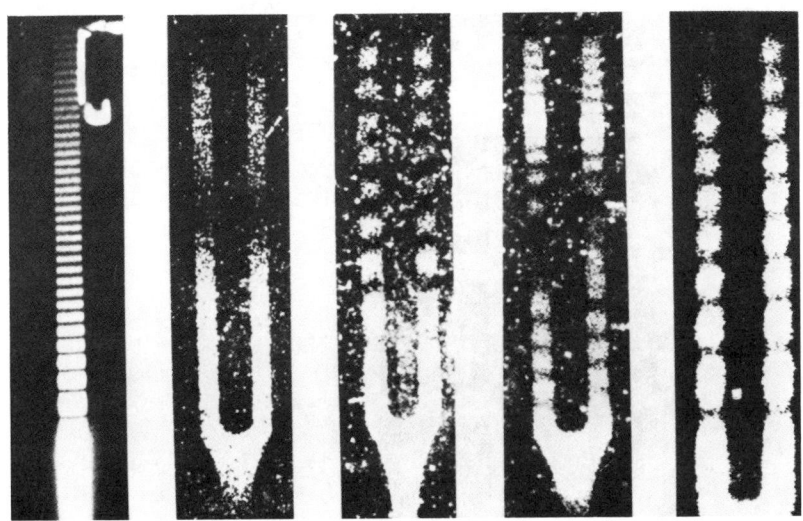

Fig.5 Fringes of equal amplitude component by various methods

A: holographic interferometry, ϑ = 433 Hz, a_1 = 6,5 μm

B: SPP, same vibration as in (A), but other view

C: SPP, higher amplitude, ϑ = 433 Hz, a_1 = 46,8 μm

D: SPP, other mode, ϑ = 2740 Hz, a_1 = 40,9 μm

E: DASPI, ϑ = 433 Hz, a_1 = 46,8 μm

Some basic experiments were accomplished with a double-pulsed ruby laser with following parameters: 10 mJ emission energy per pulse (without laser amplifier), 15 nsec single pulse length, 250–800 μsec pulse interval. Thus the laser allows to investigate vibrations with amplitudes from 5 mm*M*ϑ(s) to 800 mm*M *ϑ (s), were ϑ is the vibration frequency and M is the magnification of the SPP-image-system. For given emission energies surface areas of maximum 100 cm*cm can be illuminated by the laser. The use of a single stage laser amplifier will extend this value to about 1000 cm*cm.

Fig.6 demonstrates the application of double-pulsed SPP to an oscillating tuning-fork. The figure illustrates the two techniques of SPP displacement measurement, the treatment of Young's fringes in the fare field of the scanned specklegram and the imaging of the speckle-photo with spatial filtering in the Fourier-plane of the imaging lense. The latter will be fringes of equal component of in-plane displacement and in this case also fringes of equal amplitude of vibration.

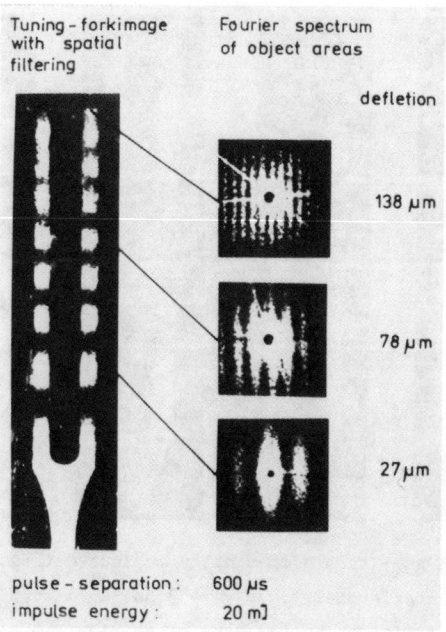

Fig.6 Oscillating tuning-fork, double-pulse specklegram

5. <u>References</u>

1   Füzessy,Z.;et al:
     Hologram interferometric measuring system for industry
     Proc. SPIE, Vol.398, 1983
2   Tiziani,H.J.:
     Vibration analysis and deformation measurement
     in:Erf,R.K. (ed.) Speckle metrology, Academic Press
     New York, 1978
3   Ennos,A.E.:
     Speckle Interferomtry
     in: Dainty,J.C. (ed.) Laser Speckle and Related Phenomena
     Springer, 1975

HOLOGRAPHIC INTERFEROMETRY METHOD FOR ASSESSMENT OF STONE SURFACE RECESSION AND ROUGHENING CAUSED BY WEATHERING AND ACID RAIN*

Prof. Cesar A. Sciammarella and Mansour A Ahmadshahi
Mechanical and Aerospace Engineering Department, Illinois Institute of Technology
3110 S. State Street, Chicago, Illinois 60616
C. Arthur Youngdahl
Argonne National Laboratory, Materials Science and Technology Division
Argonne, Illinois 60439

Environmental exposure of marble and limestone building materials produces surface recession rates of the order of tens of micrometers per year. A nondestructive damage assessment method has been developed which is capable of measuring the surface recession and roughening of stone specimens after exposure times of only a few months. Thus, studies of the effects of acid deposition on such materials are greatly facilitated.

Keywords: holographic interferometry, holographic moire, contouring, damage to momuments and statues

1. Introduction

The laser holographic moire profiling method of nondestructively evaluating minute changes in surface profiles and roughnesses produces significant measurements of the dimensional changes of marble and limestone specimen surfaces in ongoing studies of acid deposition effects on these materials.

2. Technical Approach

Stone surface material losses caused by weathering and acid rain are obtained from two types of measurements: (1) thinning or reduction of the dimensions (i.e., surface recession) and (2) rms roughness. Figure 1 illustrates the type of specimen used in this study. Recession is quantified by measuring the height of the "step" introduced between surface 2 (which is exposed to rainfall) and surfaces 1 (which are shielded from exposure). Changes in rms roughness are measured by comparing the roughness of surface 2 with the roughness of surfaces 1 and with data previously obtained from

*The work presented in this paper has been supported by the U.S. National Park Service, National Precipitation Assessment Program

surface 2. The optical technique utilized is based on the contouring holographic technique introduced by one of the authors in Ref. 1.

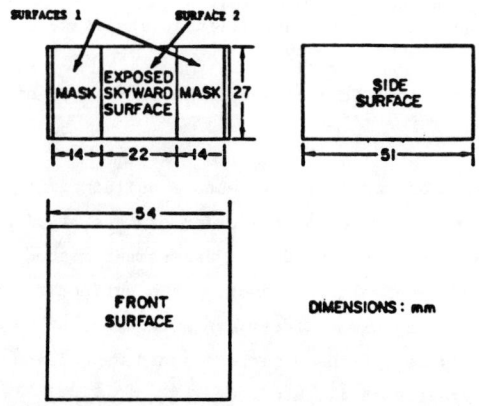

Figure 1   Schematic of field exposure specimen

The optical setup is shown in Fig. 2. The specimen is illuminated by
two beams, symmetrically inclined with respect to the specimen surface.
Two polar rotatable mirrors allow the introduction of rotations in the
direction of the illumination and the reference beams. The following steps
are required: (1) an initial hologram of the sample surface is taken
with the double illumination beam; (2) the illumination and the reference
beams are rotated, and a second exposure of the holographic plate is made;
(3) the hologram is developed; and (4) the hologram is reconstructed and
the reconstruction is recorded on film. The reconstructed image of the
sample is covered by a pattern of fringes. Two basic families of fringes
and their moire pattern are formed. The equations for each of the two
basic families are

$$I_1(x) = I_o + I \cos \phi_1(x) \quad , \tag{1}$$

$$I_2(x) = I_o + I \cos \phi_2(x) \quad . \tag{2}$$

The arguments $\phi_1(x)$ and $\phi_2(x)$ (radians) are given (Ref. 1) by the following
expressions:

$$\phi_1 = \frac{2\pi}{\lambda} \left[x(\cos \alpha\Delta\alpha - \cos \theta_R\Delta\theta_R) - h(x)\sin \alpha\Delta\alpha \right] + \phi_{1n} \quad , \tag{3}$$

$$\phi_2 = \frac{2\pi}{\lambda} \left[x(\cos \alpha\Delta\alpha + \cos \theta_R\Delta\theta_R) + h(x)\sin \alpha\Delta\alpha \right] + \phi_{2n} \quad , \tag{4}$$

where x is the coordinate of the point under consideration; α is the angle formed by the illumination beams with the direction of observation, which is the direction of the normal to the surface; $\Delta\alpha$ is the rotation of the illumination beam; θ_R is the angle formed by the reference beam with the normal to the holographic plate; $\Delta\theta_R$ is the rotation introduced in the direction of the reference beam; and h(x) is the change of height of the point of the surface with respect to a reference plane normal to the direction of observation. The first terms within the square brackets of Eqs. (3) and (4) do not depend on the relative heights of the points; the second ones do. The terms $\phi_{1n}(x)$ and $\phi_{2n}(x)$ are noise terms, or changes of phase caused by factors other than the two factors explicitly shown in Eqs. (3) and (4). The arguments of these two equations can be written

$$\phi_1(x) = 2\pi f_1 x - \beta(x) + \phi_{1n}(x) \quad , \tag{5}$$

$$\phi_2(x) = 2\pi f_2 x + \beta(x) + \phi_{2n}(x) \quad , \tag{6}$$

where

$$f_1 = \frac{\cos \alpha\Delta\alpha - \cos \theta_R\Delta\theta_R}{\lambda} \quad , \tag{7}$$

$$f_2 = \frac{\cos \alpha\Delta\alpha + \cos \theta_R\Delta\theta_R}{\lambda} \quad , \tag{8}$$

and

$$\beta(x) = \frac{2\pi}{\lambda} h(x)\sin \alpha\Delta\alpha \quad , \tag{9}$$

where $\beta(x)$ is a random phase term changing with the random variations of height of the stone surface. Consequently, the fringes can be considered as phase-modulated fringes, with the phase modulation caused by the changes of height of the surface. The term h(x) can be considered as the addition of two terms, one term corresponding to the random variable $\xi(x)$ and another term h*(x) representing the trend of the surface;

$$h(x) = h^*(x) + \xi(x) \quad . \tag{10}$$

The term h*(x) slowly varies with x, while $\xi(x)$ is assumed to satisfy the following condition:

$$\langle\xi(x)\rangle = 0 \quad , \tag{11}$$

that is, the average value of $\xi(x)$ is equal to zero. It is also assumed that the statistics of $\xi(x)$ do not change if the coordinate y is changed. This is equivalent to stating that the random changes depend on the type of stone and surface finish; in other words, $\xi(x)$ is spatially stationary.

The moire pattern produced by the two systems of fringes has an argument that depends on the difference of the arguments of the two carriers:

$$\phi_m(x) = \frac{2\pi}{\lambda} (f_1 - f_2)x - 2\beta(x) + \phi_{1n}(x) - \phi_{2n}(x) \quad . \tag{12}$$

The argument of the moire fringes contains the term $2\beta(x)$, and therefore the sensitivity of the technique has been increased by a factor of 2. The noise term contains the difference of the noise terms of the two signals. This means that all the noise contributions that are common to the two carriers have been removed. The moire fringes are additive, and consequently their contrast would be zero if not for the small contribution due to the film nonlinearity. The nonlinearity problem can be solved by optical filtering. An alternative procedure used in the present work is to filter the sampled version of the signal and multiply the filtered signals. This digital operation is effectively identical with optical filtering. A fast Fourier transform of the signal is performed, and the carriers are identified and filtered. The two carriers are multipled and another filtering process is performed to separate the moire term from other terms that are contained in the product.

The next step in the process of recovering h*(x) is to remove $\xi(x)$. To isolate the frequency, f_m, a matched filter is applied. After this step, the argument of the signal is

$$\phi_m(x) = \frac{2\pi}{\lambda} f_m x + \gamma_y(x) \quad , \tag{13}$$

where

$$\gamma_y(x) = 2\frac{2\pi}{\lambda} h^*(x) \sin \alpha\Delta\alpha \quad . \tag{14}$$

The trend of the signal is thus separated from the local variations.

Recession is calculated from the average difference between the phases of surfaces 1 and 2 (see Fig. 1) by means of the following procedure: A sinusoidal signal of the frequency $f_m = f_1 - f_2$ is generated. A cross-correlation between this signal and the moire fringes of surfaces 1 is performed, and the signal is displaced until a maximum value for the correlation is obtained. The maximum correlation indicates the best match between the phase of the reference signal and the average phase of the moire fringes of surfaces 1. The operation is repeated for surface 2, and the average phase

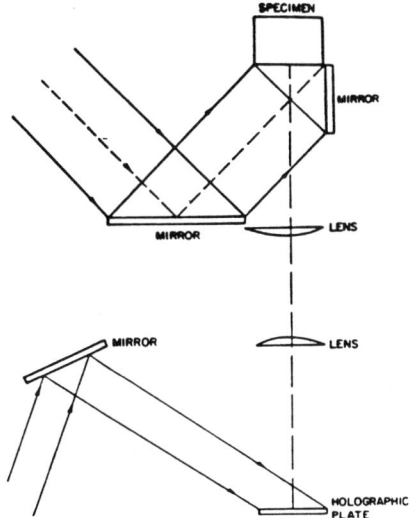

Figure 2   Optical setup

for surface 2 is obtained. The difference between the phases of surfaces 1
and 2 represents the phase jump caused by the removal of the material from
surface 2,

$$\Delta\gamma_y = \gamma_{1y} - \gamma_{2y} \quad .$$
(15)

The determination of the phases is performed in N sections and the
ensemble average is computed:

$$\overline{\Delta\gamma} = \sum_{i=1}^{i=N} \frac{\Delta\gamma_{yi}}{N} \quad .$$
(16)

From the ensemble average, the average step between the protected and un-
protected regions is computed by means of the following equation:

$$\overline{\Delta h} = \frac{\lambda}{4\pi\sin \alpha\Delta\alpha} \ \overline{\Delta\gamma} \quad .$$
(17)

To increase the number of sections included in the computation of the
ensemble average without unduly increasing the computer time, the following
procedure has been applied. As stated earlier, measurements are performed
at N locations representing the whole surface of the stone. At each location,
M contiguous sections are measured and averaged; consequently, the total
number of processed sections is NxM. Therefore, the values of $\Delta\gamma_y$ obtained
by means of Eq. (15) represent zone averages. To compute the correct
statistical values, the theory of random variables is utilized.

The roughness has been determined by applying the technique described
in Ref. 2 to obtain the phase information encoded in the fringes. In this
technique, fringes are analyzed as frequency-modulated signals. The fringes
are considered as generated by a rotating vector, whose phase information
is separated from the amplitude information by introducing an inquadrature
signal. The phase information of the moire fringes generated by the digital
filtering contains, besides the roughness values, the trend of the surface.
To remove the trend of the surface, the signal is filtered until the condition expressed in Eq. (11) is satisfied. From the values obtained by
this procedure, the rms roughness is computed.

Figure 3 shows a reconstructed fringe pattern. The pattern is enlarged
to approximately 4 times the actual sample size, and a grid is superimposed
on the image to serve as a reference frame for locating the sections to be
measured. The two lateral regions are surfaces 1 and the central portion is
surface 2. The double lines denote the areas contacted by the specimen mask
(see Fig. 1); these regions are not included in the analysis. The fringe
spacing is approximately 0.3 mm. The picture is digitized by a TV camera
which is part of the system described in Ref. 3; a schematic of this system
is shown in Fig. 4. Since the digitizing device can only digitize 512 pixels,
the digitization is performed by regions. To ensure the continuity of the
digitized regions, a pixel-matching subroutine is introduced. The electronic
noise is removed by performing 256 measurements and averaging them. The
digitization program reads 5 consecutive lines and computes the average. The
approximate distance between pixels is 5 µm. The theoretical sensitivity of
the method is approximately 0.35 µm per degree of phase angle.

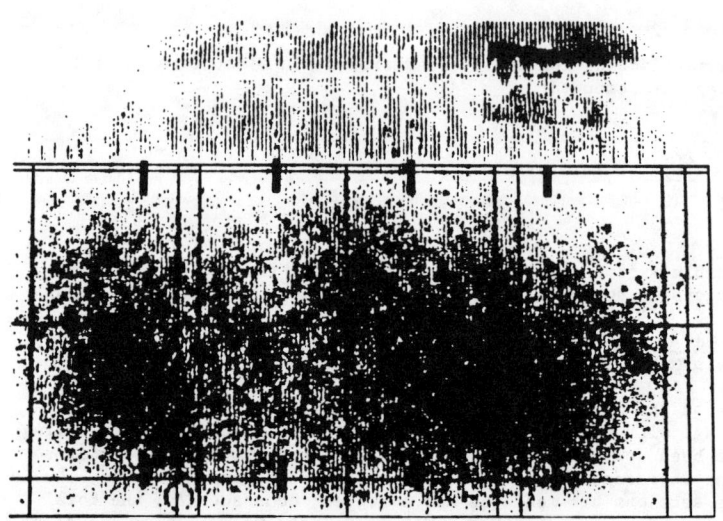

Figure 3   Reconstructed fringe pattern

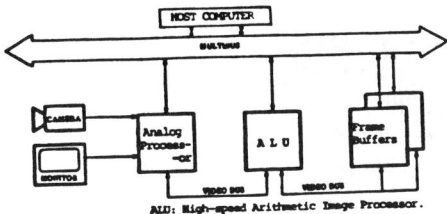

Figure 4   Schematic representation of the computer system
utilized to process the fringe pattern

Figure 5 is the flow chart for the computer program. To ensure good
pixel matching, an interpolation routine is introduced before the matching
is performed. Figure 6 shows a profile of the moire fringes from a section
of one of the tested specimens. Figure 7 shows the roughness profile
resulting from the computer analysis.

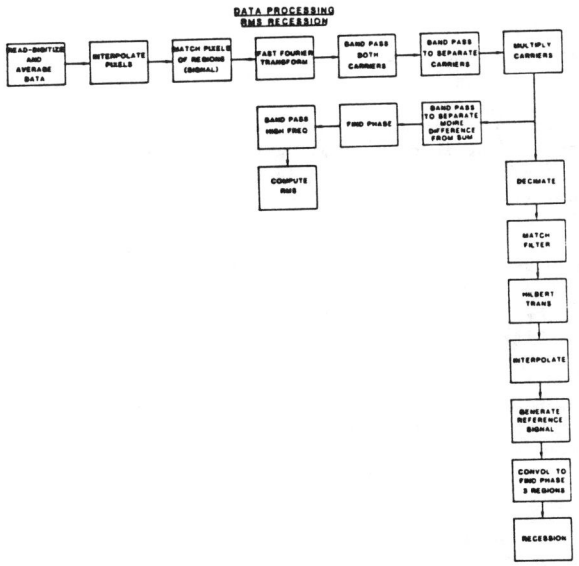

Figure 5   Flow chart of the computer program

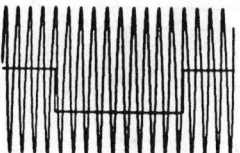

Figure 6   Moire fringe pattern
and reference signal
corresponding to a
specimen section

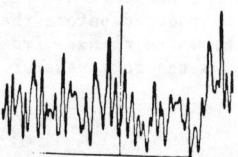

Figure 7   Roughness profile of
the section shown in
Fig. 6

3. Results and Discussion

Table 1 compares the rms roughness values obtained after 2-3 months of exposure for four marble specimens, with the optical technique described above and with a Tallysurf. The Tallysurf values were computed by analyzing the profile recorded by the instrument's analog plotter. For this reason, far fewer points are utilized in obtaining the Tallysurf rms values than is the case with the optical technique, and consequently the quantities are not strictly comparable. However, the results are given here as an indication of the kind of agreement that exists between the two methods. Table 2 shows the recession results obtained in eight specimens after the same exposure. Table 2 also gives the values measured with a profilometer on the same specimens. The profilometer readings are taken along the midline of the specimen surface, parallel to the long edges. The measured values generally fit well within the statistical estimate of the values that could be observed. All the annual samples have been remeasured after the exposures accumulated by the Fall of 1985.

Table 1   Comparison of Optical and Tallysurf Measurements[a] of
Marble Surface Roughness

| | rms Roughness (µm) | | | |
| | Protected | | Unprotected | |
Specimen	O	T	O	T
4AL2614	6.52	4.5	6.42	4.2
4PK1509	5.8	4.48	7.71	3.15
3AK1515	6.19	7.15	5.56	4.63
3AL2606	9.36	7.43	7.59	6.13

[a] O = optical method; T = Tallysurf method

Table 2   Examples of Surface Recession Results from Marble
           Specimens after Atmospheric Exposure

| Specimen | Optical Method | | | Recession by Profilometer Method (μm) |
	Recession (μm)	Std. Dev.	95% Confidence Level	
4AL2614	+4.63	± 3.69	(+3.21, +6.05)	6
4PK1509	+2.36	± 7.14	(-0.33, +5.05)	4
4PL2606	+10.5	±26.3	(+0.35, +20.6)	3
4AH3609	+4.11	±14.8	(-0.51, +11.9)	3
3AK1515	+4.20	±22.4	(-4.50, +12.7)	2
3AL2606	+4.46	± 2.44	(+3.52, +5.39)	0
3PH1120	+6.12	±12.5	(+1.40, +10.9)	4
3PM1508	-2.64	±14.1	(-10.0, +0.81)	-8

4. Conclusions

An accurate method to determine changes in depth and roughness on rough surfaces has been successfully developed and implemented. The method combines optical and digital techniques and is fully computerized. The method utilizes the moire and the holographic interferometry technologies together with a classical method of interferometry to determine depth changes on surfaces. At the same time, it makes use of important concepts and tools of the statistical communications theory.

In the current stage of development of the program, the time required to process the information from one specimen is approximately 3 hours. This time could be greatly reduced by adding new hardware components to the current system. With an array processor and an additional memory buffer, one could increase the number of sections analyzed and reduce the overall processing time.

5. References

1   Sciammarella, C. A. 1982. Holographic moire, an optical tool for
    the determination of displacements, strains, contours and slopes of
    surfaces. Opt. Eng. 21, 447-457.

2   Sciammarella, C. A., M. A. Ahmadshahi. 1984. A computer based
    method for fringe analysis. In: Proc. 1984 Society for Experimental
    Mechanics Fall Conf. on Computer-aided Testing and Model Analysis,
    Milwaukee, WI, pp. 61-69.

3   Sciammarella, C. A., M. A. Ahmadshahi. 1985. An optoelectronic
    system for fringe pattern analysis. In: Proc. 1985 Society for
    Experimental Mechanics Spring Conf. on Experimental Mechanics,
    Las Vegas, NV, pp. 861-868

INVESTIGATION OF ELASTOPLASTIC PROBLEMS WITH THE HELP
OF METHODS BASED ON HOLOGRAPHIC RECORDING OF INFORMATION

Prof.Dr.Vitalii A. Zhilkin, Department of
Theoretical Mechanics of the Novosibirsk
Institute of Railway Transport Engineers
D.Kovalchuck, 191, Novosibirsk 23, USSR

The research results of plane elastic and elastoplastic
problems of mechanics with the help of holographic interferometry and holographic moire methods based on Denisyuk hologram recording with fixture of registrate medium on the
object surface are considered.

Keywords: holographic interferometry, moire, strain

1. Introduction

Traditional schemes of holographic interferogram recording with
the registrate medium and object separated in a space have the following detects [1] :
- Strict requirements on a vibroisolation of the experimental
installation elements don't allow to use a standart testing
equipment.
- Considerable errors in a definition of displacements in plane
tangent to the specimen surface are observed.
Superposed holographic interferometers are free from these detects [2] . In the given case for a purpose of lowering requirements
of experimental installation elements vibroisolation and raising the
strain definition accuracy a photoregistrating medium is fixed on the
specimen surface and holographic interferograms are recorded in con-
coming beams according to the scheme offered by Yu.Denisyuk. In this
case a definition accuracy of displacement in the object plane is
raised owing to the greater aperture angle.

2. Superposed holographic interferometers (Fig.1)

Superposed holographic interferometer is a assembly of the object
surface section and hologram recorded in a intermediate proximity to
the object. If hologram is registrated at the metallized raster surfa-

ce it restores information peculiar to some optical methods: holographic interferometry, speckle photography, holographical moire and mirror-optical method.

When the massive specimen strain condition is studied the registrating medium fixture on its surface is realized by the mechanic and magnetic clamping (Fig.1 b). When the thinwalled constructions are studied a medium is fixed with the help of intermediate optical medium of a small shear stiffness and a piezooptical sensitivity (Fig.1 a). In the last case the photoplate attached to the specimen surface with the help of intermediate optical medium must not essentially distort the strain field.

3. Decoding of interference patterns

Displacement vector components of specimen surface points are determined in the coordinate system (x o y z) connected with the photoplate plane 2 (Fig.1). Displacements W (in the direction of Z-axis) are found by interference patterns, observed nearly normal to the specimen surface; displacements in the specimen plane U (in the direction of X-axis) and V (in the direction of Y-axis) are determined in a common treatment process of four interference patterns observed from four symmetric directions relatively to Z-axis (It is possible a conversion of the directions of viewing and a specimen surface points illumination).

3.1. Resolving equations for interferometer drawn in Fig.1 a. If interference patterns are observed in the plane (x o z) resolving equations have the following form [2]

$$\pm u \sin\tau + w(n + \sqrt{n^2 - \sin^2\tau}\,) = N_\kappa \lambda \quad , \qquad (\kappa = 1,2)$$

where n is the refractive index of intermediate piezooptic medium, N_κ - order of bright interference bands, λ is a wave length and τ is angle between the viewing direction and Z-axis.

3.2. Resolving equations for interferometer drawn in Fig.1 b are the same as in traditional methods of moire fringes.

3.3. Resolving equations for interferometer drawn in Fig.1 c. In this case only a picture of interference fringes observed by normal to the specimen surface is decoded. According to the picture it is determinated w and $\varepsilon_z = \dfrac{2w}{t}$, where t is organic glass thickness.

Such we have

$$\varepsilon_x + \varepsilon_y = - \frac{1-\mu}{\mu} \frac{2w}{t} \quad ,$$

where $\varepsilon_x, \varepsilon_y, \varepsilon_z$ are strains in the directions of x, y and z axes accordingly.

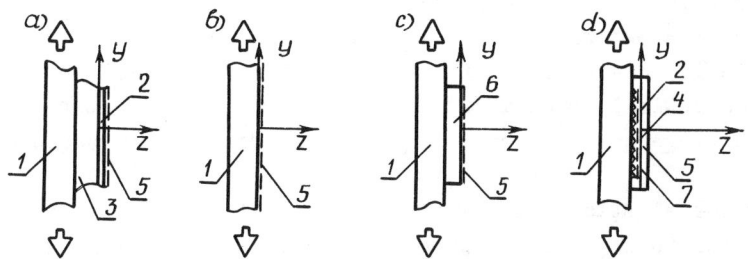

Fig. 1

Schemes of superposed holographic interferometers
1 - object, 2 - photoplate, 3 - intermediate piezooptic medium,
4 - air clearance, 5 - high solved registrating medium,
6 - piezoinsensitive organic glass, 7 - metallized reflective
grid

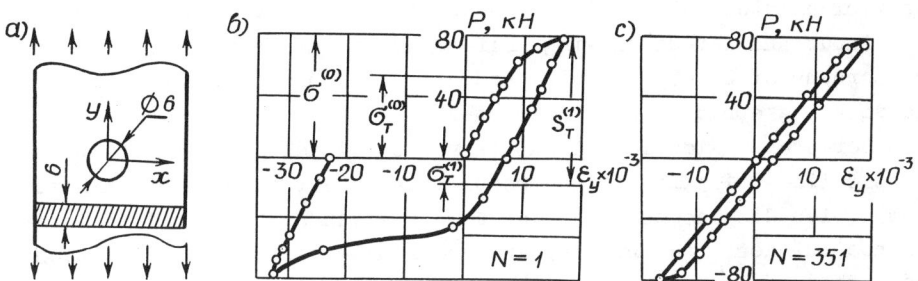

Fig. 2

a - geometric dimensions of a specimen,
b - strain diagram in the first cycle of loading (N = 1),
c - strain diagram when N = 350

109

3. 4. Resolving equations for interferometer drawn in Fig.1 d [3] .
When interference fringes are observed in the plane (x o z) we have
$$\pm\,u\sin T + w(1-\cos T) + \xi\,\frac{\partial w}{\partial x} = N_\kappa \Lambda \qquad (\kappa = 1,2)\quad,$$
where $\xi = H\,\mathrm{tg}\,T$, H-is the air gap thickness between a raster and a photoplate; $\frac{\partial w}{\partial x}$ is mean derivative on basis between points of beam entry
and exit from the photoplate.

Holographic moire method is less sensitive to displacement w
than holographic interferometer method. Under small raster frequencies
(small angle T) a scheme of shift interferometer is realized. Equations are essentially symplified when H approximates to zero.

Strain range measured by considered interferometers accounts for
$6\cdot 10^{-5}\ldots 2\cdot 10^{-3}$, therefore with the help of them it is possible to
study only the elastic problems of mechanics of deformed medium. A study of the plastic deformation process is only possible in a combination with other optic methods of the lower sensitivity either using
a device of controlled phase shift or using staged methods of specimen loading and interference friges recording. In the last case for
studying deformation kinetics of specimen (D16П), tested on tension
before its destruction it is necessary to record about 200 interferograms. It is clear that a solution of the such problems is related with
elaboration of automated output methods of experimental information
and with a creation of data computer processing algorithm. This problem is solved with the help of the calculating complex or a base of
the minicomputer "CM-4". It includes the devices for optical information input and for displaying half-toned and graphic information on
the screens of black and white and colour TV and storages on magnetic disks, graphic chart and printer [4] .

For a approximation of interference fringes order function the
elastic lines of circular and rectilinear statically indeterminate
beams (pier displacements of which were given proportional to interference fringes order number) and the finite-element approximation
and method of boundary integral equations are used.

4. Experimental results
4.1. The plotting of deformation diagram for maximum stressed point
with a hole under low-cycle loading. Holographic interferometer method is used. Specimens were made of sheet material   DI6T   .
Their geometric dimensions are given in Fig.2. Soft loading of specimen by sign-variable load is carried out on a testing machine
УМЭ – 10 ТМ. Maximum load does not exceed 80 kH. The diagrams

110

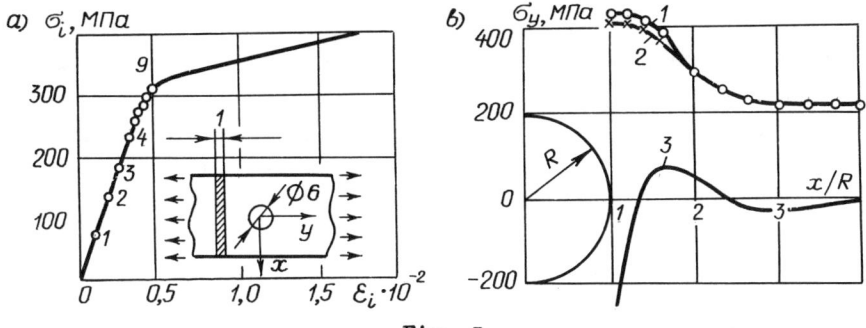

Fig. 3

a - geometric dimensions of a specimen and its strain diagram,
b - diagram (σ_y) in a critical section: 1 - a theory of
small elasto-plastic strain; 3 - residual stress, 2 - a theory
of flowing

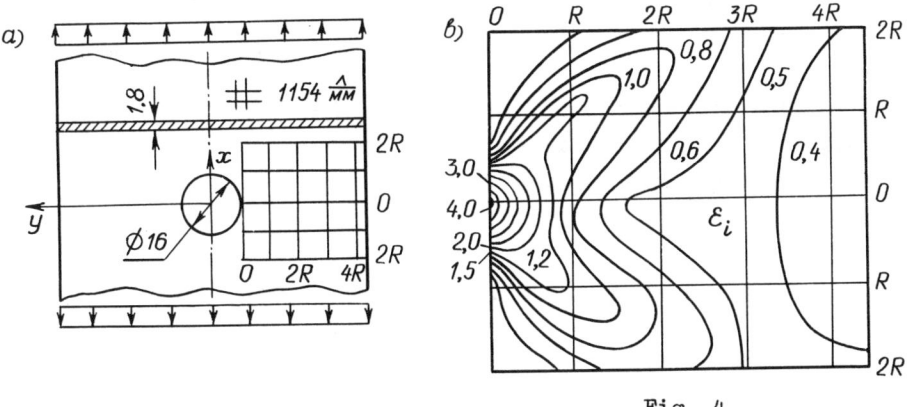

Fig. 4

a - geometric dimensions of a
specimen and grid region within which interference fringes
decoding is made; b - strain
intensity level lines in region
near a hole; c - interference
fringes observed in the first
diffraction order (+1): 1 - in
a clearance, 2 - on a reflection

111

drawn in Fig.3 are plotted according to the results of 22 hologram treatment.

Analysis of curves given in Fig.2 b and 2 c shows that alloy DI6T is hardening in a process. A diagram of material point deformation corresponds to a elastic performance of material. The well-known relation of Masing is fulfilled with error no more than 1 %.

4.2. Analysis of strained state of alloy plates (DI6T) with the central circular hole.

4.2.1. Using of holographic interferometer method (Fig.3). Points on deformation diagram correspond to the stages of loading and double-exposed hologram registation. Quantitative decoding of interferograms is carried out for five stages of specimen loading.

4.2.2. Using of holographic moire method (Fig.4). Orthogonal raster with frequency of 1154 lines/mm is marked on the specimen surface. A specimen strain state is studied within of the grid region, contained 483 joints in which an intensity of strain $\mathcal{E}_i$ under loads from 40,6 to 44 kH is determined.

It is supposed that Poisson's ratio is equal 0.5.

5. References

1. Вест Ч. Голографическая интерферометрия. М., Мир, 1982. 504 с.
2. Жилкин В.А., Герасимов С.И. О возможности изучения деформированного состояния изделий с помощью накладного интерферометра. - Журнал технической физики, 1982, т.52, № 10, с.2079-2085.
3. Жилкин В.А., Зиновьев В.Б. Расшифровка интерференционных картин в методе голографического муара. - Журнал технической физики, 1986, т.56, вып.I, с. II3-II9.
4. Герасимов С.И., Гужов В.Н., Жилкин В.А., Козачок А.Г. Автоматизация обработки интерференционных картин при исследовании полей деформаций. - Заводская лаборатория, 1985, т.5I, № 4, с. 77-80

Moiré and optoelectronic methods

OPTICAL METHOD OF STRAIN MEASUREMENTS
APPLICATION TO STUDY BIAXIAL TENSION SPECIMENS

Dr. F. Brémand - Prof. A. Lagarde

Laboratoire de Mécanique des Solides - Equipe de Recherche Associée au C.N.R.S
- 40, Avenue du Recteur Pineau - 86022 Poitiers Cedex France

The method uses a coherent light diffracted by an orthogonal grating engraved on the surface of the specimen. This method has the convenience of giving, in large as small deformations, over a small measuring area, the orientations and the values of the principal extensions as well as the rotation of the rigid body. A simulation of a known deformation field enabled us to test the method. With a measuring base (less than a square millimeter), we used it to compare two cross-shaped tension specimens. It is shown one shape introduces a biaxial stress state in a small region around the central point.

Keywords : strain measurement - biaxial tension specimen

1. Introduction

For a long time researchers have shown interests in the measurements of large deformations on the surface of an object. They use several techniques. Three gauge rosettes enable the access to the three parameters of Mohr circle of deformations with a good linearity and a good sensitivity for up to 20 %. However they present the inconvenience of not being able to resist successive alternating strains and in addition the measurement base is quite large.

The moiré methods are more difficult to use in the case of large displacements and are not able to measure strain greater than 30 % and angle greater than 30°. The more useful method in the large deformation field is the grid method which consists of engraving a series of orthogonal lines or a group of circles on the surface of the specimen. The use of circles give directly the orientations and the values of the principal strains, but the dispersion on the results may go up to 15 %. The solution we are proposing is based on the use of two grating of parallel orthogonal lines (10 lines per mm) marked on the surface of the specimen of which the photographic film is being analysed by the diffraction procedure. This method has already been applied in the case of small deformations [1].

We applied it to measure strains in the central part of two crossed-shape specimens loaded in a biaxial tension test. The comparison between these two shapes is done and we show only one leads at the existence of a biaxial stress state in the central point.

Principle of the method [2 - 3 - 4]

We consider the case of planes deformations on the surface of the specimen. Let us suppose an initial square which is defined in a referential $0,X,Y$ by four points : $O(o,o)$; $A(o,p)$; $B(p,p)$; $C(p,o)$ where p is the length of its sides (fig. 1).

We suppose this square is transformed in a parallelogramme O', A', B', C' in the deformed state such that :

- the vector $O'A'$ has components $OA' = (a_2 \cos \alpha_2, a_2 \sin \alpha_2)$
- the vector $O'C'$ has components $OC' = (a_1 \cos \alpha_1, a_1 \sin \alpha_1)$

Therefore an interior point $M(X,Y)$ is transformed in $m(x,y)$ by the following transformation :

$$x = \frac{a_1}{p} \cos \alpha_1 \, X + \frac{a_2}{p} \cos \alpha_2 \, Y$$

$$y = \frac{a_1}{p} \sin \alpha_1 \, X + \frac{a_2}{p} \sin \alpha_2 \, Y$$

This analytical transformation allows us to determine the gradient of the transformation tensor $\overline{\overline{F}}$ by the matrix :

$$F = \begin{bmatrix} \frac{a_1}{p} \cos \alpha_1 & \frac{a_2}{p} \cos \alpha_2 \\ \frac{a_1}{p} \sin \alpha_1 & \frac{a_2}{p} \sin \alpha_2 \end{bmatrix}$$

The Cauchy Green's right tensor $\overline{\overline{C}} = {}^{t}\overline{\overline{F}}\overline{\overline{F}}$ and Cauchy Green's left tensor $\overline{\overline{c}} = \overline{\overline{F}}{}^{t}\overline{\overline{F}}$ have respectively the following matrix :

$$C = \begin{bmatrix} (\frac{a_1}{p})^2 & \frac{a_1 a_2}{p^2} \cos(\alpha_2 - \alpha_1) \\ \frac{a_1 a_2}{p^2} \cos(\alpha_2 - \alpha_1) & (\frac{a_2}{p})^2 \end{bmatrix} \quad ;$$

$$C = \begin{bmatrix} (\frac{a_1}{p} \cos\alpha_1)^2 + (\frac{a_2}{p} \cos\alpha_2)^2 & (\frac{a_1}{p})^2 \cos\alpha_1 \sin\alpha_1 + (\frac{a_2}{p})^2 \cos\alpha_2 \sin\alpha_2 \\ (\frac{a_1}{p})^2 \cos\alpha_1 \sin\alpha_1 + (\frac{a_2}{p})^2 \cos\alpha_2 \sin\alpha_2 & (\frac{a_1}{p} \sin\alpha_1)^2 + (\frac{a_2}{p} \sin\alpha_2)^2 \end{bmatrix}$$

Now we suppose that the square previously looked for, is obtained from an orthogonal grid composed of two gratings of parallel lines of pitch p. These gratings could be either engraved or printed or stick on the surface of a studied specimen. If we assume the grid perfectly follows the displacement in each of the point of the model, we can visualise the deformations of each small initial square.

In accordance with the figure 2, we write the following relations where β is given by $\beta = \pi/2 - (\alpha_2 - \alpha_1)$

$$\cos \quad = \frac{P_2}{a_2} = \frac{P_1}{a_1}$$

Then it is possible to get the components of $\overline{\overline{C}}$ and $\overline{c}$ in any base by measuring the pitches p_1 and p_2 and the orientations of the two deformed gratings initially orthogonal and of the same pitch p. We will always take the base resulting from the directions of the two initial families of lines :

It is evident a diagonalisation made on $\overline{\overline{C}}$ leads at the knowledge of proper values and proper vectors of $\overline{\overline{C}}$. One can easily obtain the magnitude of the principal strains and the orientation γ' of the proper vectors of $\overline{\overline{C}}$ in accordance with one axis of the referential. The angle γ' represents the direction of the pure principal strains. In fact generally, it is different of γ which visualises the orientation of the principal directions of the strain tensor, and we know the difference $\gamma - \gamma'$ is the rotation of the rigid solid R :

$$\gamma = \gamma' + R.$$

Using the polar decomposition of $\overline{\overline{F}}$ we show the relation :

$$\text{tg } R = \frac{a_2 \cos\alpha_2 - a_1 \sin\alpha_1}{a_2 \sin\alpha_2 + a_1 \cos\alpha_1}$$

Thus we get the orientation and the value of the principal extensions and the rotation of rigid solid from the knowledge of four parameters (two pitches p_1 and p_2 and two angles α_1 and α_2). These values are obtained using the procedure of diffraction on photographic negatives representing the deformed state of the studied gratings.

The diffraction phenomena of a parallel beam of a coherent light through a plane grating is well known (5,6). The hypothesis made in the case of a phenomenon of Fraunhofer's diffraction (infinite diffraction giving regularly spaced points) allow the determination of the pitch of the grating knowing the wavelength λ of the radiation, the distance L between the screen (E) and the

117

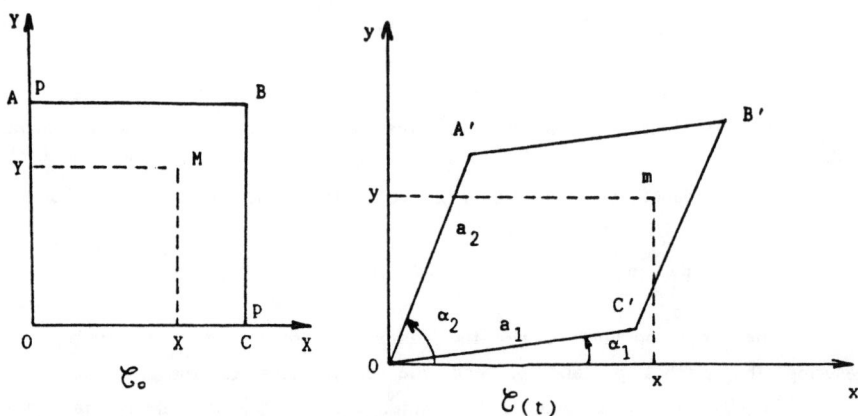

Fig. 1 Deformations of a square

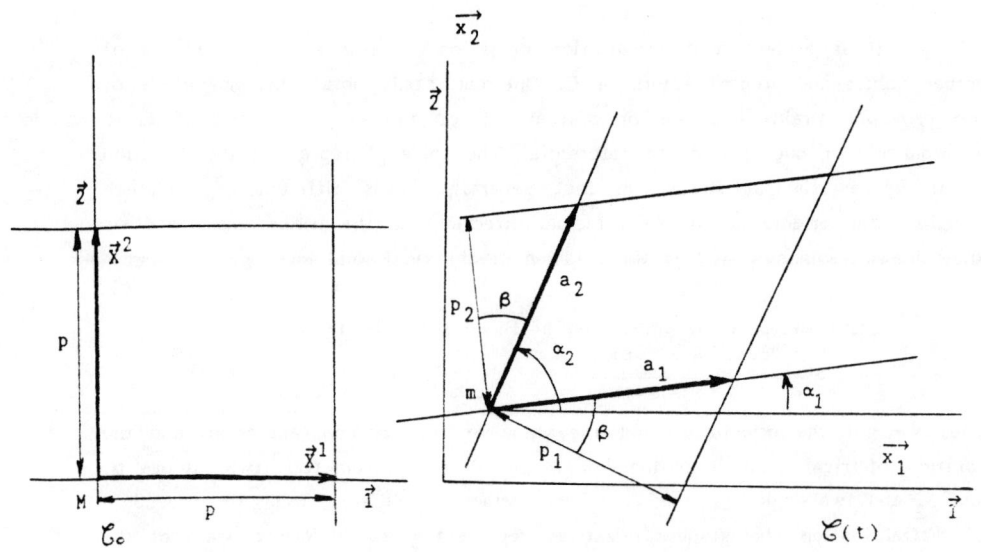

Fig. 2 Deformations of an orthogonal grating

photographic film, the distance d between two consecutive points of diffraction

$$p = \frac{\lambda L}{d}$$

This relation assumes small angles of diffraction, in other words a large value of L with respect to d. When this hypothesis is not verified we use the relation :

$$p = \frac{\lambda_m}{Arctg \frac{d^m}{L}}$$

We have represented on figure 3 the diffraction image of a grating of parallel crossing lines. We notice that the directions formed by the diffraction points are perpendicular to the orientation of the family of corresponding lines. It is now easy to describe P_1 ; P_2 ; α_1 ; α_2 as functions of d_1^m ; d_2^m ; δ_1 ; δ_2 :

$$P_1 = \frac{\lambda m}{Arctg \frac{d_1^m}{L}} \qquad P_2 = \frac{\lambda m}{Arctg \frac{d_2^m}{L}}$$

$$\delta_1 = \delta_1 - \frac{\pi}{2} \qquad \delta_2 = \delta_2 + \frac{\pi}{2}$$

The polar coordinates of these points are numerically read by using a digital table. In order to minimise the uncertainties over the four parameters the center of each point is take three times. A numerical analysis is done on a micro computer and it gives the statistical analysis of the data and computes the strain values.

2. Simulation [2 - 3 - 4]

In order to test the validity of this measuring method we have made a simulation of an homogeneous strain field. Then, in the deformed state, we consider a grating of pitch p, which is composed with two same families of parallel lines but inclined one with respect to another with an angle $\pi/2 - \delta$ (figure 4). We suppose these two families are initially perpendicular. The transformation from the initial state $(X_1 ; X_2)$ to the final state $(x_1 ; x_2)$ has the following expression :

$$x_1 = \frac{1}{\cos \delta} X_1 + tg \, \delta \, X_2$$

$$x_2 = X_2$$

We deduce the expression of the gradient of transformation tensor :

$$F = \begin{bmatrix} \frac{1}{\cos \delta} & tg \, \delta \\ 0 & 1 \end{bmatrix}$$

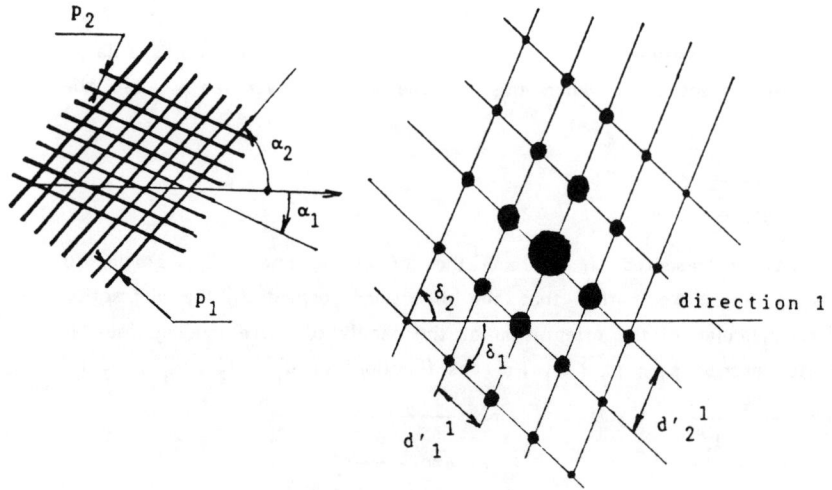

Fig. 3 Diffraction image of a grating

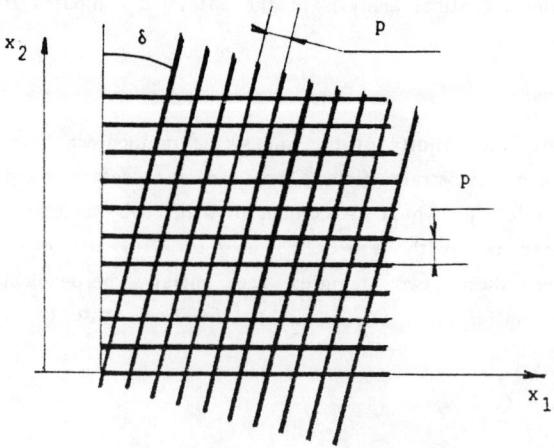

Fig. 4 Geometry of the studied grating

and the tensors $\overline{\overline{C}}$ and $\overline{\overline{c}}$ have as expression :

$$C = \frac{1}{\cos^2 \delta} \begin{bmatrix} 1 + \sin \delta & 0 \\ 0 & 1 - \sin \delta \end{bmatrix} \qquad c = \begin{bmatrix} 1 + \sin \delta & 0 \\ 0 & 1 - \sin \delta \end{bmatrix}$$

Since $\overline{\overline{E}} = \frac{1}{2}(\overline{\overline{C}} - \overline{\overline{I}})$ and $\overline{\overline{e}} = \frac{1}{2}(\overline{\overline{1}} - \overline{\overline{c}})$ where $\overline{\overline{E}}$ and $\overline{\overline{e}}$ are the Green Lagrange's tensor and Euler Almansi's tensor respectively, one can easily write

$$E_1 = \frac{1}{2} \sin \left(\frac{1 + \sin \delta}{\cos^2 \delta}\right) \qquad\qquad e_1 = -\frac{1}{2} \sin \delta$$

$$E_2 = \frac{1}{2} \sin \left(\frac{1 - \sin \delta}{\cos^2 \delta}\right) \qquad\qquad e_2 = \frac{1}{2} \sin \delta$$

Considering the physical representation of the transformation, we know that the principal directions in the spatial representation are diagonals of the lozenges.

Hence $\quad \gamma = \frac{\delta}{2} + \pi/4$

The diagonalisation of C leads us to the value of γ'

$\gamma' = \pi/4$

From there we deduce

$R = \delta/2$

We note that the rotation is equal to the variation of the orientation of a diagonal.

The used gratings are identical (p = 0.042 mm). Four tests were done with values of equal to - 36.5° ; - 27.5° ; 27.5° and 36.5°, the distance L being equal to 1095 mm while the wavelength of the laser beam being equal to $632.8*10^{-6}$ mm. The experimental results (figures 5 and 6) compare very well with the theoretical curves. It is worth noting that the gratings were of excellent quality and the diffraction points were circular with a good contrast.

Biaxial tension test [4]

We used two croos-shaped specimens (figure 7) made from a TM60A urethane. In order to limit the influence between the two loading directions the arms of one shape are made of thin strips from the molding [7 - 8]. In addition, 10 lines orthogonal gratings (pitch p = 0.1 mm) were engraved at the base of the mould and are reproduced on the molded specimens. The tension apparatus has been designed to enable independant loadings in the two directions. The forces are transmitted by means of a nylon thread passing over pulleys mounted on rolling bearings.

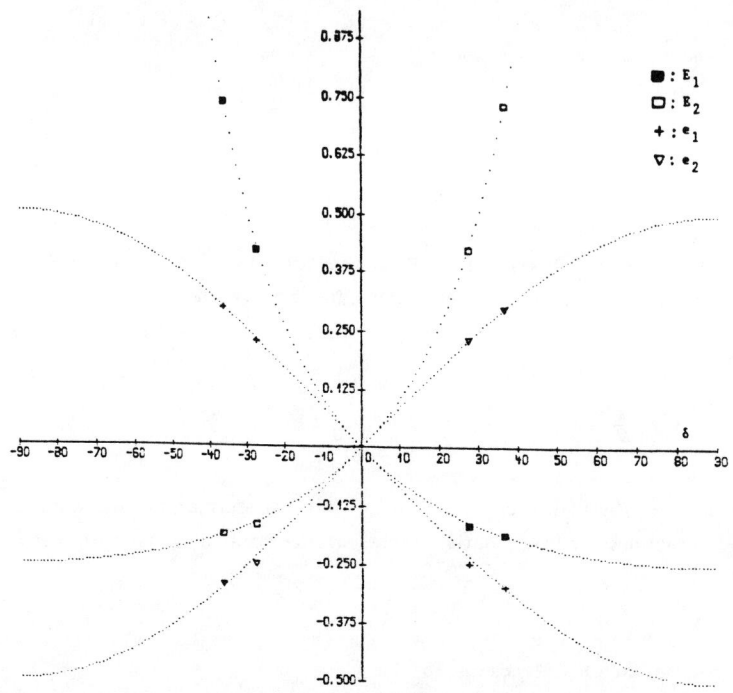

Fig. 5 Principal lagrangian strains E_1, E_2
Principal eulerian strains e_1, e_2

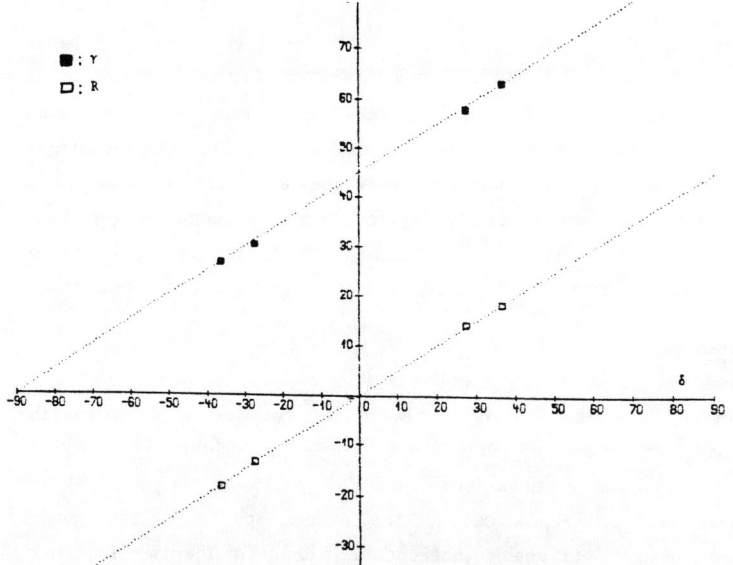

Fig. 6 Orientation of the principal strain and rigid body rotation

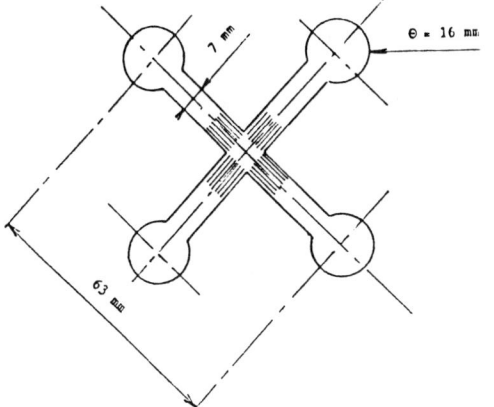

Fig. 7   Geometry of the tension specimen B

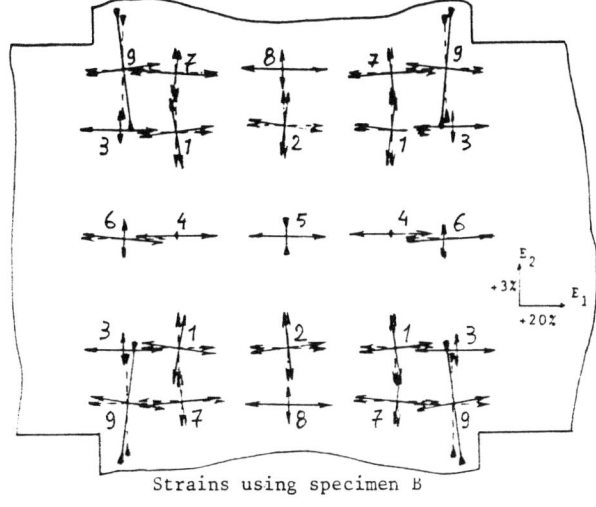

Strains using specimen B

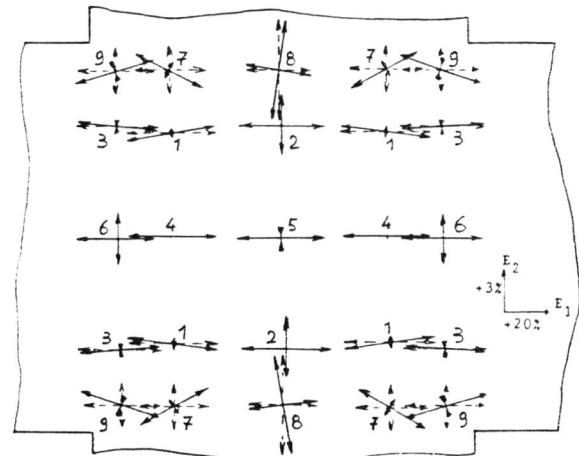

Strains using specimen A

Fig. 8   Comparison between the two specimens

In first we studied strains on all the central part of each shape. Let us call : specimen - A the classical cross-shaped specimen and specimen - B the shape made with lamellaes. Figure 8 shows the comparison of these two tests where the loading was F_1 = 4 kg and F_2 = 2 kg. In fact we measured strains in nine points and we used symmetry conditions to extend the values anywhere. The black lines represent the principal deformations. On can note E_2 is smaller than E_1 ($E_2 \simeq \pm$ 2 %, $E_1 \simeq$ 20 %). We note also, a biaxial tension state exists in a little region around the central point 5. Further more, the principal and loading directions correspond better for the specimen B than for A in the corner 3, 1, 7, 9.

In a second time we measured strain and calculated stresses in the central point 5 of the two specimens. These experimental data were compared with theoritical values get from a given mechanical behaviour law. Then we supposed the material was incompressible, hyperelastic of neohookean type, and one can easily obtain the following relationship between stress and strain tensor :

$$\overline{\overline{T}} = p \, \overline{\overline{1}} + G \, \overline{\overline{c}}$$

where

$\overline{\overline{T}}$ is the Cauchy's stress tensor.

p represents an hydrostatic pressure function of a point.

G is the modulus of rigidity in sliding.

Using the incompressibility equation ($c_1 \, c_2 \, c_3$ = 1) one can write

$$(\sigma_1 - \sigma_2) = G \, (c_1 - c_2) \qquad (1)$$

$$\sigma_1 = G \, (c_1 - \frac{1}{c_1 \, c_2}) \qquad (2)$$

$$\sigma_2 = G \, (c_2 - \frac{1}{c_1 \, c_2}) \qquad (3)$$

The first relationship shows the linearity between the difference of the principal stresses and strains (figure 9). But the values of G are different and depend of the type of the test and the shape of the specimen. Nevertheless we remark G is the same in the biaxial tension with specimen B and uniaxial tension with specimen A. From the equations 2 and 3 and from this last value of G we computed c_1 and c_2 in function of σ_1 for a given σ_2 (figure 10). Then we note the perfect correspondance between experimental points and theoretical curves a and d . Hence we deduce the biaxial tension test done on specimen B generalizes the uniaxial test done on specimen A but not biaxial test with A. To confirm this we made birefringence measurements in the central point 5. Then we suppose Maxwell's laws are verified even in the finite deformation field. So we can write the linear relationship between N (the fringe order) and $\sigma_1 - \sigma_2$ such that :

$$N - \frac{Ce}{\lambda} (\sigma_1 - \sigma_2) \qquad (4)$$

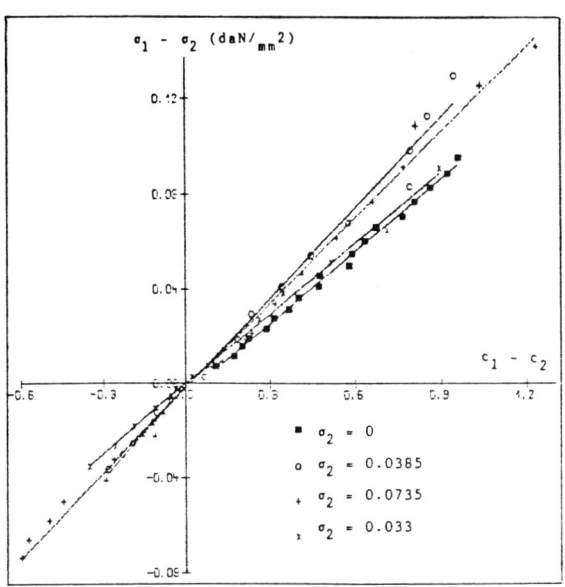

Fig. 9   Linearity between stresses and strain

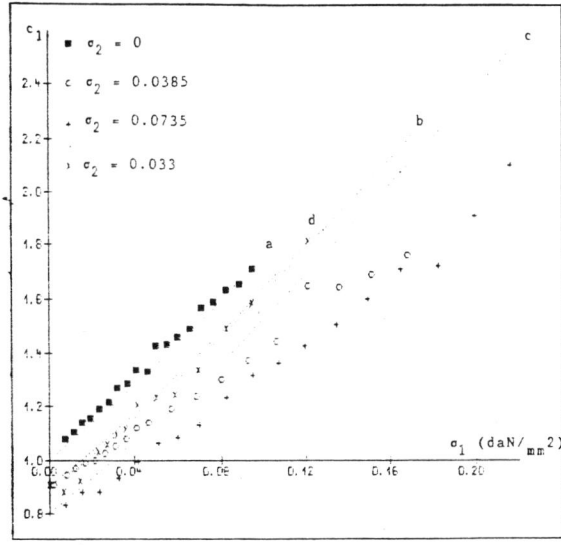

Fig. 10   Strain c_1 versus σ_1 (σ_2 given) in the central point

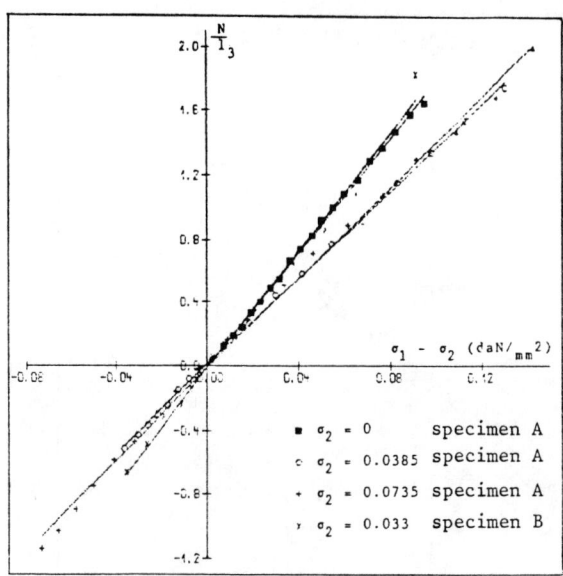

Fig. 11 Fringe order – principal stresses

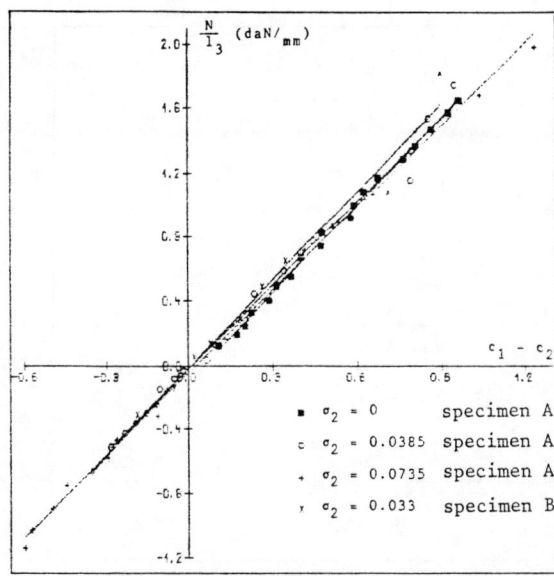

Fig. 12 Fringe order – principal strains

where   C is the photoelastic constant

             e is the thickness of the specimen

             λ is the wavelength of the laser beam.

From (4) we obtained figure 11 which always shows a comparison between experimental and theoretical data. Again we can see a difference on the proportionality coefficient showing the specimen A is not adapted in biaxial tests. Yet if we plot (figure 12) the fringe order versus the principal strains, there is superposition of every curves. It is easy to understand. Alone this last figure is withtout stresses concluding : the problem is on the determination of the stresses which are not very well biaxial using the shape A, because there is dependance between the loading directions.

3. Conclusion

Our measuring method leads at the knowledge of the orientation and magnitude of the principal strains and of the rigid solid rotation on a small region. It allowed to measure strains in the central part of two cross-shaped tension specimens using a 10 lines/mm orthogonal grating. The different results show only the shape done with lamellaes give a biaxial stress state when loading in biaxial tension test.

4. References

1   Sevenhuijsen P.J. : "The development of a laser grating method for the measurement on strain distribution in plane opaque surfaces." 6th international conference on experimental stress analysis, 1978 München.

2   Brémand F. et Lagarde A. : "Méthode optique de mesure des déformations utilisant le phénomène de diffraction". CRAS Paris, t. 303, Série II, n° 7, 1986.

3   Brémand F. et Lagarde A. : "Optical method of strain meeasurement. Application to study of circular bending of beam in the large strain range" VIIIth International Conference on Experimental Stress Analysis Delft May 12-16, 1986.

4   Brémand F. et Lagarde A. : "A new method of optical strain measurement with applications" SEM Spring conference on Experimental Mechanics New-Orleans June 8-13, 1986.

5   Bruhat J. : Cours de physique générale-Optique. 6ème édition revue et complétée par A. Kastler, Masson et Cie 1965.

6   Françon M., Maréchal A. : Diffraction - Structures des Images. Masson et Cie 1970.

7   Mönch E. and Galster D. : Brit.J.Applied Phys. 14 810, 1963.

8   Hayhurst D.R. : "Experimental testing techniques for high temperature" Seventh international Conference on experimental stress analysis Haifa Israël 23-27 August 1982

APPLICATION OF MOIRE WITH COHERENT AND NOT COHERENT LIGHT TO DYNAMIC MEASUREMENTS OF DEFORMATION UP TO 10^3 s^{-1} STRAIN-RATE

Dr. Carlo Albertini, Dr.Ing. Mario Montagnani, Mr. Erminio Pizzinato
Commission of the European Communities
Joint Research Centre - Ispra Establishment
21020 Ispra (Va)   Italy

Professor Dr.Ir. Pieter Boone, Mr. Gustave van der Steen
Rijksuniversiteit - Gent, Laboratorium Soete, Belgium

Professor Dr.Ing. Luciano Pirodda
University of Cagliari, Italy

Moiré techniques with coherent and not coherent light have been successfully used in measuring deformation parameters during tensile tests at
strain-rate up to 10^3 s^{-1} performed with Hopkinson bar devices. The
deforming grating was always formed on a photoresist support while
the reference grating was supported by stripping film or formed by
coherent light interference. Fringes were recorded by a fast camera at
framing rates up to 10^5 s^{-1}.

Keywords: Moiré fringes, fast camera, high strain-rate

1. Introduction

The determination of the mechanical properties of materials at high strain-rate
up to 10^5 s^{-1} is requested for the calculation of mechanical structures able to
withstand dynamic loads, as well as for the optimization of the thermomechanical
transformation processes of metals. At such very high strain-rates, the load distribution in the measuring equipment and in the specimen is made under the condition
of wave propagation. Therefore, all the loading systems, the measuring equipment
and the design of the specimen must be properly chosen in order to govern in detail
the stress distribution by wave propagation. The split Hopkinson bar [1] is the main
piece of equipment developed following such criteria and it is supposed to achieve
a homogeneous stress distribution by successive reflections of the elasto-plastic
waves propagating in a specimen which is short in comparison with the length of the
pulse. Therefore, specimens - a few millimetres long - are used, on which the stress
and strain distributions must, consequently, be carefully checked. Unfortunately,
the use of very small specimens renders the effects of the restraints critical in
dynamic tensile tests, impeding lateral deformation, due to the stiffness of the
connections between gauge length and loading equipment.

IMEKO

These crucial points have been verified by comparing measurements of the full field strain distribution on small and large specimens of austenitic stainless steel AISI 316 performed by the Moiré technique, with coherent and not coherent light during tensile tests at strain-rates up to 10^3 s^{-1}.

2. Moiré techniques applied to dynamic tensile tests on small specimens

The small specimens in question have a gauge length of 5 mm (Fig. 1) and a cylindrical or rectangular cross-section; they are tested in the modified Hopkinson's bar, sketched in Fig. 2 and described in [2]. The measurement of the full field strain distribution over the short gauge length has to be performed over the three characteristic deformation phases of metals tested in tension. These phases are: elastic, uniform plastic and plastic with necking, respectively. The three deformation phases are characterized by very different values of the particle displacement, the measurement of which requires different Moiré techniques, as described below.

2.1 Displacement measurements in the elastic field

In order to measure the elastic displacement over the full field, a method having a sensitivity of about a micron is needed. The interferometric Moiré technique, realized with coherent light [3,4], has been applied with the experimental set-up as shown in Fig. 3. A grating is fabricated on or attached to the specimen

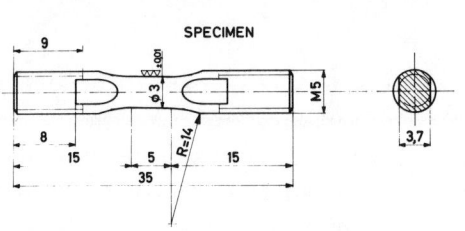

Fig. 1 Specimen

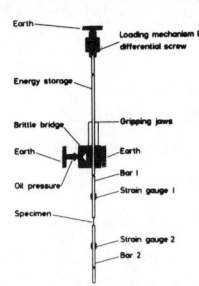

Fig. 2 Modified Hopkinson bar with
prestressed bar loading device

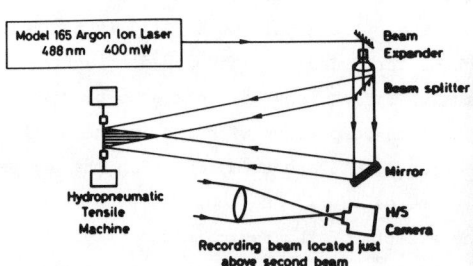

Fig. 3 Moiré interferometry.
Optical set-up

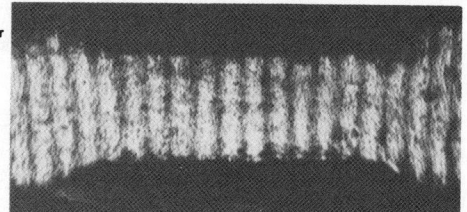

Fig. 4 Moiré fringes with coherent
light on specimen 10x15 mm

surface and it deforms with the surface during a test. The two coherent light beams
incident on the grating are diffracted by it and the interference of the diffracted beams produces a set of fringes on the surface which are recorded by a high
speed camera. Up to now [5] we have produced gratings of $\sim$ 450 lines per mm and we
were able to record displacement fields (Fig. 4) at a frame frequency of 3500 per
second using a high speed camera. This frame frequency is too low for Hopkinson's
bar experiments in the elastic field where a frame frequency of at least 10^5 per
second is required because of the short rise time of the loading pulse (about 30 μs).

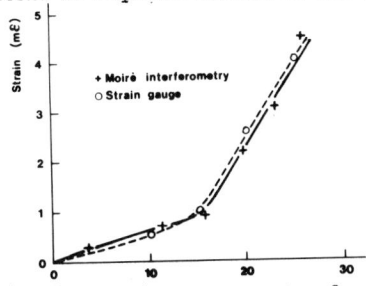

Fig. 5 Dynamic measurements of
        strains

Nevertheless, in a test performed at an
average strain-rate of 1 s^{-1} and with a
grating of $\sim$ 450 lines per mm, it was
possible to check that the strain measurement was in good agreement with the
one obtained from an electrical strain-
gauge fixed on the reverse side of the
specimen, as shown in Fig. 5 [5].

2.2 Displacement measurements in a uniform plastic field

    In this case, due to the larger displacement to be measured, we have used a
deforming grating of 40 lines per mm, reproduced on photoresist Kodak KPR, distributed on the surface of the specimen. The reference grating was reproduced on a
stripping film [5] (Gevaert 0825 Gevalith film) which was superposed on the deforming
grating having in-between a thin film of Kodak photoflow, which ensured contempo-
raneously the smooth adherence and the relative sliding of the reference grating over
the deforming one. Illumination was made with white light. By this experimental set-
up it was possible to record the displacement field over the gauge length at a frame
frequency of 10^5 per second, using a Cordin model 377 high speed camera, as shown
in Fig. 6, during a tensile test in which the specimen was deformed for 50% in
500 μs ($\dot{\varepsilon}$ = 1000 s^{-1}). This figure, taken before the beginning of necking, also
shows a displacement gradient from the extremities towards the centre of the specimen which is caused by the stiffness of the connections between gauge length and the
loading half bars, which impede lateral deformation.

    In order to discriminate between the effects due to lateral restraints and those
due to wave propagation, it would be necessary to record the displacement field at a
frame frequency higher than 10^6 per second; that means at an interframe time lower
than the travel time of the wave along the gauge length. Up to now we did not have
a fast camera permitting such a frame rate so that we decided to observe the displacement distribution by the same Moiré technique at lower strain-rate, where no
wave propagation effects exist, with a frame frequency of about 10 frames per minute.
The results are shown in Fig. 7. The comparison of Fig. 6 with Fig. 7 suggests that

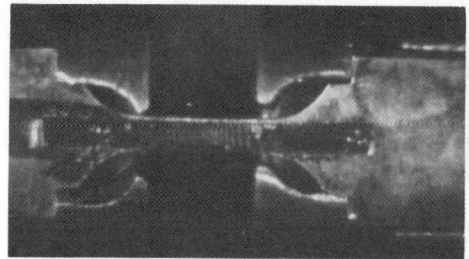

Fig. 6 Moiré fringes with non
coherent light.
Hopkinson bar test

Fig. 7 Moiré fringes with non
coherent light. Static test

probably the effect of the lateral restraint is predominant with respect to the wave
propagation effect in determining the observed displacement gradient. In order to
better analyse the effect of the lateral restraint, the contemporaneous observation
of the displacement fields along the two orthogonal axes (Fig. 8) was also successfully tried using the same technique, for a test at low strain-rate (10^{-2} s^{-1}).

2.3 Displacement measurements in the plastic field during necking

For the recording of the large displacements immediately before and during
necking, due to the small size of the specimens, a grating of four lines per mm was
directly filmed at a frame frequency of 10^5 per second without previous interference
with a reference grating, as shown in Fig. 9. This type of record has been used in
order to calculate the true stress - strain diagram during necking following the
Bridgman analysis, which requires the measurement of both the diameter of the
smallest specimen cross-section and the diameter of the circle which copes with the
profile of the necked area. These measurements can be taken from Fig. 9. Furthermore,
from the record it is possible to check the validity of the Bridgman analysis [7] for
the material under investigation by comparing the true strain calculated from the
diameter measurements and the longitudinal strain measured in the neck, from the
photographic record. These investigations are in progress.

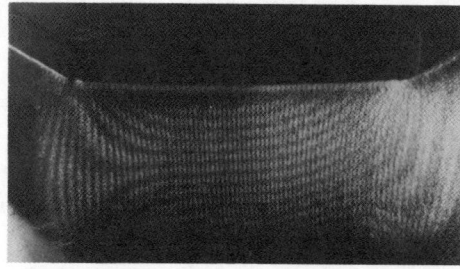

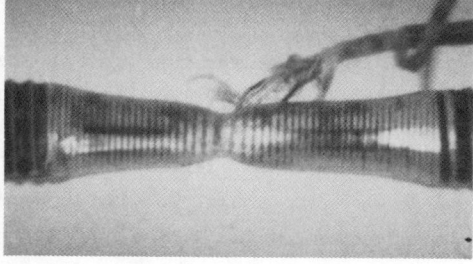

Fig. 8 Biaxial Moiré fringes with non
coherent light

Fig. 9 Grating of 4 l/mm and necking.
Hopkinson bar test

3. Moiré techniques applied to dynamic tensile tests on large specimens

In order to check the existence of size and shape effects on the dynamic mechanical properties of materials, tensile tests have been performed on large specimens
of cylindrical and rectangular cross-sections, where the displacement field was
measured directly on the specimen by the Moiré technique.

3.1 Check of the effect of size at high strain-rate on large cylindrical specimens

Specimens, 10 times longer and 100 times larger in cross-section than as shown
in Fig. 1, have been tested using a 5 MN large dynamic test facility. The displacement field has been measured by the same Moiré technique as described in 2.2, with
a grating of 2 lines per mm and a frame frequency of 4000 per second (Fig. 10), as
well as by discrete electrical strain gauges. A displacement gradient is also present along the specimen from the extremities towards the centre of the gauge length,
which confirms the lateral restraint given by the connections between gauge length
and test apparatus. These results are also under more detailed evaluation at present.

3.2 Check of the effect of shape at high strain-rate

Specimens of rectangular cross-sections with a width of 80 mm and a thickness
varying from 10 to 30 mm, have been tested. We used the same Moiré technique as
described in 2.2, based on the use of stripping films. In this case the displacement field has been measured both over the width side and over the thickness side
of the specimen, as shown in Fig. 11, which allowed us to obtain detailed information about the flow mode of the material.

Fig.10 Moiré fringes with non coherent
        light of a large cylindrical
        specimen. Dynamic test

Fig.11 Moiré fringes with non coherent
        light of a large flat specimen.
        Dynamic test

4. Automatic evaluation of full-field displacement, strain and strain-rate

A computer software package has been developed which permits us the evaluation,
point-by-point, of the displacement, strain and strain-rate values, as shown in
Fig. 12. Up to now we have digitized the coordinates of the Moiré fringes by a
hand-activated magnetic table. This procedure represents the main bottle-neck for
an efficient use of the Moiré techniques in measuring mechanical parameters during
dynamic materials testing. We are now trying to automatize the digitizing read-out
procedure by using a solid-state electronic video camera controlled by a personal
computer.

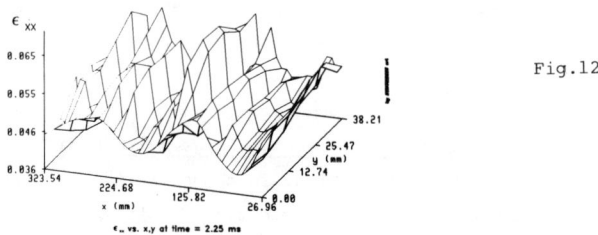

GRAPHIC OUTPUT FROM MARISA CODE
HORIZONTAL STRAIN
SAMPLE CASE MARISA

Fig.12

5. Conclusions

Moiré techniques, based on stripping films and not coherent light, have been successfully employed in measuring deformation parameters during tensile tests at high strain rates up to 10^3 s^{-1} on very small and very large specimens. These measurements gave, in a first approximation, an indication of the important influence of lateral restraint during high strain-rate testing, due to the imperative need of using short specimens in order to average instantaneously the loading by wave propagation. This indication is needed for the correct evaluation of deformation parameters and for developing more efficient specimens and test equipment.

Interferometric Moiré with coherent light needs further development in order to be applied at very high strain-rates.

6. References

1   Davies, R.M.: A critical study of the Hopkinson pressure bar. Phil. Trans. Roy. Soc., London, Ser. A, 240, 375 (1948).

2   Albertini, C., Montagnani, M.: Dynamic material properties of several steels for fast breeder reactor safety analysis. EUR 5787 EN.

3   Boone, P.M.: Strain 5(2), 89 (1962).

4   Pirodda, L.: Strain analysis by grating interferometry. Optics and Lasers in Engineering 5 (1984) 7-28.

5   Boone, P.M., Vinckier, A.G., Denys, R.M., Sys, W.M., Deleu, E.N.: Application of specimen-grid Moiré techniques in large scale steel testing. Optical Engineering, July/August 1982, Vol. 21, Note 615.

6   Griffiths, L.J., Lu, H., Pizzinato, E.V.: A full-field, high sensitivity dynamic testing technique using Moiré interferometry. Technical Note EUR I.07.01.82.193.

7   Bridgman P.W., Studies in large plastic flow and fracture, Mac Graw-Hill, 1952

OPTO-ELECTRONIC MEASUREMENTS USING DIFFRACTION OF LIGHT
IN DYNAMIC EXPERIMENTS

Milan Držík, CSc
Institute of Construction and Architecture
of the Slovak Academy of Sciences
842 20 Bratislava
Dúbravská 9
Czechoslovakia

An optical **method** is presented to measure the dynamic
strains on the deformable bodies. The light passing through
a thin slit is sensing by photo-electric detector and by
this way the variations of the slit opening may be recorded
immediatelly on the oscilloscope screen. The sensitivity
of such displacement measurement is comparable with
wavelenghts of light. An attention has been payed to the
questions of rigid body motion as well as to the output
linearity of the measurement system.

Keywords: displacement measurement, photo-electric
recording, diffraction of light

1. Introduction

The effect of the diffraction of light ranks among the optical
principles applicable to the dimension and/or deformation
measurements. Despite the fact that this idea is experimentally
simple to materialize the technique has been little used by stress
analysts as yet. At this occasion we can mention the papers of
Pryor, North and Hageniers [1-3] and others [4] . This little
utilization is probably due to the fact that the use of
diffractographic method as a measurement tool for strains
determination has a serious limitations
i/ very small sensitivity for the slit opening variations
ii/ inaccurate subjectively influenced manual identification
of the relative positions of the intensity maxima
In order to suppress these disadvantages a photo-electric recording
of the intensity in a diffraction pattern may be used [5] .
Moreover, this photo-electric approach offers a possibillity to
work at the time variable dynamical conditions.

2. Theoretical background

To illustrate the behavior of the light intensity behind the slit in an opaque screen, we calculate the Fraunhofer diffraction pattern. We assume the illumination to be a normally incident plane-wave field of amplitude $U(x)$. We also assume $z \gg w$ (where z and w are the distance slit - observation plane and width of the slit respectively) so that Fraunhofer conditions are satisfied. The transmittance function $t(x)$ of the slit is equal

$$t(x) = rect\left(\frac{x}{w}\right) = \begin{cases} 1 & -w/2 \leq x \leq w/2 \\ 0 & |x| > w/2 \end{cases} \tag{1}$$

Thus using the diffraction integral [6] we obtain the amplitude at the observation plane

$$U(x_0) = U(x)\, w\, sinc\left(\frac{w\,x_0}{\lambda z}\right) \tag{2}$$

where the function

$$sinc\left(\frac{w\,x_0}{\lambda z}\right) = sin\left(\frac{w\,x_0}{\lambda z}\right) \Big/ \left(\frac{w\,x_0}{\lambda z}\right). \tag{3}$$

So, the intensity distribution registered by the quadratic detector

$$I(x_0) = |U(x_0)|^2 \left(\frac{w\,x_0}{\lambda z}\right)\, sinc\left(\frac{w\,x_0}{\lambda z}\right). \tag{4}$$

This is the well known form of irradiance with the expressive central peak of maximum intensity and the small alternating maxima and minima which are vanishing on both sides of the pattern. As it can be seen from Eq.4, the irradiance varies as the square of width of the slit. If the effective area of the photo-electric sensor (with linear characteristic and point-like dimensions) is placed to the maximum intensity the output electric signal depends quadratically on the measured value, too. So, the response of the system to the slit opening displacement is nonlinear and an electronic correction circuit would be needed in this case.

More simple and reliable way how to secure the output to be linear, is to use an integral recording of the light intensity in a diffraction pattern as a whole. To do this, the diverging "fan" of light is collected by a wide-open collimating lens. By this way the image of the slit is projected onto the monocrystal of the photo-diode. In such arrangement a contribution of the high order maxima is negligible and the photo-diode is illuminated practically by all of light passing through the slit.

However, the central problem of such opto-electronic setup is how to achieve the system to be invariant with regard to relative motions of optical components (namely the slit) and fixed laser beam. This is very important in dynamical conditions considering an outside vibrational influences as well as possible large rigid body motion of the measured object at the phase of dynamic loading. We can neglect small rotational motions ($\sim \pm 5^\circ$) around both the axises - rotational of the laser beam and parallel to slit edges. In the latter case an apparent width of the slit varies as a cosine factor of the rotational angle consequently the light irradiance behind the slit do not change more than $\sim 1\%$.

Much more dangereous is a shift of the slit in a direction perpendicular to its edges outside the laser beam center. Usually the CW laser resonators emitte the TEM_{00} mode radiation so that the Gaussian transverse irradiance profile is present. Any displacement of the narrow slit in such case gives rise a considerable false output signal [7] . Therefore, an essential condition is to create at the plane of the slit a transversally constant irradiance distribution. To do this and simultaneously preserve the simplicity of the optical system we have payed attention to the spherical abberation of the simple thin positive lens. Spherical abberation appears when marginal rays (passing through the lens near the edge) from a point source come to a focus at different distances from those which pass through the central zone of the lens (Fig.1). Here is the focal region shown in meridional section. The foci of both

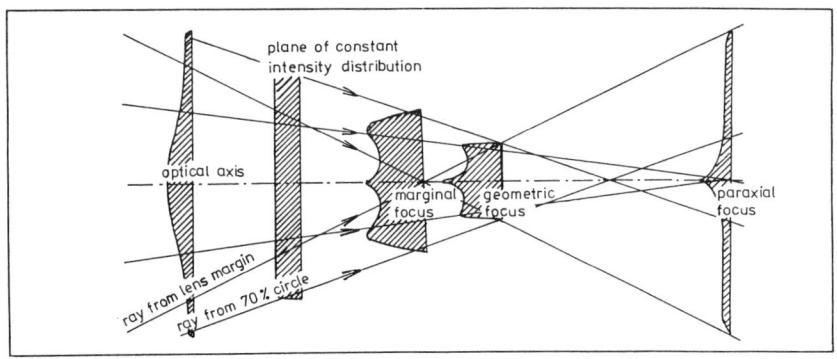

Fig.1 Detail of focal region of the rays focused by simple thin positive lens with spherical aberration

paraxial and marginal rays as well as the plane of minimum geometric blur - geometric focus are drawn. The irradiance distributions at the characteristic planes are shown provided that the Gaussian laser beam is concentrated in a pinhole. As it can be seen in some distance in front of the point where marginal rays are concentrated a constant level of irradiance across the beam may be accepted. This is the plane of best position for the narrow slit used as a measuring tool. To search this position theoretically it is a troublesome task because of many real characteristics of the optical system under consideration are needed to include. So the preliminary experimental measurement was carried out by scanning of the laser beam using photosensor screened by a pinhole diaphragm. By this way in any particular case optimum parameters such as the mutual distances of components, an aperture number and focal distance of lens as well as the parameters of pinhole objective may be chosen.

Another specific problem connected with the photo-electric sensing is a high frequency noise of CW laser light. The typical frequency of the noise by which the light intensity is irregularly modulated lies in the region around 100 kHz. Its value may exceed $\pm$ 10 % of the maximum intensity and an output signal may be completely hidden in the noise. Therefore, it is necessary to use a differential electronic scheme for the separation of the usefull signal. We have developed a simple electrical bridge with both sensing photo-diodes and resistors (Fig.2). The bridge includes three

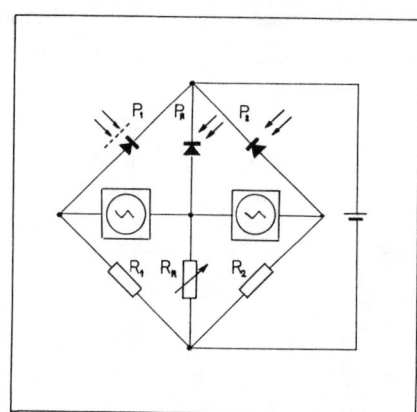

branches one of which serves as a reference for the information sensing detectors. This reference photo-diode P_R is illuminated directly by a laser light separate on a beamsplitter and aimed at the photo-diode. Both the information channels are mutually independent, thus two processes are simultaneously tracked. Cut-off frequency of the whole circuit including sensors and resistors was approximatelly 100 MHz, much more as it is usually needed for the research in mechanics.

Fig.2 Scheme of electrical bridge for compensation of laser light noise

3. Experimental arrangement

The arrangement shown in Fig.3 has been used in which a CW He-Ne 60 mW laser was the source of light and GaAs photo-diode was the reciever of radiation. By a simple lens a light diverging from the point source was focused onto the slit. The positive lens with an aperture number 1 : 3,2 was chosen and this was sufficient to create a constantly illuminated area with a diameter of about 3 mm. An aperture of the slit was created by the two sharp edges of razor blades. These were mounted using an adhesive to a steel specimen on both the sides of crack and in the second case on the opposite sides along the diameter of the ring-shaped dynamometer. The widths of the slits were precisely adjusted by means of diffraction pattern

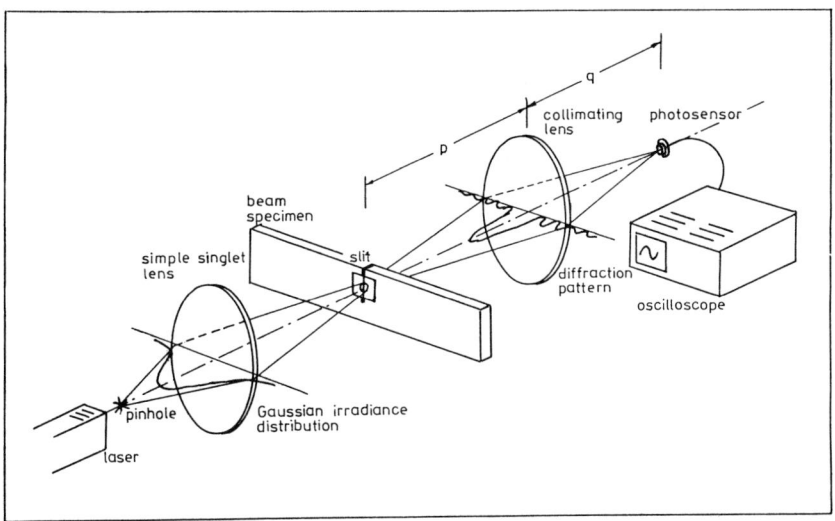

Fig.3 Schematic diagram of the experimental arrangement

projected onto the testing screen. Because of two slits were simultaneously measured - one for registration of crack opening displacement and the other for dynamometer - both the slits were adjusted to the same value of the width. Thus, the reference branch of bridge serves for both the channels. Alternativelly, the bridge may be balanced by using variable beamsplitter in front of photodiode P_1 (or P_2). For the sake of simplicity only the crack opening displacement measurement is schematically drawn in Fig.3.

If the two independent measuring channels are needed the beamsplitters are included in the optical scheme. These are dividing the laser beam to three beams one of which illuminates the reference photo-diode.

4. Conclusions

An optical non-contact method for measuring strains using diffraction of light was developed. Photo-electric sensing of strains has advantages over other techniques - including resistance strain gages - first of all in hostile conditions such as temperature variations, humidity and dynamic vibrations. The proposed arrangement is shown to have a sufficient sensitivity, good linearity as well as dynamic stability. In the experiments time dependencies of load vs. crack opening displacement were recorded with resolving power of μs. It is believed it may be effective in a number of applications.

5. References

1 Pryor, T.R., North, W.P.T.: The diffractographic strain gage, Exp. Mech. 11(12), 565-568, 1971
2 Pryor, T.R., Hageniers, O.L., North, W.P.T.: Displacement measurement along a line by the diffractographic method, Exp. Mech. 12(8), 384-386, 1972
3 Pryor, T.R., Hageniers, O.L., North, W.P.T.: Diffractography vs. holography - a stress analyst´s comparison, Exp. Mech. 13(5), 220-224, 1973
4 Keprt, J., Zemánek, Z.: Bezkontaktní měření průměrú lineárních útvarú užitím difrakce světla a moiré techniky, JMO 24(1), 3-6,1979
5 Schubert, S.: Laser-optical measurement of the deformation of surfaces due to dynamic loading, Proc. 10th Congress Imeko, Vol. 4 Praha 1985
6 Goodman, J.W.: Introduction to Fourier optics, McGraw-Hill Inc., 1968
7 Klimkov, J.M.: Prikladnaja lazernaja optika (in Russian), Mashinostrojenije, 1985

THE DEVELOPMENT AND APPLICATION OF

A THERMOELASTIC TECHNIQUE FOR STRESS ANALYSIS

Professor Peter Stanley, M.A.,M.Sc.,Ph.D.,

Department of Engineering, University of Manchester,

Simon Engineering Laboratories, Oxford Road,

MANCHESTER M13 9PL, England

A highly sensitive infra-red radiometric system (the
SPATE equipment) has been developed that can respond
to the stress-induced temperature changes in solids.
The principal features of the equipment are described
and its use as a non-contact full-field stress analysis
device is illustrated with reference to a number of
quantitative stress analysis studies.

Keywords: infra-red radiometry, thermoelasticity, stress analysis

1. Introduction

Stress changes in a solid body give rise to small temperature changes [1]
which result, in turn, in the emission of infra-red radiation from the surface of
the solid. The measurement of the infra-red radiant flux by the means of a linear
detection system is the basis of the thermoelastic technique for the quantitative
determination of the stress changes. Using such a system, the relationship
between the received signal S and the changes in the principal surface stresses
(σ_1, σ_2 and σ_3), assuming adiabatic elastic conditions, is

$$\sigma_1 + \sigma_2 + \sigma_3 = AS \tag{1}$$

in which the constant A is a "calibration factor" dependent on the characteristics
of the system and the properties of the material [2].

1st IMEKO TC15 Conference, 4th Danubia-Adria Conference, 25th Czechoslovak
Conference on Experimental Stress Analysis, Plsen, Czechoslovakia,
May 25-28, 1987

 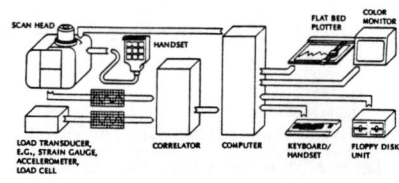

Fig.1 The SPATE equipment   Fig.2   SPATE schematic

The measuring system (the SPATE equipment - Stress Pattern Aanalysis by the
measurement of Thermal Emission - made by SIRA Ltd., England) is shown in Fig.1
and, in block diagram form, in Fig.2. It consists essentially of (i) a scanning
head and detector unit, (ii) an analogue signal processing unit and (iii) a
digital control, display and storage unit. The detector is a lead/tin/telluride
cell, approximately 100 μm in width, cooled by liquid nitrogen and operating over
the 8-14 μm wavelength band. A two-mirror raster-type scanning system is
incorporated so that a selected area of the test specimen or component can be
studied. The correlation of the received signals, in frequency, magnitude and
phase, with a reference signal derived from a transducer in the loading system or
on the component, is an important feature of the equipment.

In practice a selected surface area is scanned, point-by-point, while the
specimen or component under investigation is subjected to cyclic loading. The
received signals are stored in digital form and can be displayed on a high-
resolution video monitor as a 16-colour linearly sub-divided stress contour map or
produced, in hard-copy, as a line plot using an x-y plotter. There is a wide
range of choice in the size of the scanned area, the number of sampling points and
the sampling time per point. The resolution limit is 0.6 x 0.6mm. The
temperature sensitivity of the equipment is such that stress changes of the order
of 1 N/mm^2 in steel and 0.4 N/mm^2 in aluminium can be resolved. Further important
practical details are given in references [2,3].

Calibration studies [2] using simple specimens (e.g. beams and discs) have
confirmed the linearity and consistency of the response of the SPATE equipment and
A values (see equation (1)) have been obtained for several materials. Further
work on plates and discs under in-plane loading and on a range of other test
pieces and components, both metallic and non-metallic, has been described in the
literature, [3,4,5,6]. Two recent applications are outlined here to illustrate
the characteristics of the technique.

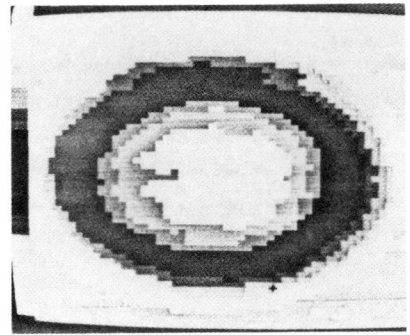

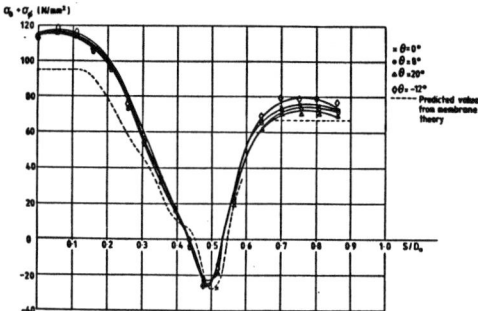

Fig.3 SPATE scan of knuckle region   Fig.4   Stress sum distributions

2. Model pressure vessel end-closure

Semi-ellipsoidal and torispherical end-closures are used for a very wide range
of cylindrical pressure vessels, and it is known [7] that for such ends, with a
thickness to diameter ratio of 0.002 or less, subjected to internal pressure, the
governing failure mode is buckling due to the circumferential compressive stresses
developed in the end. The stress distribution in an end-closure, and hence the
tendency to buckle, is very dependent upon the shape of the meridional profile of
the end; other end shapes may be stronger than similar semi-ellipsoidal or
torispherical ends. A "freely formed" end has been studied in this context.

A small closed cylindrical pressure vessel (76mm diameter, 165mm long, 0.4mm
thick) was formed from two pressed cylindrical steel "cups" joined together by a
soldered lap joint. The lower end of the vessel was hemispherical and contained a
pressure connection; the upper end was initially flat and, prior to testing, was
"freely formed" into a "dished" shape by the application of a steadily increasing
internal pressure in the vessel. Attention was concentrated on the "knuckle"
region of the "freely formed" end (i.e. the region of the end adjacent to the
cylinder) where high stresses usually occur and where buckles develop in
torispherical ends [7]. A SPATE scan of this region, with the vessel subjected to
a cyclic internal pressure, is shown in Fig.3. Stress sum distributions, obtained
from the stored data, along four meridional lines from the centre of the end,
through the knuckle and into the cylindrical portion of the vessel are shown in
Fig.4. These results show clearly that there is a predominant compressive stress
in the knuckle region of the end and it was concluded that buckling is a possible
failure mode in ends of this type.

A theoretical stress sum distribution, derived from membrane theory using
curvature values obtained from detailed end-profile measurements, is also shown in
Fig.4. The theoretical curve corresponds closely with the experimental curves in
the knuckle region. This leads to the important additional conclusion that the

143

bending stresses in the knuckle of the "freely formed" end are small compared with the membrane stresses, contrasting markedly with the stress system in conventional ends.

These novel observations are of considerable value to pressure vessel designers and analysts. It is noteworthy that the thermoelastic study upon which they are based was completed in a few hours, with minimal preparatory effort.

3. Crack-tip stress fields

The analysis and characterisation of the stress field around a crack-tip are topics of major importance in fracture mechanics. A general relationship [8] between the SPATE signal from a point (coordinates r, θ) in the vicinity of a crack-tip and the mode I and mode II stress intensity factors (K_I and K_{II} respectively) has been derived from the Westergaard equations in the form,

$$AS = 2K_I \cos(\theta/2)/\sqrt{2\pi r} - 2K_{II} \sin(\theta/2)/\sqrt{2\pi r} \qquad (2)$$

Using this equation, K_I and K_{II} can be derived from SPATE data for either single-mode or mixed-mode loading. Graphical techniques for this purpose have been developed and applied [8]; work on mode I loading is outlined here.

For pure mode I loading (i.e. ($K_{II} = 0$) it can be shown from equation (2) that the signal maximum S_{max} in a line scan through the crack-tip region, parallel to the crack and distance y from the crack, is related to the value of y by the expression

$$y = \frac{3\sqrt{3} \ K_I^2}{4\pi A^2} \ \frac{1}{S_{max}^2} \qquad (3)$$

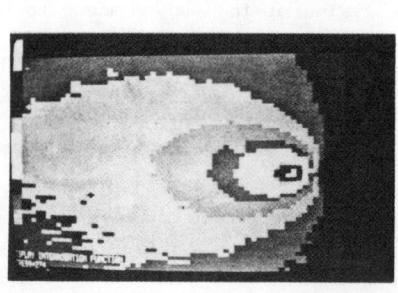

Fig.5   SPATE scan of crack-tip region

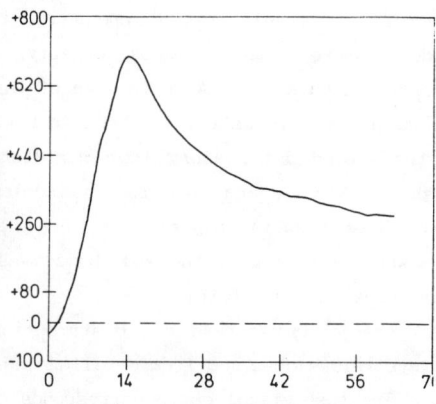

Fig.6   Line scan parallel to crack

144

K_I is therefore readily derived from the gradient of the linear portion of a plot of y versus $1/S_{max}^2$ obtained from a series of line scans parallel to the crack.

The mode I specimen was a flat steel plate (300mm x 300mm x 1.6mm) with a small central hole from which two cracks had been propagated by controlled fatigue loading before the test. A "black and white" version of the SPATE scan over the tip-region of one of these non-propagating cracks is shown in Fig.5, and a typical line scan parallel to the crack is shown in Fig.6.

The plot of y versus $1.S_{max}^2$ obtained from a series of such scans both above and below the crack is given in Fig.7. There are clearly defined linear portions in the graph. (Non-linear behaviour at the crack tip is associated with local plasticity.) Non-dimensional K_I values derived from the gradient of such plots for several stable crack lengths were on average 5% smaller than corresponding theoretical values. Alternative analyses of the SPATE signals gave equally satisfactory results [8].

This mode I work was extended to demonstrate the use of the SPATE technique in propagating crack studies. The governing relationship for the growth rate of a propagating crack is

$$da/dN = C(\Delta K_I)^m \qquad (4)$$

where a is the crack length, N is the number of cycles, and C and m are material constants. A second similar mode I test specimen was subjected to a cyclic load sufficient to cause controlled propagation of the two cracks. Fig.8 is a "black and white" version of the SPATE scan around one of the propagating crack-tips. K_I and a were determined for a particular crack length from a line scan along the line of the crack (i.e. $\theta = 0$).

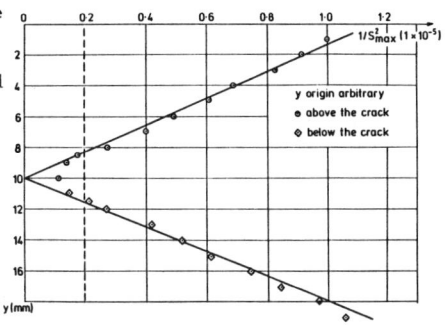

Fig.7 y versus $1/S_{max}^2$

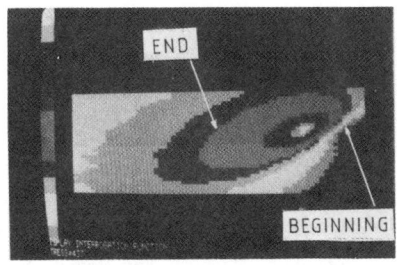

.Fig.8 SPATE scan of propagating crack

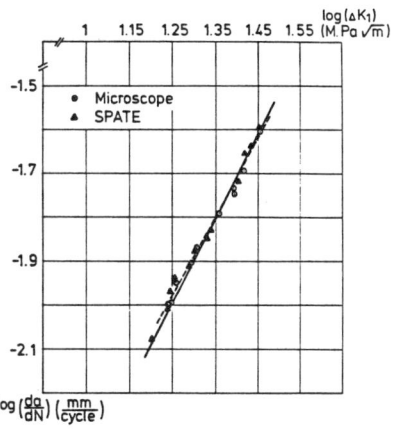

Fig.9 log(da/dN) versus $\log(\Delta K_I)$

145

Values of C and m (equation (4)) were obtained from the intercept and gradient of a log-log plot of da/dN versus ΔK_I - see Fig.9 - constructed from a series of such scans and were consistent with independently determined values.

The considerably reduced time and effort required for this work, compared with that required in alternative experimental approaches, is again noteworthy.

4. Closing remarks

The SPATE technique has been applied successfully to a diverse range of stress analysis problems and it is now widely recognised as an invaluable new tool. There are about 50 equipments in use world-wide; this number is growing steadily. Important advances can be expected as experience grows and as the very considerable scope for further development is exploited.

5. References

1   Thompson, W. (Lord Kelvin): On the dynamical theory of heat. Trans.Roy.Soc. Edinburgh, 20, pp 261-283, 1853.

2   Stanley, P. and Chan, W.K.: Quantitative stress analysis by means of the thermoelastic effect. J.Strain Analysis, 20, No.3, pp 129-137, 1985.

3   Stanley, P. and Chan, W.K.: SPATE studies of the stress distributions in steel plates and rings under in-plane loading. Proc. SEM Spring Conference on Exptl.Mechs., Las Vegas, U.S.A., June 1985, pp 747-757. (Also to be published in the January 1987 issue of Experimental Mechanics.)

4   Stanley, P. and Chan, W.K.: Stress studies in composite cylinders based on measurement of infra-red emissions due to cyclic loading. Proc.Int.Symp. on Mechs. of Polymer Composites, (Czechoslovak Academy of Science), Prague, Czechoslovakia, April 1986, pp 266-273.

5   Stanley, P. and Chan, W.K.: Mode II crack studies using the SPATE technique. Proc. SEM Spring Conf. on Exptl.Mechs., New Orleans, U.S.A., June 1986, pp 916-923,

6   Stanley, P. : An appraisal of a new infrared-based stress analysis technique, Optical Engineering, 26, No.1, pp 075-080, 1987.

7   Stanley, P. and Campbell, T.D. : Very thin torispherical pressure vessel ends under internal pressure : strains, deformations and buckling behaviour. J.Strain Analysis, 16, No.3, pp 187-203, 1981.

8   Stanley, P. and Chan, W.K.: The determination of stress intensity factors and crack-tip velocities from thermoelastic infra-red emissions. Proc.Int.Conf. on Fatigue of Engrg.Matls. and Structures, Vol.1 (paper C262/86), (I.Mech.E.), Sheffield, England, September 1986, pp 105-114

146

THEORETICAL AND EXPERIMENTAL STUDY OF BOND IN REINFORCED CONCRETE

Dr. Piet Stroeven - Dr. Jan Bien[x]
Stevin Laboratory, Faculty of Civil Engineering, Delft University of
Technology, Stevinweg 4, 2628 CN Delft, The Netherlands
[x]Institute of Civil Engineering, Technical University of Wroclaw, Poland

An analytical and an experimental approach to bond problems
are presented. The analytical method is based on Mindlin's
strain nuclei technique. In the experiments use is made of
two-dimensional pull-out specimens. Surface deformations
are visualized by means of holographic interferometry.

Keywords: bond, Mindlin analysis, holographic interferometry

1. Introduction

The reinforcing effect of steel fibres in concrete depends domi-
nantly on the bond properties of the steel-matrix interface. This
holds also for some other types of fibres. The bond phenomenon also
governs the length of lap splices of conventional steel reinforcement
in concrete. A better insight into the distribution of the stresses a-
long fibre or bar length and its effect upon the gradual debonding of
the interface under increasing loadings would therefore be of utmost
importance for more accurate designing. A thorough understanding of the
debonding phenomenon in terms of interphase morphology would also allow
for systematic improvements of the bond characteristics and, hence, of
the mechanical properties of the composite.

A research strategy towards the bond problem in reinforced concrete should therefore at least consider the following questions:
a. what is the stress-strain distribution in a simple pull-out specimen
   and what will be the changes induced by increased loadings?
b. how does this effect interfacial debonding of a bar or fibre?
c. how does interphase morphology influece the mechanism of debonding?
d. are these characteristics reflected by the mechanical tests?
These questions have been treated in a wide-ranging research program.
We will deal here exclusively with questions a. and b. To solve the

Int.Conf. Meas. Stat. Dyn. Par. Struct. Mat., Plzen May 25-28, 1987

problem stated in a. use is made of Mindlin's solution for a concentrated load in the interior of a homogeneous, isotropic half-space. By
superimposing a large series of such loads, the pull-out problem can be
simulated quite accurately. In the tests we employed two-dimensional
specimens in which the debonding front between a steel strip and the
mortar matrix was visualized by means of holographic interferometry.

2. Analytical approach

The strain nuclei technique is based on Mindlin loads. The situation of a single load is shown in Fig.1. All stress and deformation

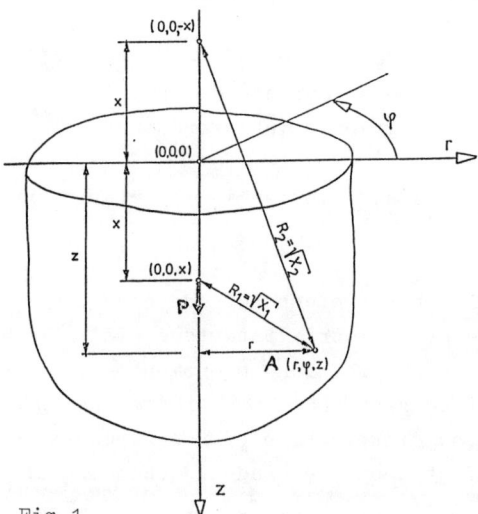

components are expressed in terms
of the location of the Mindlin load
$P(0,0,x)$ and the radial distances
from a point A to P and its mirrored position $P^X(0,0,-x)$ with respect
to the free surface, indicated by
R_1 and R_2. For the relevant formulae, see (1).

The embedded steel element is hypo-
thetically sectioned. A concentrated load P_i is subjected to the
centre of gravity of element i. P_i
is in equilibrium with the loads acting at the end sections. Hence,
$P_i = F_i - F_{i+1} = -T_i$ (F is load in bar).
T_i is the shear load at the external surface of the bar; it is spread

Fig.1

Sinle force P in a half-space

uniformly over this surface area. The array of loads $P_i (i=1,..,n)$ leads
upon substitution to the final stress-strain state. Compatibility in
longitudinal direction yields, by equating the strains in concrete and
steel, a system of n equations with n unknown loads $P_i (i=1,..,n)$. After
solving for these loads, all stress and deformation components can be
calculated. During debonding a mixed boundary value problem is obtained. In that case a maximum value of the shear resistance can be introduced. Basically, the approach stays the same, however.

Optimization of this analytical method should be pursued. Instabi-
lities occur near the place where the bar protrudes from the matrix.
These are due to the discretization. A single result is shown in Fig.2.
Interfacial shear stress is before debonding occurs hyperbolically distributed along the embedded bar, with a maximum value close to the con-

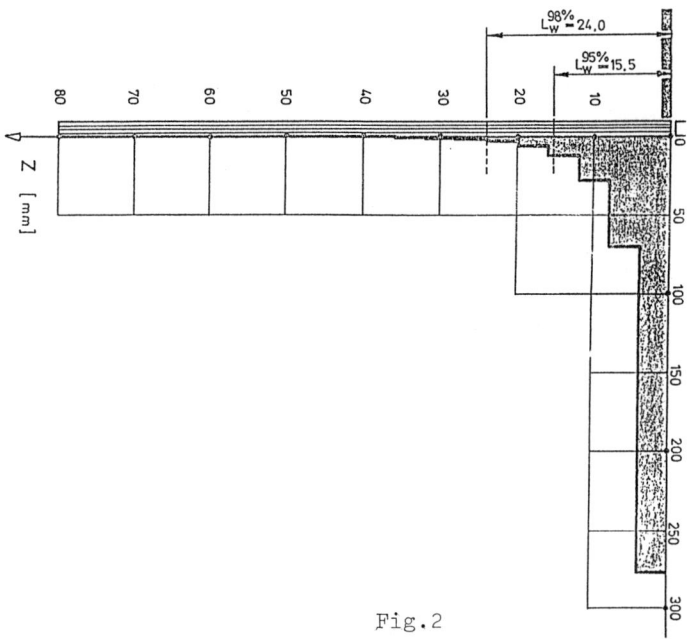

Fig.2

Shear stresses at steel concrete interface

crete surface. The length of the bar significantly contributing to the
stress transfer amounts to 15 bar diameters. This is independent from
bar length. This effective length of the bar decreases hyperbolically
with an increase in Young's modulus of the concrete and increases
slightly with an increase of Poisson's ratio of the concrete. For
further results, see (2).

3. Experimental approach

A large number of pull-out specimens have been tested during the
past years. The test series encompassed single steel fibres, reported
on in (3) and steel wires, 4 mm in diameter, tested according to RILEM
standards, reported on in (4). They serve as reference, as indicated
in the introduction under d. The latter tests were also evaluated in
microstructural terms. Samples were prepared from the matrix-wire interphase and polished. When studied under the microscope they revealed
the interphase layer and the slip line. For "smooth" surfaces, this
slip line followed the interface between steel and concrete. However,
for the "rough" surfaces (i.e. with a surface roughness surpassing RMS-
values of about 1.5 to 2 µm) the slip line was located inside the more
porous interphase layer having a thickness of 20 to 40 µm. These features of interphase morphology agree with data in the literature (5).

In the reported test series it was the objective to visualize the debonding front along the bar length. To that end, a two-dimensional set-up was chosen. The pull-out specimens consisted of steel strips with a thickness of 3 mm, embedded in rectangular mortar specimens with a cross-section of 40x40 mm. Three values of the surface roughness were selected, one close to the transition point - where interface debonding changed into interphase sliding - and two higher values (i.e. R_a (=RMS-)values of 1.80, 5.45 and 8.75 µm, respectively). Before testing, a surface layer was removed from the specimens by sawing, so that the resulting specimen thickness coincided with the strip width, i.e. 20 mm. Fig.3 shows the specimen configuration.

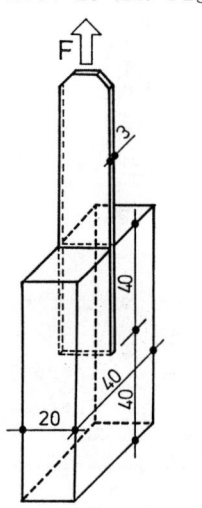

Fig.3 Specimen dimensions

In the case of two preliminary test series of 2x9 specimens for extremal values of R_a, pouring was accomplished with the steel strip in a horizontal position. Next, three series of 6 specimens per indicated value of R_a were casted with the cross-section of the steel strip in a vertical position. Pull-out testing was executed with a displacement-controlled, self reacting system. Loading was performed incrementally, so that holograms could be made for each load increment.

The surface displacement field could be visualized by means of a double exposure holographic interferometric technique. A 8 mm Argon laser (type LEXEL) was applied as a light source. The holograms were prepared on AGFA GEVAERT 10E56 (NAN) plates. In the final test series, the relative displacement, δ, between the top surface of the mortar and the steel strip was additionally recorded.

4. Experimental results

In the preliminary test series, the different bonding conditions on both sides of the strip could easily be observed (6); debonding was in some occasions even completed over the embedded length before any debonding could be observed on the other side of the strip. In the final test series, the debonding front could also be detected accurately

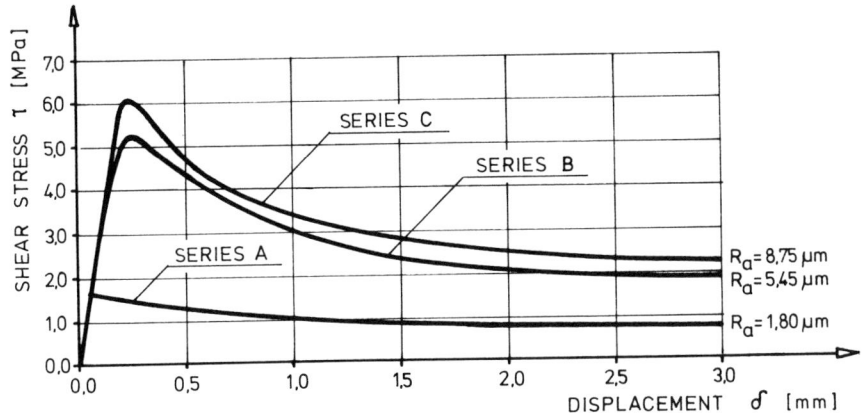

Fig.4   Average shear-displacement curves (the slip-stick phenomenon is omitted) for each of the test series

(sensitivity being governed by fringe spacing). As in conventional pull-out tests, we have observed a characteristic difference in the shear stress-displacement curves, shown in Fig.4.

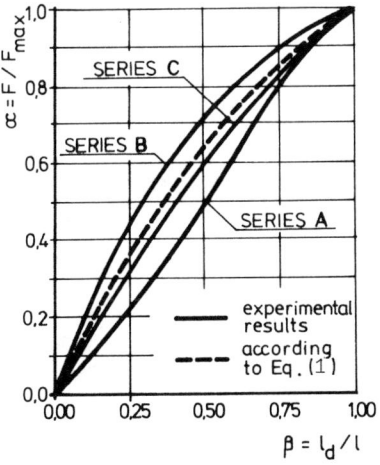

Fig.5   Debonding under load

By expressing the debonded surface fraction, β ($=l_d/l$, l_d and l being the debonded and total bar length, respectively), in terms of the load intensity, α ($=F/F_{max}$, F and F_{max} being the load and maximum load, respectively), it could be shown that the three composites varied reasonable close around the parabolic approximation:

$$\alpha = \beta(1.6 - 0.6\beta) \qquad\qquad -1-$$

As can be observed in Fig.5, the "smooth" surface yields relatively high values for the debonded fraction - given a certain load intensity - the "rough" surfaces score on the average low.

The interferograms demonstrate that along most of the ascending branch of the shear stress-displacement curve we deal with mixed boundary conditions along the embedded bar surface (see, Fig.6). Debonding starts already at 15 to 20% of ultimate loading. A more sensitive approach would probably have revealed even lower values.

5. Concluding remarks

The various approaches to the bond problem have to be combined. This will definitely give a better insight into the bond phenomenon. To that end, the H.I. results will be elaborated further to yield local information. Data on the stress distribution along the embedded interface, obtained in the analytical approach, would additionally allow for a more detailed analysis of the results obtained sofar.

6. References

1   Mindlin,R.D.: Force at a point of the interior of a semi-infinite solid. Physics, 7, May? 1936
2   Bień,J.-Grady,S.: Discrete model of interaction between concrete and steel bar subjected to pull-out loadings. Stev.Rep.1-86-12,Delft
3   Stroeven,P.-et al: Pull-out tests of steel fibres. Proc.RILEM Symp. Sheffield, U.K., pp.345-353, The Constr.Press, 1978
4   Stroeven,P.-de Wind,G.: Structural and mechanical aspects of debonding of a steel bar from a cementitious matrix. Proc.Conf.Bond in concrete, Paisley, Scotland, pp.40-50, Appl.Sc.Publ.,London, 1982
5   Barnes,B.D.-Diamond,S.-Dolch,W.L.: The contact zone between Portland cement and glass "aggregate" surfaces.Cem.Conc.Res.,8, 1978
6   Kassová,D.: Mechanics of debonding in pull-out testing. Stev.Rep. August 1985, Delft
7   Bien,J.: Holographic interferometry study of steel-concrete bond in pull-out testing, Stev.Rep.1-86-9, Delft

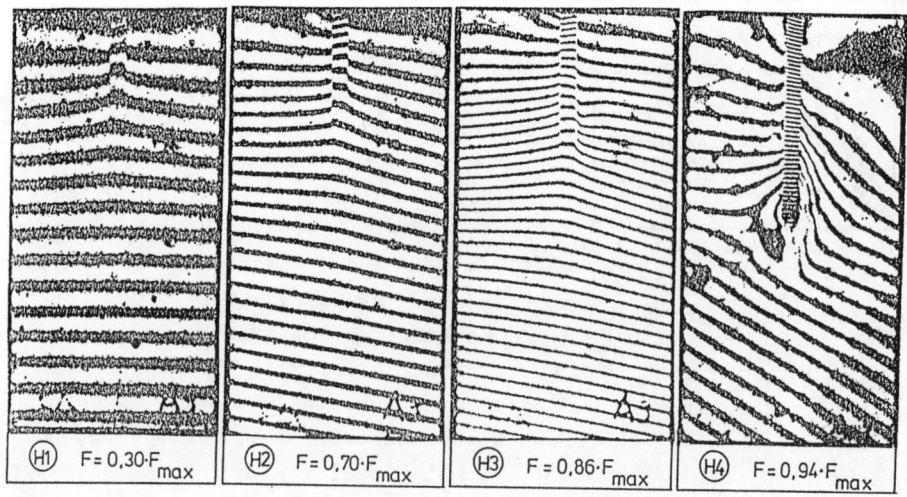

Fig.6 Successive interferograms revealing gradual debonding

Application of strain gauges and
other electromechanical transducers

THE DRIFT AND CREEP BEHAVIOUR
OF ENCAPSULATED WELDED STRAIN GAGES
AT 300 °C OVER A PERIOD OF SEVERAL YEARS

Dipl.-Ing. Claus Amberg
Bundesanstalt für Materialforschung und -prüfung (BAM)
Unter den Eichen 87, D-1000 Berlin 45

This 3rd part of a test report gives an estimation of the strain
gages (s.g.) errors obtained during the long-term supervision of
the hot liner within a prestressed concrete pressure vessel (PCPV)
at a constant temperature of 300 °C by comparing the measuring
results to those of long-term laboratory measurements of the
drift and creep of the strain gages: After the first 10 days the
creep of the PtW-s.g. SG 425 decreasingly declines compared to
the drift of the NiCr-s.g. SG 125, and, therefore the SG 425
are much better qualified for long-term supervision at and
above 300 °C.

Keywords: high temperature - weldable strain gages - long-term
measurement - static strain measurement - supervision

1. INTRODUCTION

The 1st and 2nd part [1, 2] of the report show that these s.g.
are well suited for long-term supervision of concrete structures
over a period of more than 10 years at temperatures up to 120 °C as
well as for long-term supervision of the expansion restraint of the
hot liner within a PVPV [3, 4] at temperatures ranging from room
temperature up to 260 °C. These two parts of the report also give
information on the construction and number of s.g., the measured
area in the vessel, the arrangement of the s.g. at the liner and the
installation techniques (also on active and dummy s.g.), the hyster-
esis (especially of the PtW-s.g.) and the typical values of the
measured liner strains at 300 °C.

Int. Conf. Measurement of Static and Dynamic Parameters of Structures
and Materials, IMEKO-TC 15, May, 25-28, 1987, Plzen

155

This 3rd part of the trilogy deals with the suitability of these s.g. to determine the strain variations of the measured liner area at a constant temperature of 300 °C.

Fig. 1 shows typical strain indications of the tangential, axial and dummy s.g. at the liner during 3 consecutive temperature cycles of the PCPV: The indications of the NiCr-s.g. SG 125 changed to a greater extent than those of the PtW-s.g. SG 425, nearly independently of the measuring direction: From these measurements the liner strain changes cannot be calculated without knowledge of the long-term behaviour of the s.g. at 300 °C.

2. LABORATORY INVESTIGATIONS

The long-term behaviour of the s.g. types SG 125 and SG 425 at high temperatures was investigated in the BAM laboratory on liner material in furnaces (drift) and in a 4-point-bending-beam-s.g.-calibrator (drift and creep) by means of a few (lost) s.g. specimens.

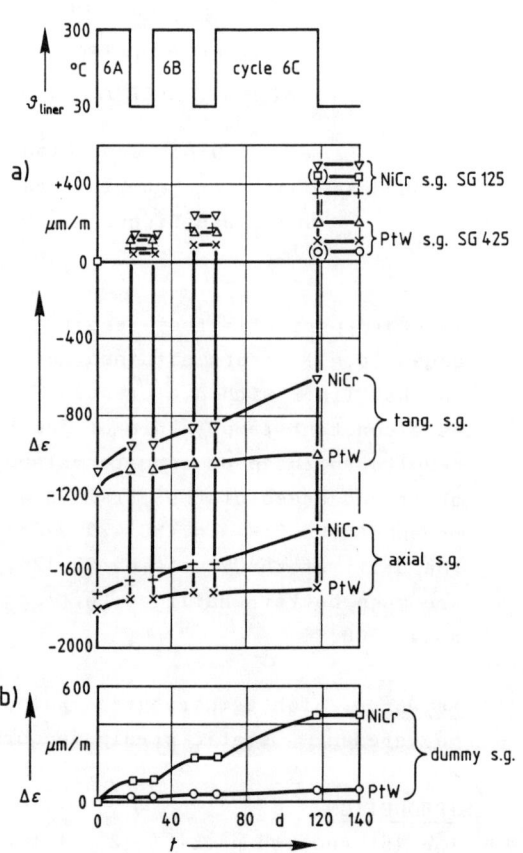

Fig. 1 Typical strain indications of tangential, axial and dummy s.g. at the liner during 3 consecutive temperature cycles of the PCPV up to 300 °C

Table 1 Time - dependent drift rate of the s.g. SG125 and creep rate of the s.g. SG425 at 300 °c within different periods of time (from laboratory measurements)

time at 300 °C	0-10	0-25	10-25	0-50	25-50	0-100	50-100	0-250	100-250	days
cycle No.	Σ1-5		6A		6B		6C		7	
drift rate: SG125	11^+_-9	$8,5^+_-6.$	7^+_-4	7^+_-4	$5^+_-2,5$	5^+_-3	4^+_-2	4^+_-2	$2,8^+_-1$	$\frac{\mu m}{md}$
creep rate SG425	10^+_-8	5^+_-4	$2^+_-1,7$	$3^+_-2,5$	$0,7^+_-4$	$1,8^+_-1,4$	$0,6^+_-0,4$	$0,9^+_-0,7$	$0,3^+_-0,2$	$\frac{\mu m}{md}$

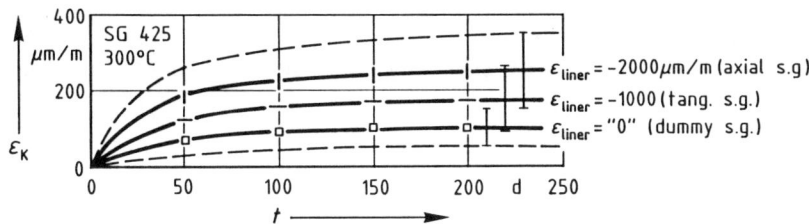

Fig. 2 The creep behaviour of the PtW-s.g. SG425 at 300 °C on liner material

The drift of the NiCr-s.g. SG 125 at 300 °C is rather strong; their creep, however, is negligible. The drift rate (per day) is specimen-dependent between 2 and 20 μm/md within the first 10 days and between 2 and 6 μm/md after e.g. 100 days [6, 7], see Table 1. From our measurements and from literature [7] a drift range of the s.g. SG 125 can be calculated. It is indicated in Fig. 3a as upper and lower drift estimation.

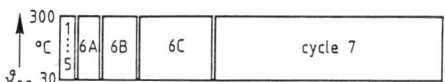

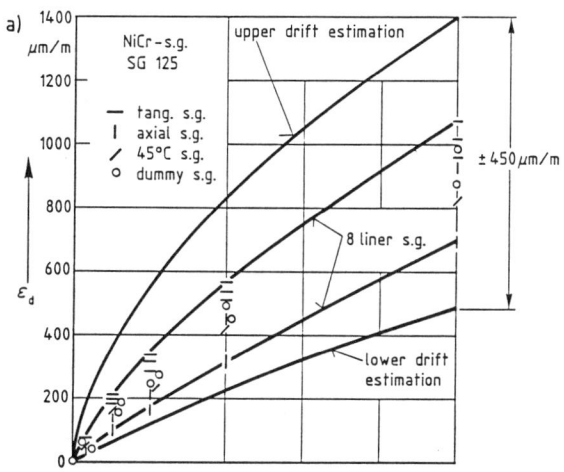

The PtW-s.g. SG 425 are creeping, if extended or compressed, opposite to their loading direction [8]. Their drift at 300 °C can be neglected. Some of the s.g. specimens were kept on liner

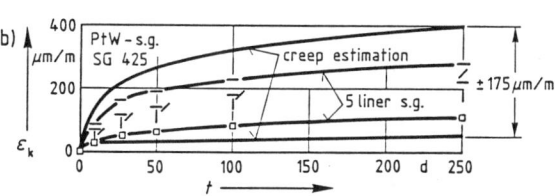

Fig. 3 Lined-up s.g. indications during the 300 °C periods of the liner as well as drift and creep estimations resulting from laboratory investigations

material at 300 °C and at strain levels of ε = 0, -1000 and -2000 μm/m. This corresponds to the conditions of the dummy s.g., the tangential s.g. and the axial s.g. at 300 °C on the liner [2]. The noticed behaviour (Fig. 2) can be explained by creep of the s.g. [8, 9]. The creep of the s.g. specimens differs greatly even under the same

157

load; therefore, the whole range between the dashed curves of Fig. 2 is marked into Fig. 3b to show the estimated creep of the s.g. at the liner.

3. LINER MEASUREMENTS

In Fig. 3 the s.g. indications measured during the 300 °C periods of the liner are summarized; at the same time the 300 °C phases of the cycles 1 to 5 could be condensed to 10 days. Since the s.g. indications are within the drift or creep estimation limits, it can be gathered from Fig. 3, that the changes of liner strain during the 300 °C periods within 250 days did not exceed

$\pm$ 450 µm/m, regarding the NiCr-s.g. and

$\pm$ 175 µm/m, regarding the PtW-s.g.

For shorter 300 °C periods more restricted error limits can be stated, if Fig. 3 and Table 1 are used:

If the s.g. indications during the 300 °C periods of the temperature cycles 6A, 6B and 6C (see Fig. 1) are regarded individually, and if the corresponding values of Table 1 are used as drift respectively creep estimation, the error limits are essentially restricted (Fig. 4), because they are the corresponding curve parts from Fig. 3, composed at the beginning of each 300 °C period. Since the s.g. indications during these 300 °C periods were again within the error limits (Fig. 4), it is possible to conclude: the changes of the liner strain were within these more restricted limits.

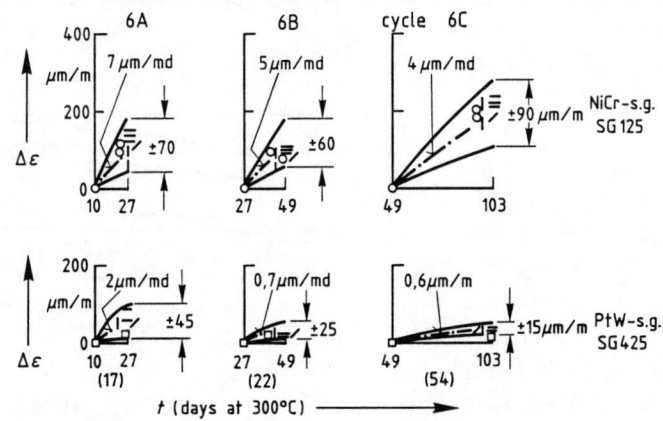

Fig. 4 Strain indications of the liner s.g. during the separate temperature cycles of Fig. 1 as well as drift and creep estimations in accordance with Table 1 and Fig. 3

158

Due to the fact, that the drift and creep curves of the s.g. become constantly flatter, even lower error estimations (per day) can be given for strain measurements after e.g. 100 days, carried out over an extremely long period of time (Fig. 5).

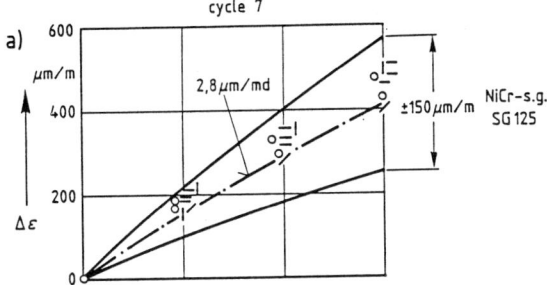

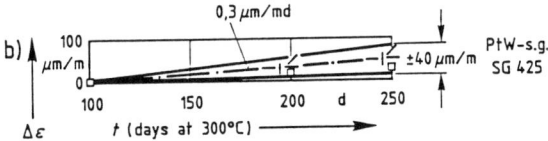

Fig. 5   Long-term measuring error of the s.g. after 100 days at a constant temperatur of 300 °C:
a) NiCr-s.g. SG125, with the drift correction of Table 1
b) PtW-s.g. SG425, nearly without any correction

4. SUITABILITY OF THE STRAIN GAGES

Fig. 3:   within the first 10 days at 300 °C no remarkable difference between the drift of the NiCr-s.g. and the creep of the PtW-s.g. is to be expected.

The Figs. 3, 4 and 5 show: after the first 10 days at 300 °C the creep of the PtW-s.g. SG 425 becomes increasingly slower compared to the drift of the NiCr-s.g. SG 125. For long-term measurements of static strains at 300 °C can be stated:

- the s.g. SG 125 are possibly suitable, but only if the comparatively big drift is corrected in accordance with Table 1 (see incated values of drift rate in Fig. 4 and 5),
- the s.g. SG 425 are well suited, even without any creep correction. In this case long-term errors which correspond approximately to those given in Fig. 4 and 5 are to be expected.

ACKNOWLEDGEMENTS

The author wants to express his appreciation to Mr. N. Mayer from
the Bundesanstalt für Materialforschung und -prüfung (BAM) Berlin,
for many useful discussions on the whole test work and to Mr. J.
Specht from the BAM for the precise preparation and installation
of the data recording device and to Mr. R. Scheiber from the Öster-
reichisches Forschungszentrum Seibersdorf (ÖFZS) for the computing
of the s.g. data obtained from the temperature cycles of the vessel.

REFERENCES

1   Amberg, C. and Czaika, N.: On long term strain measurements
    in prestressed concrete structures at temperatures up to 120 °C
    with encapsulated weldable strain gages. Prepr. IMEKO '85,
    April 22-26, 1985, Praha, Vol. 3, pp. 78-85.

2   Amberg, C. Strain Measurements at Variable Temperatures up to
    300 °C within a Prestressed Concrete Pressure Vessel by means
    of Encapsulated Welded Strain Gages. Proc. VIIIth Int. Conf.
    on Experimental Stress Analysis, Amsterdam, May 12-16, 1986,
    pp. 589-597.

3   Fritz, K. and Német, J.: A PCPV with Elastic Hot Liner, BNES,
    (Brit. Nuclear Energy Soc.) Bristol, 1982, Gas Cooled Reactors
    Today, 8p.

4   Witt, A., Zemann, H. Patry and Weissbacher, L.: Experimental
    Verification of a PCPV Concept with Hot Liner, BNES, Bristol,
    1982, Gas Cooled Reactors Today, 8p.

5   Scheiber, R.: Simulation of Thermal Processes in a Prestressed
    Concrete Pressure Vessel. Conf. Applied Modelling and Simulation
    1982, Paris, Proc. IASTD, ISTN-No: 0-88986-034-3, pp.258-261.

6   Amberg, C. and Czaika, N.: Über das langzeitige Drift- und Kriech-
    verhalten gekoppelter, aufschweissbarer Dehnungsmesstreifen mit
    NiCr-Messdraht bei Temperaturen bis 320 °C, VDI-Berichte Nr.339,
    1981, pp. 105-111.

7   Böhm, W., Hofstötter, P., Rasche, N. and Weichsel, J.: Prak-
    tischer Einsatz gekoppelter Hochtemperatur - Dehnungsmesstreifen
    bis 315 °C. VGB Kraftwerkstechnik 61, H. 6, June 1981, pp. 502-
    509.

8   Amberg C. and Czaika, N.: On Long-Term Behaviour of Encapsulated
    Weldable Half Bridge Strain gages with PtW-Wires at Temperatures
    up to 550 °C. Proc. VII Int. Conf. on Experimental Stress Analyses,
    Haifa, August 23-27, 1982, pp. 237-255.

9   Andreae, G. and Niessen, G.: Über Möglichkeiten und Grenzen der
    Hochtemperatur - Dehnungsmesstreifen mit Platin-Wolfram-Leiter,
    Teil I und II. Materialprüfung Band 27 (1985), No. 11 pp. 344-
    346 and No. 12 pp. 381-383.

DYNAMIC FORCES AND IMPEDANCE MEASUREMENT IN STATICALLY
LOADED STRUCTURES

Prof.Dr. Mikchail Genkin-Dr.Vladimir Tikchonov-Dr.Vladlen Yablonskij
Mechanical Engineering Research Institute of the Academy of Science
Griboedov Str. 4, 101830 Moscow Centre, USSR

A method is investigated to measure relatively low dynamic
forces and impedance in statically loaded structures by
means of thin elastic layers and built-in piezoelectric
transducers. The elastisity theory equations being solved
for two regions of the layer a pressure distribution is
achieved and condition minimizing pressure nonuniformity
determined. Computation and experimental results are shown.

Keywords: force measurement, elastomer thin layer

1. Introduction

A measurement of small dynamic forces in presence of relatively
high static loads is actual problem in vibroacoustics of mechanical
structures. For example, resilient mounts of heavy energetic machines
carry static loads equal to tens of tons whereas vibration forces may
be 10^3- 10^4 times less than those. Such ratios exclude utilizing of
ordinary tensoresistive gauges from vibroacoustic force measurement.
In this case only piezoelectric transducers ensure required rigidity
and sensibility simultaneously. Force measurement in resilient mounts
gives the best estimation of vibroisolation efficiency in accordance
with its definition. However, one can't install ordinary piezoelectric transducers in heavy loaded mechanical chains because of fragi-
lity of ceramics. Besides, such transducers are usually intended for
unilateral loads measurements at laboratory tests. In real structures
the dynamic loads are three-dimensional ones and require the corresponding devices for their measurement. Force vector components can
also be measured by the group of one-dimensional dynamometers having
relatively low stiffness in "unmeasured" directions. In this way force flows may be distributed between different dynamometers. An axial

IMEKO 5th TC7 Symposium, Plzen, Czechoslovakia, May 25-28, 1987 stiffness required is usually 10-20 times higher than one of mechanical chain in the same direction.

Dynamometers with stiffness anisotropy are also very useful in measurement of impedance, dynamic stiffness and other characteristics of mechanical structures. The dynamic compliance in i-th point of the structure under the action of k-th generalized force excitation is defined as:

$$y_{ik} = u_i / F_k \;\Big|\; F_l = 0,\; l \neq k$$

All the other forces, except for F_k , including transverse reactions in the excitation point should be equal to 0. This condition is realized in dynamometers with one-dimensional stiffness.

In multi-point vibration test systems a flexible beam or a string is used for transverse forces isolation. But in practice an unavoidable bending of the beam results in considerable lowering of the highest frequency limit.

The problems assigned can be successfully resolved by means of dynamometers which contain not only force transducers but also one or several layers of liquid or elastic material carrying a dynamic load.

2. A force measuring device with an elastomer thin layer

A force tranducer being plunged in material or in contact with it is actually working as a pressure transducer. As an example, take a dynamometer, which consist of an elastic layer acting upon a tensoresistor [1], or a device, containing a liquid cavity hermetically sealed and a membrane tensoresistor [2] .

A dynamometer with a thin rubber layer and a piezoelement (Fig.1, [3]) has the advantage of high sensitivity and signal-to-noise ratio in spite of partial loading of the sensor. Other useful characteristics will be shown later. A schematic view of the dynamometer with one elas-

tic layer is shown in Fig.1. Upper plate 1 is subjected to external loading being passed to low base 3 through elastic layer 2. If the layer is uniform, the preassure passed through is proportional to the force applied. A low base is provided with a tapped hole with a

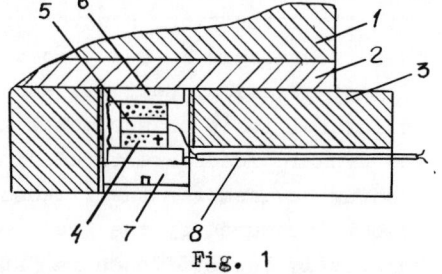

Fig. 1

thread , where two plates 4 of piezoceramics with signal electrode 5, ground electrode 6 and supporting plug 7 are arranged. A signal generated is lead out through a canal in low base 3 by means of output cable 8. If lateral displacement is limited, shear loads in the layer may

be negleted therefore even a dynamometer with relatively high cross
sensitivity of the piezoelement measures an axial force only.

The most interesting construction contains one or more thin elastomer layers (of rubber or rubber-like material) with thickness-to-
diameter ratio less than 0.03-0.05. Such layers guarantee maximum axial stiffness and don't distort dynamics of the structure given.

As mentioned above, relatively small transverse stiffness and
sensitivity of the dynamometer make it to be useful for measuring force vector components. A static shear displacement doesn't change its
main characteristic considerably.

An elastomer layer being fastened with metal plates together
shows poor compressibility. A normal load results in a quasi-hydrosta-
tic pressure. It exceeds tangent stresses anywhere but in small boundary region. Compression loads up to 50-200 MPa are available.

The ratio of static compression layer stiffness $c_n = E_n S/h$ to shear
stiffness $c_\tau = G_\tau S/h$ depends on correlation of geometric parameters of
the layer [4] :

$$c_\tau / c_n = G_\tau / E_n = \alpha + a\zeta^2 \quad ,$$

where a is a numerical coefficient (a= 8/3 for a cylinder), α = G/K is
the shear-to-volume compression moduli ratio, ζ is the layer thickness
-to-dimension ratio (ζ = h/d for a cylinder diam.d), G_τ is the layer shear modulus ($G_\tau = G(1- 0.23\zeta)$) and E_n is the layer compression
modulus. Taking into account a considerable difference of the elastic
moduli of thin layers (ζ = 0.001- 0.05) the compression-to-shear
stiffnesses ratio rises up to 10^3-10^4.

The polarized built-in piezoelement measuring the force contacts
a small part of the layer. Therefore it carries a small part of static
and dynamic loads and is well protected from shocks and moment loads.
High compression loads being permissible
can reduce the layer dimension and a load appeared localize on a small area. A
layer thickness being reduced compression
in contact area becomes more uniforms.

3. Basic equations

Stress distribution analysis is developed according to the model shown in
(Fig. 2). Two regions of the layer are
considered: one of cylindrical form
has a diam. $2R_1$ and another one is
of a ring form with diam $2R_2$, which
is in contact with the cylinder (a).

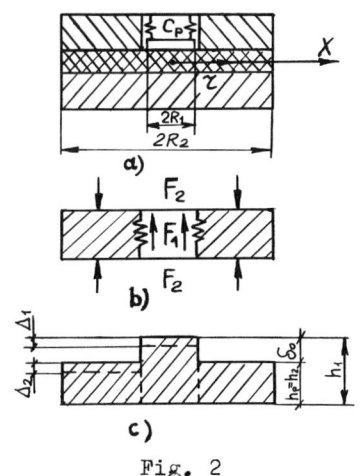

Fig. 2

163

A thread stiffness is taken into account. The "piston model" is used for calculation. Axial forces being applied to the piezoelement (F_1), to the ring layer (F_2) and to the resulting force (F) are considered (b). In general the inner cylinder h and the outer ring $h = h$ stiffnesses are different (c).

The elasticity theory problem is being solved under the following assumptions:
- main stresses are equal to hydrostatic pressure: $\sigma_{11}=\sigma_{22}=\sigma_{33}=\sigma$;
- pressure distribution varies only in one direction: $\sigma = \sigma(r)$;
- being written in cylindical coordinates the displacements changement law satisfies such condition: $\partial u_r/\partial z \gg \partial u_z/\partial r$.

It is shown that an equilibrium equation along the full thickness of the layer for different coaxial regions results in the following equation:

$$\Delta_r \sigma - k^2\sigma = q \qquad , \qquad (1)$$

where Δ_r is a Laplace operator being written in the middle layer cross-section coordinates: $k^2=12\alpha/h^2$; $q= k^2\sigma_o$ - are asymptotic expressions of the coefficients for small values of ζ ; $\sigma_o=K\Delta/h$ is the pressure corresponding to one- dimensional compression; Δ is the axial displacement of the upper plate relative to the base.

A solution of the boundary problem

$$\Delta_r = \frac{d^2}{dr^2} + \frac{1}{r}\frac{d}{dr} , \qquad \sigma(R)= 0 \qquad (2)$$

will be as follows:

$$\sigma(r)= \sigma_o \left[\frac{I_o(kr)}{I_o(kR)} - 1 \right] .$$

Non-zero stress- and strain tensor components are expressed by means of a hydrostatic pressure: $\sigma_z=\sigma_r=\sigma_\varphi=\sigma(r), \tau_{zr}= [(h/2) - z]\sigma'(r)$.

For an equal-thickness layer an approximate estimation of the pressure distribution can be evaluated by the following formula:

$$\eta = F_1/ \beta^2 F, \qquad \beta^2= S_1/ S= (R_1/ R_2)^2. \qquad (3)$$

A layer thickness decreases both the parameter $kR_2= 2R_2\sqrt{3\alpha}/ h$ enlarges and attenuation hydrostatic pressure field nonuniformity reduces ($\eta \to 1$).

In general, for layer of various thickness a joint deformation of the inner cylinder with thickness h_1 and area $S_1=\pi R_1^2$ and the outer ring with thickness h_2, area $S_2=\tau(R_2^2-R_1^2)$ is considered. In the place of S_1 the boundary problem (1) should be solved with the operator (2) and by boundary contact pressure set: $\sigma_1(R_1)= \sigma$.

The solution is:

$$\sigma_1(r)= \frac{\sigma +K\varepsilon_1}{I_o(k_1R_1)} I_o(k_1r) - K_1 , \qquad (4)$$

where $k_1 = 2\sqrt{3\alpha}/h_1$, $\varepsilon_1 = \Delta_1/h_1$; Δ_1 is a mutual axial displacement of
the layer boundary surfaces under the piezoelement.

In the place of S_2

$$\sigma_2(r) = C_1 I_o(k_2 r) + C_2 K_o(k_2 r) - K \varepsilon_2 \,, \tag{5}$$

where $k_2 = 2\sqrt{3\alpha}/h_2$, $\varepsilon_2 = \Delta_2/h_2$, Δ_2 is a mutual axial displacement
of the upper plate and the base. The parameters C_1 and C_2 can be found
by means of two boundary conditions: $\sigma_2(R_1) = \sigma$, $\sigma_2(R_2) = 0$.

As a result, the equation system

$$\sigma_1{}'(R_1) = \sigma_2{}'(R_1), \quad \Delta_1 = \Delta_2 + \delta_s - \delta_r, \quad F = F_1 + F_2,$$
$$F_1 = c_r \delta_r = -2\pi \int_0^{R_1} \sigma_1(r)rdr, \quad F_2 = -2\pi \int_0^{R_2} \sigma_2(r)rdr \tag{6}$$

and the equations (4) and (5) should be solved together.

There are two problems to be solved. The first one is to determine the stress and strain state with an assumption of the supporting
plug being preliminary tied. And the second one is to change the state
determinedunder static and dynamic loading.

If $F = 0$ and the supporting plug displacement δ_s, the parameters
δ_r, Δ_1, Δ_2, σ should be found from equation (4)-(6) and coefficient
nonuniformity should be determined from expression (3). In the second
problem the load F and condition $\delta_s = 0$ are considered to be set. Both
solutions should be superimposed.

In computation algorithm a harmonic dinamic load is considered to
be small relative to the static one. At every step of deformation
growth a small excitation relative to the state achived with linear
relation of parameters is given. Generally speaking, both problems are
nonlinear from the geometrical and physical point of view because of
sensitivity of the characteristics to the layer thickness variation
and of elastic moduli dependence on hydrostatic pressure.

4. Computation and experimental results

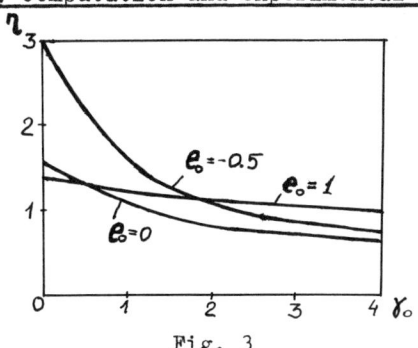

Fig. 3

Fig. 3 shows a nonuniformity
coefficient (η) change as a
function of an argument $\gamma_o = KS_1/$
$/c_r h_1$, where c_r is a thread stiffness, for various values of parameter $e_o = (h_1 - h_2)/h_2$.

These curves show the thicknesses relation at which a pressure nonuniformity is minimum.
An amplitude response of

165

dynamometer charge sensitivity (dB re zero level 12.6 pK/N) is shown in Fig. 4. A set of 40 mm diameter contained 21 rubber layers of 0.3 mm thick and 20 steel plates of 0.1 mm thick. A piezoelement had two plates of 1 mm thick and 8 mm diameter.

A static load of 5 kN was excited by a dynamic force of 12 N in the range of 5 - 500 c/s.

The signal investigated (curve 1) with an error being less than 1 dB was proportional to control transducer signal (curve 2).

Irregularity of the response in 100- 500 c/s band is caused by a test equipment resonance. Its frequency depends on stiffnesses of control and test dynamometers.

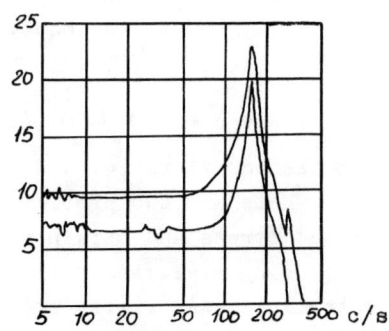

Fig. 4

5. Conclusion

The methods discussed above concern force and impedance measurement at great and small static loads and can be widely used in vibration tests, in dyagnostics and in exploitation control of various mechanical constructions thus enlarging the possibilities of the ordinary "cinematic" vibrometry.

6. References

1 Pat. France. No. 2,236,171. Cl. GO1, L1/20, 1975
2 Pat. USA No. 3,130,382. Cl. 338_5, 1964
3 Genkin,M.D.-Yelezov,V.G.-Yablonskij,V.V.: Metody upravlyajemoi vibrozashchity mashyn.Nauka, Moscow, 1985
4 Tikchonov,V.A.- Yakovlev,N.G.: Primeneniye tonkosloinych resinometallicheskikch elementov dlya vibrozashchitnykch sistem. "Kolebaniya i vibroakusticheskaya aktivnostj mashyn i konstruktsij". Nauka, Moscow, 1986, pp. 33-42

EVALUATION OF SURFACE ROUGHNESS USING A NEW, SIMPLE, HANDY DEVICE

Prof. Dr. Hanna Abdel Misseh Hanna , Ph. D.
Head of Mechanical Department . Faculty of Engineering . Zagazig . EGYPT .
12 El Magd st. Roxy , Heliopolis , Cairo
Egypt

A simple mechanical device has been designed to evaluate surface roughness value . Design of this device is based on the known relation between the static coefficient of friction for steel/steel surfaces and their roughness values . Different machine surfaces (milled, ground and lapped) were used for the device calibration . The surface roughness parameters for (Ra, Rp, and Rt) were measured using a Talysurf-6 with incorporated computer. Relations between the roughness parameters and static coefficient of friction were determined . The statistical analysis, of results obtained for 1000 specimens, showed a good correlation between values determined by the device and those measured on the Talysurf-6 .

Keywords: Surface roughness, coefficient of friction, Talysurf-6

1. Introduction

The object of this study is to design a simple industrial device for evaluating the surface roughness; being similar to limit gauges used for the inspection of dimensions . Generally, the surface roughness devices are expensive and very complicated in design and even in operation. Many works were elaborated for the surface roughness evaluation through hydraulic or pneumatic systems|1|which need special precautions, but all fail to find a clear and definite relation. Besides these trials, many studies|2,3|were elaborated to find a relation between the static coefficient of friction and the surface roughness parameters. Definite relations were found for steel/steel milled, ground and lapped surfaces |4-6|.
Based on such relations, it is thought to design a friction meter having

a reference smooth surface of hardened steel to operate on the measured surfaces. The operating forces were thought to be magnetic, electric or mechanical ones. The latter method is chosen using normal and tangential action springs.

2. Principle of Operation

The designed friction meter, shown Fig. 1, when it is manually loaded to the test specimen, the sliding blocks 2 will slide vertically in the frame and will be additionally loaded by the springs 4. The self aligned ball bearings 3 will act to keep the disc in line contact with the test specimen. The locking pin 6 will prevent the disk from rotary relative motion on the test specimen. The levered adaptor 7 will be in contact to the electronic switch 15. Just before measuring, the pin 6 will be pulled out and the disc will be prevented from rotation by friction only. Thus when rotating the handle 8, the piston 9 will act to compress the springs 10 against the amplifying plate 11 fastened to the disc 1.

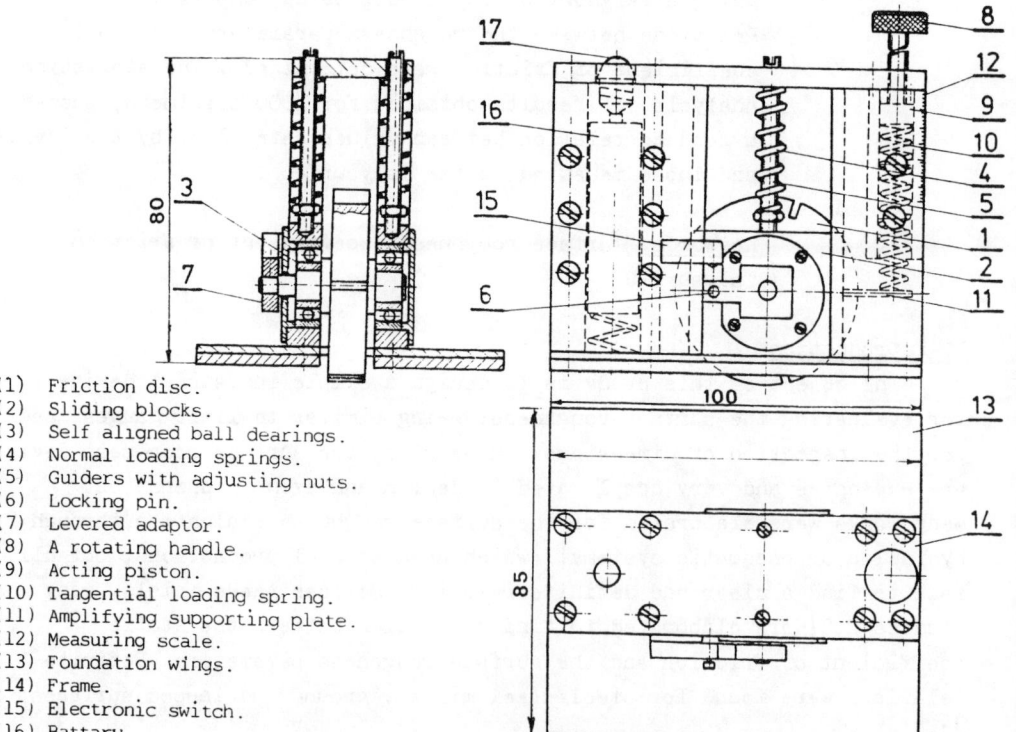

(1) Friction disc.
(2) Sliding blocks.
(3) Self aligned ball dearings.
(4) Normal loading springs.
(5) Guiders with adjusting nuts.
(6) Locking pin.
(7) Levered adaptor.
(8) A rotating handle.
(9) Acting piston.
(10) Tangential loading spring.
(11) Amplifying supporting plate.
(12) Measuring scale.
(13) Foundation wings.
(14) Frame.
(15) Electronic switch.
(16) Battary.
(17) Lamp.

Fig. 1 Assembly Drawing of The Friction Meter

So, a tangential moment will be created to overcome the frictional moment between the disc and specimen. When a small relative motion occurs, i.e. when the created tangential moment reaches the frictional one, the liberated lever adaptor 7 will rotate to switch-on the electronic switch and the lamp. The measured displacement of piston on the scale 12 represents the compression on the spring 10 and consequenttly the friction moment.

3. Relation between Coefficient of Friction and Device Reading

Figure 2 simplifies the operation of the friction meter. The normal load

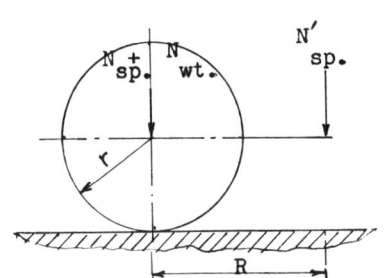

Fig. 2 Loading Scheme for the Device

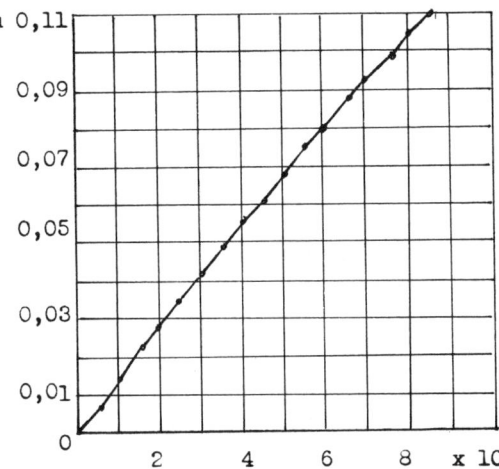

Fig. 3 The reading of the measuring device x(mm) and the coefficient of friction μ

N'_{Total} will be expressed by :

$$N_{Total} = N_{Spring} + N_{Weight} + N'_{Spring}$$

where N_{Spring} is the normal load of springs 4 , N_{Weight} is the weight of disc 1 and the sliding blocks 2 with the ball bearings 3 . N'_{Spring} is the tangential load created by compressing the spring 10 to the scale reading x. Thus, the friction force can be expressed as :

$T_{Total} = N_{Total} \cdot \mu$; and the frictional moment will be :

$M_{Friction} = T_{Total} \cdot r$ while the tangential acting moment will be :

$M_{tang.} = N'_{spring} \cdot R$

So, at equilibrium, when the disc starts to rotate;

$M_{tang.} = M_{friction}$, from which the device reading, shown Fig. 3, can be expressed as :

$$x = \frac{\mu \cdot r}{K \cdot R} (K . \Delta L + K'. x + N_{weight})$$

$$= f(\frac{\mu}{1 - c})$$

where μ is the static coefficient of friction between the disc and test specimens, and c is constant.

4. Test Specimens

Material of specimens is plain carbon steel, the mechanical properties of which are : Ultimate tensile strength = 42 Kp/mm^2; Yield point= 25. Kp/mm^2 and the Brinell hardness number = 120.
Four machining and finishing processes covering the most common practice in metal cutting were used, such as milling, grinding and lapping with different cutting conditions.
Specimens used are parallelograms of rectangular plain surfaces (100x 100x10) mm and composition of which is (0.5% S_i , 0.5% M_n , 0.2% C)

5. Surface Roughness

An indicative relation is searched for static coefficient of friction and the various surface roughness parameters|7|represented as :
a) Center Line average values Ra ;
b) Depth of smoothness   Rp ;
c) Maximum peak to valley height Rt.
It has been claimed by Rubert, M.P.|8|that the depth of smoothness Rp or the maximum profile height is an indication of the shape of crests and is a significant parameter when considering frictional characteristics of a surface . Halling|9|emphasizes the significance of distribution and frequency of occurence of the peaks.

6. Results

1000 surfaces have been measured on the Talysurf-6 ten times each and the average values were considered for all measured parameters, or implicitely calculated through the incorporated computer of the Talysurf-6. The measured surfaces were classified into 4 groups, according to method and conditions of machining as :
a) Milled surfaces.
b) Rough ground surfaces.
c) Fine ground surfaces.
d) Lapped surfaces.
Each specimen has been measured by the friction meter 10 times at different positions of its surface and statistically analysed to have the mean value of each and the standard deviation defining the precision of the devices.
It is found that the maximum value of precision σ_{max} is within 10% from

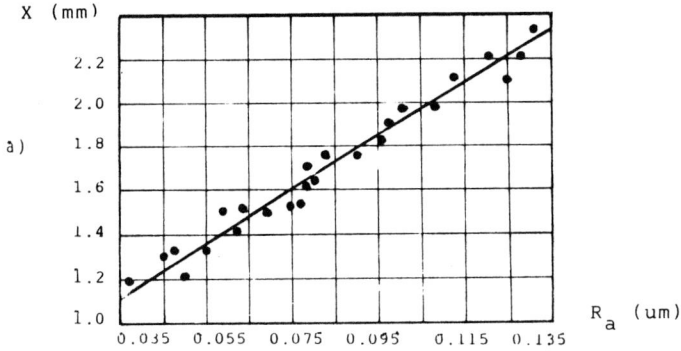

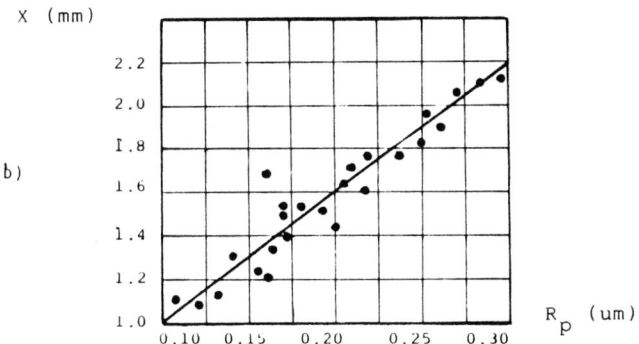

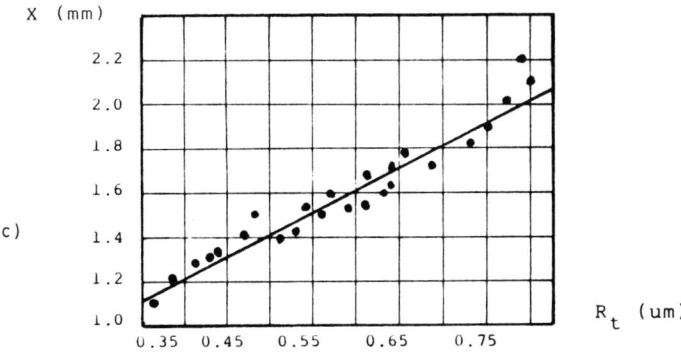

Fig. 4   Regression Curves for the $(X - R_a, R_p \& R_t)$
Relations for the Lapped Surfaces

the main value, a result considered satisfactory for the device. Results were regressed through 3 main formulae : Linear, exponential and powered relations. It was found that the linear relation is the most significant one, having the maximum correlation coefficient. Figure 4 represents the linearly regressed relation between the device scale reading and the significant surface roughness parameters Ra, Rp and Rt, for a lapped surface.

7. Conclusion

The designed measuring device , based on mechanical operating nature is an easy, simple and an industrial device for evaluation of surface roughness of products. It is similar to the limit gauges used for inspection of the products dimensions. The statistical analysis of the obtained readings on the device shows an average value of the estimated mean square within 7% for 10000 readings taken on 1000 specimens. The correlation coefficient calculated for the regression relation between the scale reading of the device and different surface roughness parameters for each cutting conditions has been found to be within a range of 70-98%. Results seems to be very satisfactory for the measuring device. For the collected overall readings, the correlation coefficient is found to be within 72% and 79% which is a satisfactory result.

8. References

1 Halling, J. " The Principle Of Tribology ", Macmillan Press, 1975.
2 Buckley, D.H., " The Metal To Metal Interface And Its Effects On Adhesion And Friction ", J. Coll. Int. Sc., 1968.
3 Ghabrial, S.R., " Effect Of Progressive Lapping On Smoothness Index For Spark Erroded Surfaces ", Faculty Of Eng. Bulletin, Ain Shams University, 1977.
4 Takahashi, T., " On The Coefficient Of Friction Under High Pressures ", Trans. Of The JSME, Vol. 17, 1949, Japan.
5 Furey, M.J., " Surface Roughness Effects On Metallic Contact And Friction ", ASLE Trans., Vol. 6, 1963.
6 Buckley, D.H., " The Metal To Metal Interface And Its Effects On Adhesion And Friction ", J. Coll. Int. Sc., 1977.
7 Tsukada, T. And Anno, Y. " An Analysis Of The Deformation Of Contacting Rough Surfaces ",[1st] And [3rd] Reports Bulletin, ISME, Vol. 15, 1975.
8 Rubert, M.P.,"Measuring Surface Roughness", The Prod. Eng., March 1960.
9 Halling, J., "The Specification Of Surface Quality", The Prod. Eng. May 1972

VIBRATION AMPLITUDE TESTING BY ELECTROACOUSTIC MEANS

Dr.Ing.Aleš Boleslav,C.Sc. - Ing.Karel Antropius,C.Sc.[x]
Research Institute of Radiocommunications, Novodvorská 994, Prague 4
[x]Faculty of Civil Engineering, TU Prague, Thákurova 7, Prague 6, ČSSR

A method of indicating the vibration amplitude by means of a
miniature pressure microphone is presented, and a simple way
of keeping the constant distance between the acoustic input
and the body surface without damping the vibration is described. The testing experiment results were checked up by holographic interferometry. Amplitude measurement by a magnetodynamic pick-up was also tested, but its mechanical impedance
was found to be much higher, and vibration damping occured.

Keywords: electroacoustic sensors, amplitude indication

1. Introduction

Vibration amplitude measurements during resonant mode shape investigations have been commonly carried out by various coherent optics
methods, and particularly by the time-average holographic interferometry.
The optical methods fail, however, at hidden parts of investigated surfaces which cannot be lighted and simultaneously observed even by making
use of mirrors. This is the case, e.g. of some radial-bladed engine im-
pellers, where the small and the large blades alternate, and the small
ones are very close to, and almost fully hidden behind the large ones /1/.

Consequently, complementary methods have been searched for to measure the vibration amplitude in hardly observable places. The main requirement is that the measuring technique must not affect the vibration
itself. It follows that the mechanical /or acoustical/ impedance of a
measuring sensor must be substantially lower than the impedance of the
vibrating body in the measurement spot. The method should allow for a
quick scanning of surface investigated, and so the sensor dimensions
should be minimized, and its position hand-controlled during measurement.
It is also desirable that the sensors used are simple, and that they
can be easily set up by adapting the commercially available common
elements.

2. Vibration amplitude indication by a pressure microphone

Vibration amplitude in direction normal to the surface can be approximately determined by measuring the acoustic pressure in acoustic near-field, since this pressure predominantly results from vibration of surface in the close vicinity of microphone acoustic input /2,3/. It follows from the theory, that the acoustic pressure is proportional to the vibration amplitude, and also proportional to the square of vibration frequency.

Measuring in near-field means that the distance of the measuring point from the surface must be much less than the wavelength of the acoustic signal. It is important for stable measurement sensitivity that this distance is kept constant. The dimension of the microphone acoustic input should be as small as possible, and both this dimension, and the distance from the surface should be less than the spacing of nodes and antinodes.

To test this method we have made use of a small cilindrical /length 10 mm, Ø 10 mm/ electret pressure microphone /type EM 60/ with built-in integrated preamplifier. This microphone, a cross-section of which is shown in fig. 1a, is most often used as a built-in part of portable tape-recorders. The diameter of the acoustic input is 3 mm. The microphone front side was covered by a thin aluminium cap with drilled openning Ø 3 mm. Three cylinder-shaped miniature compliant stops /see fig. 1b/ made of polyurethane foam were stuck on this cap to ensure the constant distance from the measured surface during manual scanning. The exact dimensions of the stops are not vital. To ease the handling of the microphone it may be connected to a light strip holder. The voltage on the output of the microphone preamplifier is proportional to the measured acoustic pressure. Its sensitivity is approx. 8 mV/Pa.

3. Experimental testing by comparing with holographic interferograms

The applicability of the pressure microphone described above for vibration amplitude determination was experimentally tested by comparing the courses of the measured acoustic pressure to the course of vibration amplitude found by the holographic time-average method.

The vibrating object was a steel cantilever plate beam 120 mm long, 30 mm wide, and 3 mm thick, fixed at the lower end. It was excited to resonances by two piezoelectric transducers /approx. 3 x 10 x 0.5 mm/ stuck at the bottom of its back side, and fed by AC voltage in the range from 1 to 6 kHz. At resonances, time-average holographic interferograms were recorded in a conventional holographic off-axes setup. The geometry of the

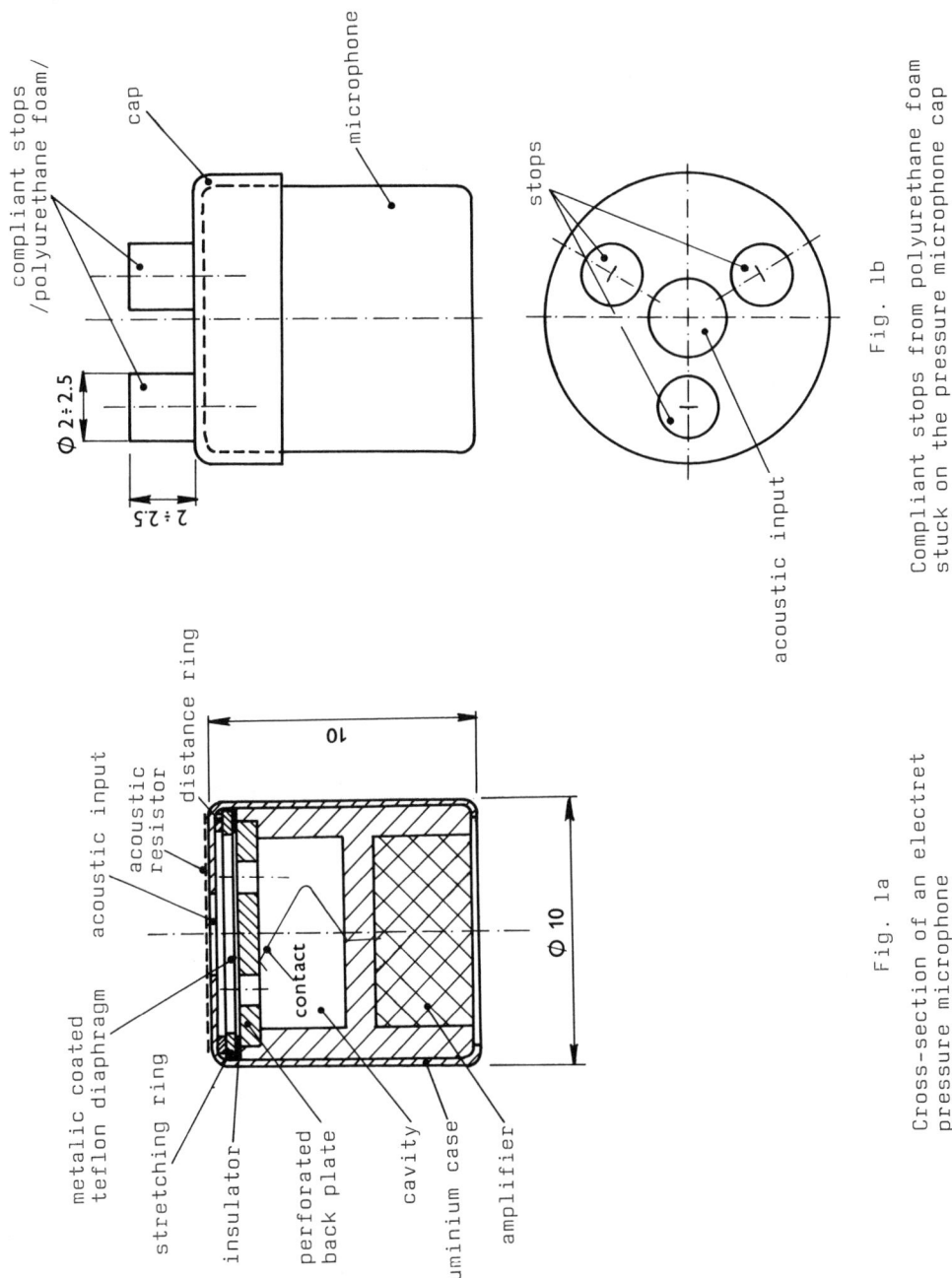

compliant stops
/polyurethane foam/

cap

microphone

Ø 2÷2.5

2÷2.5

Fig. 1b

Compliant stops from polyurethane foam
stuck on the pressure microphone cap

stops

acoustic input

metalic coated
teflon diaphragm

acoustic input

acoustic
resistor

distance ring

stretching ring

insulator

perforated
back plate

contact

cavity

aluminium case

amplifier

10

Ø 10

Fig. 1a

Cross-section of an electret
pressure microphone

175

setup was such that only the amplitude component normal to the beam surface could be recorded.

After a hologram had been recorded, the beam at the same vibration was scanned by the microphone along the vertical axis of symmetry. The values of output voltage from the microphone, and the vibration amplitude courses read from the interferograms were plotted along the beam length. For two resonant modes at the frequencies 1102 Hz, and 3060 Hz the interferograms and the plots are shown in figs 2a, and 2b.

Firstly, it was found that a light contact of the polyurethane foam stops with the vibrating beam surface, i.e. the contact without applying an unnecessary force on the microphone, affected the amplitude only very little, even when the microphone was touching the beam directly in antinodes. In case of the lowest measured resonant frequency 1102 Hz the amplitude decay was only approx. 10 %, at higher frequency resonances the decay was already negligible.

The experiments have shown that except in the neighbourhood of surface edges the courses of measured acoustic pressure fairly well correspond to the course of vibration amplitude. The agreement is the better the greater is the ratio of nodes and antinodes spacing to the diameter of microphone acoustic input. It was found that when the spacing of antinodes is relatively small than at the intermediate node the microphone output voltage decreases to a certain minimum value only but does not reach the zero value. The position of the nodes, however, can be even then well localized.

When the microphon output signals were compared for surface spots vibrating with the same amplitude but at different frequencies, the measured values were, within tolerance of 25 %, in ratio of frequency squares.

4. Use of a magnetodynamic transducer for amplitude determination

Some pilot tests have been also carried out to find when the magnetodynamic transducers can be used to indicate the vibration amplitude. For the tests a magnetodynamic stereophonic pick-up of the type 2102 /TESLA/, made for Hi-Fi signal reproduction from gramophone records was used. The speed sensitivity of this pick-up is approx. 1.4 mV per 10 mm/sec. At 1 kHz this speed corresponds to the amplitude 2.25 µm, and taking into account that voltages as low as 10 µV can be measured with sufficient accuracy, it follows, that at 1 kHz it should be possible to indicate amplitudes as low as 16 nm, which is much smaller value than could be detected by optical interferometric methods. Since the pick-up

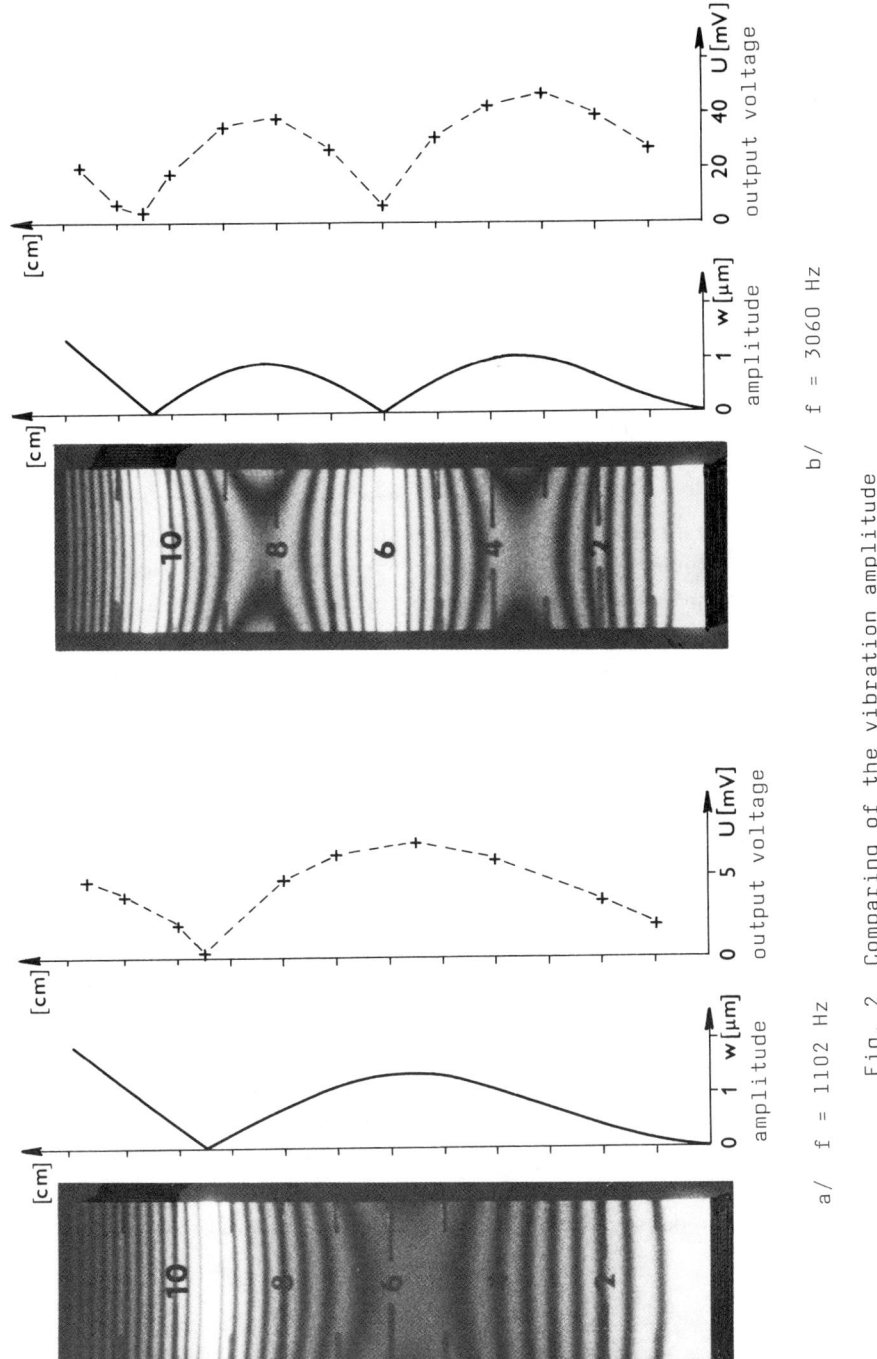

a/ f = 1102 Hz

b/ f = 3060 Hz

Fig. 2   Comparing of the vibration amplitude
          with the microphone output voltage courses

responds to the speed, its amplitude sensitivity still increases proportionally to the vibration frequency.

The main purpose of our testing was to compare the vibration damping by the contact of the pick-up stylus with damping by the pressure microphone. An experiment on the same testing beam has shown that at the lowest resonant frequency 1102 Hz, and at the same place where the contact of microphone reduced the emplitude by 10 % only as reported above, the touch of the pick-up stylus resulted in more than 50 % decrease of the vibration amplitude. At higher frequency resonances the influence was not so high, it is clear, however, that the mechanical impedance of the pick-up is much higher than that of polyurethane foam stops at the microphone.

5. Conclusions

A miniature pressure microphone can be made use of to determine vibration amplitudes, and at frequencies higher than 1 kHz to do so with at least the same sensitivity as the optical interferometric methods. If equiped with polyurethane foam stops mechanical impedance of the microphone in contact with vibrating surface is so low that in most cases resonant vibrations are not affected. Proportionality of the output signal, and the amplitude magnitude under the centre of the microphone input is met the better the smaller is the amplitude gradient i.e. the greater is the spacing of nodes and antinodes in the neighbourhood of this input.

A magnetodynamic pick-up may be also used for this purpose but its impedance is much higher, and care must be taken not to damp the resonant vibrations.

6. References

1 Antropius,K. - Paslerová,A.: Holographic Mode Shape Recording of a Radial-bladed Impeller M 602 /in Czech/. Res.rep. no. KSM 84/86, Faculty of Civil Eng., TU Prague, March 1986.
2 Merhaut et al.: Příručka elektroakustiky, SNTL, Prague 1964.
3 Keele,D.B.Jr.: Low-Frequency Loudspeaker Assesment by Nearfield Sound-Pressure Measurement. Journ.of the AES, April 1974

FEM OPTIMIZATION OF THE SHAPE OF PROFILED MEMBRANES UTILIZED AT STRAIN GAUGE INSTRUMENTED TRANSDUCERS

Dan-Mihai Ştefănescu, M.Sc.E.E.
Strength of Materials Chair, Polytechnic Institute of Bucharest
Splaiul Independenţei 313, 77206 Bucureşti, Romania

Transducers with strain gauges bonded on profiled membrane flexible elements are widely used for the electrical measurement of mechanical quantities. Analytical computation of these membranes is complicated while the classical design, based on nomograms, is rather approximate. In the paper an analysis is suggested by the finite element method, which permits the estimation of profiled membrane behaviour for various material and dimensional parameters, hence the shape optimization of these elastic structures.

Keywords: finite element method, strain gauge, transducer

1. Introduction

The profiled membrane (with well defined thickness variation along the radius) is an elastic element with small height and easily machined. As shown by Rohrbach and Andreae [1] it provides an increased radial sensitivity ε_r with respect to the simple membrane (Fig. 1).

Analytical computation of this profiled membrane is complicated while the classical design, based on nomograms, is rather approximate. A more rigorous calculation will be carried out using the finite element method (FEM) which has been applied with very good results at the study of rib-stiffened membrane [3] and of other elastic elements having more sophisticated shapes [2].

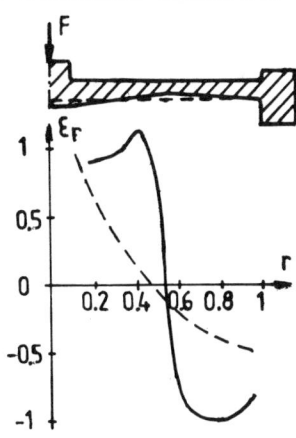

Fig. 1   Radial strain diagrams for simple and profiled membranes

IMEKO 1st TC15 Conference, Plzeň, Czechoslovakia, May 25-28 1987

2. Computation Model

The model of a profiled membrane is presented in Figure 2. It offers many variants by establishing the dimensions of the central protrusion, the membrane thickness and the exterior rim.

Due to the axial symmetry it is sufficient to model half of the structure. The mesh scheme contains 119 axisymmetrical finite elements and 165 nodes. The membrane itself is composed by three layers, each being generated by two rows of finite elements located both sides of the inflexion circle.

The concentrated load is applied on the central protrusion, at the central node no. 6. Nodes 1-6 are guided on the vertical. The rim of the studied elastic element is supported at nodes 147, 156 and 165.

The calculation using the computer program SAP IV give as results the node displacements and stresses from finite elements. For strain gauge instrumented transducers it is necessary to calculate the strains of the elastic element. In order to process the large volume of data, a proprietary [4] postprocessing computer program for radial strains ε_r is used.

It follows that taking advantage of the modern computer science one can establish the optimum shape of the transducer elastic body.

3. Numerical Results

Let consider the chosen profiled membrane model and the nominal load F_t = 5000N, the modulus of elasticity E = 2.1 X 10^5 N/mm^2 and the Poisson's ratio ν = 0.3. It results the diagram of radial strains ε_r on the lower membrane side presented in Figure 2. The corectness of dimensioning is proved by the maximum strain obtained: about 1000 μm/m.

It is interesting to mention that the computation model allows the simulation of a machining error of the elastic element, for example by moving the node 29 with 0.2 mm in other eight locations placed on sort of "wind arrows". The results presented in Table 1 show that the inflexion circle doesn't change the position. It remains slightly displaced towards the rim, near the node 75, due to the larger thickness from the middle of the profiled membrane. Conversely, in the zones of strain gauge locations, the values of ε_r differ from one case to another, proving the complexity and the accuracy of a study performed using the finite element method.

Based on the computation model, an experimental model has been

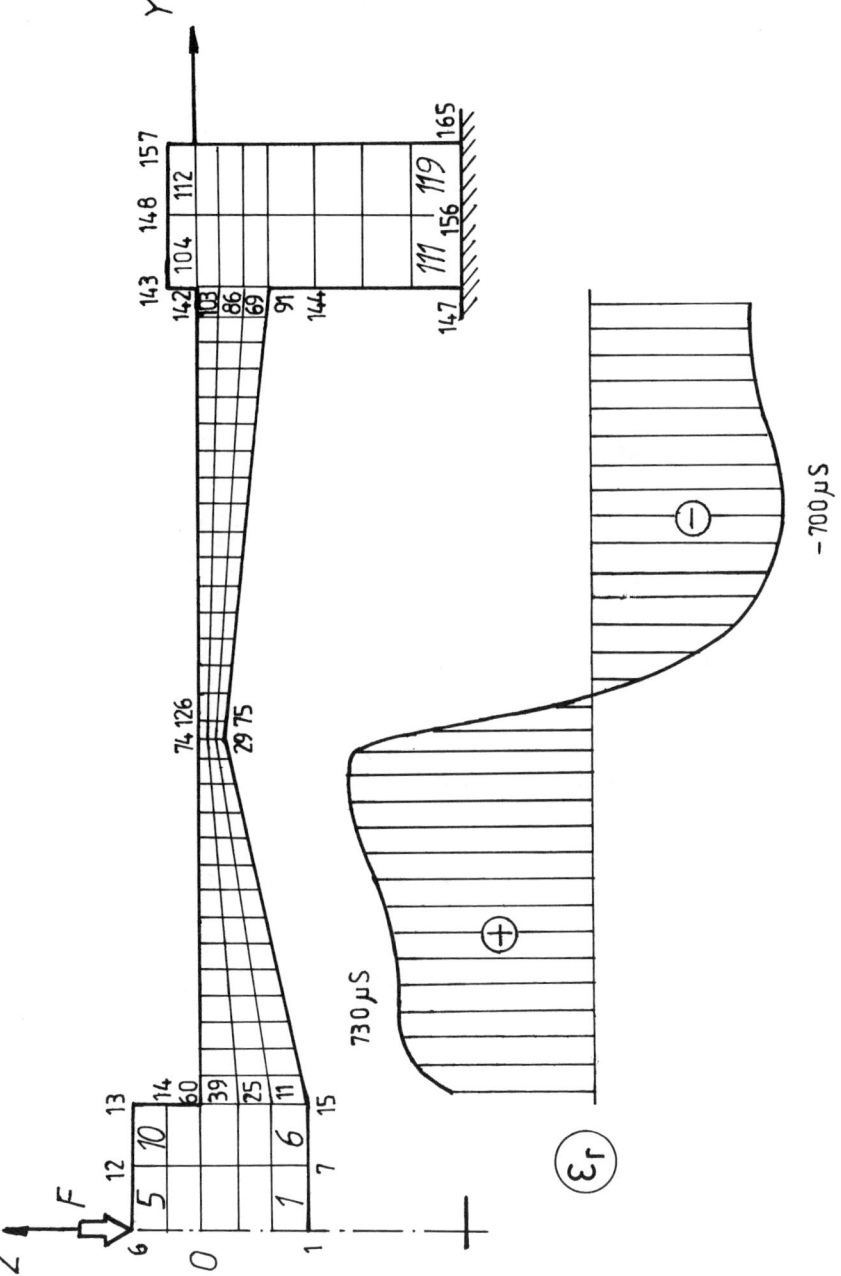

Fig. 2

Computation model and ε_r diagram on the lower side of the profiled membrane

Table 1

Diagram (top right):
```
      B   C   D
   A ←(29)→ E
      H   G   F
```

Node	Radial strains [μm/m]									
	29	A	B	C	D	E	F	G	H	
15	559	560	561	560	562	563	563	560	558	
16	672	673	677	679	680	675	665	664	663	
17	729	734	746	745	743	731	716	715	717	positive
18	730	736	750	749	748	732	711	712	713	radial
19	733	740	762	760	757	735	709	709	712	strains
20	750	756	781	780	775	751	715	717	720	
21	770	777	812	808	799	769	728	730	734	
22	795	803	849	845	837	793	744	745	750	
23	827	836	888	884	878	824	760	765	769	
24	854	869	941	934	928	852	779	781	789	
25	891	908	993	987	981	889	798	802	811	
26	919	942	1061	1046	1036	911	802	806	820	
27	923	945	1068	1055	1052	915	783	790	803	
28	924	960	1122	1105	1078	899	752	768	794	
29	600	582	624	659	703	630	521	486	462	
75	140	101	43	102	161	178	172	130	101	inflexion
76	-181	-196	-284	-258	-224	-150	-114	-131	-147	circle
77	-395	-414	-513	-497	-482	-386	-305	-315	-337	
78	-533	-549	-650	-635	-621	-523	-443	-497	-454	
79	-620	-634	-723	-718	-715	-619	-527	-533	-537	
80	-675	-682	-763	-761	-759	-667	-585	-587	-591	
81	-697	-708	-782	-777	-775	-698	-622	-620	-627	negative
82	-705	-715	-776	-775	-774	-703	-637	-637	-640	radial
83	-702	-712	-763	-762	-762	-704	-643	-642	-645	strains
84	-694	-701	-746	-745	-741	-690	-641	-640	-643	
85	-675	-684	-719	-718	-716	-676	-633	-632	-637	
86	-655	-664	-695	-691	-691	-654	-621	-620	-623	
87	-634	-642	-665	-663	-661	-634	-607	-605	-608	
88	-610	-617	-635	-634	-630	-608	-587	-585	-589	
89	-571	-579	-592	-589	-587	-571	-555	-554	-558	
91	-606	-615	-625	-622	-619	-605	-593	-593	-597	

made. The scheme showing the location of the four or eight strain gauges and their connection in the Wheatstone bridge is presented in Figure 3.

The results of the FEM analysis from the point of view of strain sensitivity on this profiled membrane are in good agreement with those experimentally obtained.

Such profiled membranes are successfully used at force measurement in wind tunnel balances or testing machines for springs.

4. Conclusions

Elastic elements of complex shape utilized in transducer design have complicated load - deflection relationships. The finite element method, the most versatile technique of structural analysis, permits both the global approach of the problem and the detailed investiga-

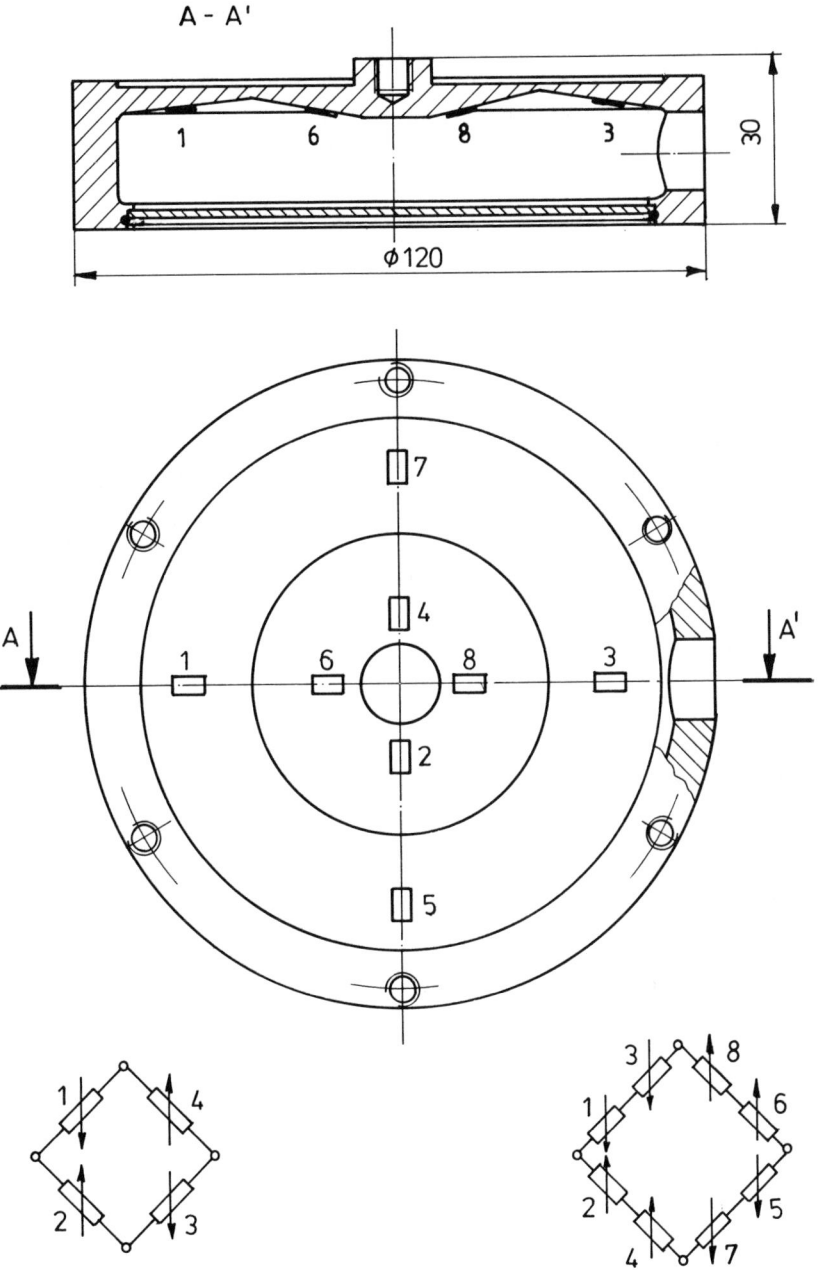

Fig. 3
Experimental model and schemes for
the bridge connection of strain gauges

tion in any point of structure; it is a rapid computation technique
allowing a graphical and/or tabular presentation of results. This
numerical method validates the approximations done in the analytical
computation and, at the same time, increases the efficiency of classical experimental stress analysis techniques, avoiding, for instance, the wrong location of strain gauges.

A structural analysis program for static and dynamic response
of linear systems (SAP IV) and a proprietary postprocessing program for strain calculation (SV 001) permit the parametric determination of design characteristics and the establishment of the
optimal shape of the profiled membranes utilized in strain gauge
transducer design.

5. References

1 Rohrbach,C.-Andreae,G.: Eine neuartige flache Präzisionskraftmeß-
   dose einfacher Bauart mit Dehnungsmeßstreifen. Materialprüf., 3,
   No.8, pp.300-304, Aug.1961.

2 Ştefănescu,D.M.: Untersuchung des Verformungszustands eines elas-
   tischen Rohrelementes mit Spalten, anwendbar beim Bau der Kraft-
   aufnehmer mit DMS. Rev.Roum.Sci.Techn.-Méc.Appl., 29, No.5,
   pp.519-533, Sept.-Oct. 1984.

3 Ştefănescu,D.M.: Comparative Study of the Sensitivity of Various
   Measuring Techniques on the Model of a Stiffened Membrane Analysed
   by the Finite Element Method. Sensoren - Technologie und Anwendung,
   Bad Nauheim, FRG, March 1986.

4 Ştefănescu,D.M.: Sensitivity Analysis by FEM of the Strain Gauge
   Instrumented Rib-stiffened Membranes Utilized in Transducer Design.
   Mathematical Methods in Engineering, Karlovy Vary, Czechoslovakia,
   Dec.2-5, 1986

NEW INFORMATION ON APPLICATION OF SEMICONDUCTOR STRAIN GAUGE SENSORS

Ing. Jiří Lukas, CSc
Aeronautical Research and Test Institute
Beranových 130, 199 05 Praha 9 - Letňany
Czechoslovakia

The demand for up-to-date, accurate and reliable transducers of mechanical quantities in industrial practice is still topical. The new physical principles are being found and, by applying the known ones, the relevant parameters of transducers are being improved. This is the case even in application of the silicon diaphragms with diffused strain gauges. One of the problems relating to this area is the temperature dependence of the strain gauge behaviour. The potential solution of this problem is discussed in the presented report.

Keywords: diffused strain gauges

1. Introduction

The up-to-date technology of microelectronics finds its application even in the field of electric measurement of non-electric quantities. A case in point is a silicon diaphragm on which are formed the diffused strain gauges. Their configuration on the diaphragm allows to indicate the state of stress of the diaphragm affected by a mechanical quantity to be measured. Such elements, manufactured by TESLA Concern Enterprise in Czechoslovakia, are very suitable, for example, for construction of pressure transducers. Certain limitation in their application represents the effect of temperature on the electric resistance and on the strain gauge sensitivity. The report discusses new information on potential elimination of this disadvantage, obtained by mathematical analysis of the circuit configurations used. Two types of circuits will be discussed.

2. Constant-voltage supply bridge circuit

Fig. 1 depicts two types of circuits, A and B, in which the compensation of the measuring signal for temperature effect is realized by applying element R_6 with a linear and/or a non-linear temperature dependence of resistance, diffused strain gauge R_5 and adjusting resistor R_{b1}.

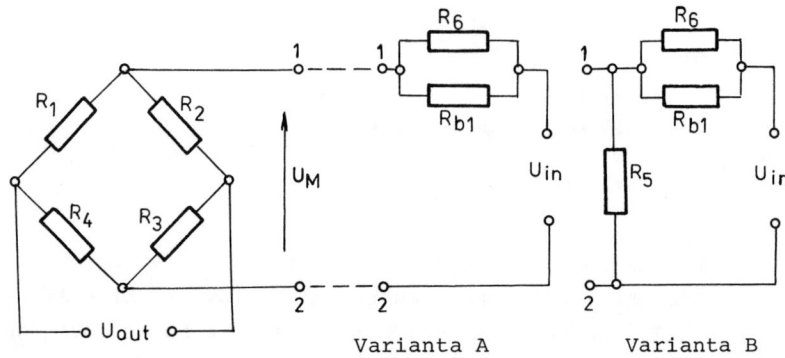

Fig. 1

Constant-voltage supply bridge circuit

If there is used the Type A circuit applying linear element R_6 with a temperature coefficient of resistance of 0.0225 Ohm/Ohm $^\circ C$, and a strain gauge resistance of 500 Ohms with a temperature coefficient of resistance of 0.0015 Ohm/Ohm $^\circ C$, then the temperature change in input bridge voltage U_M can be described as follows:

a/ The temperature dependence is strongly non-linear, particularly for higher differences in temperature.

b/ For lower differences in temperature, the temperature dependence pursued for different values of resistance R_{b1} has its extreme within a wide range of compensating element R_6; however, only for lower values of the latter resistor, the temperature dependence is slightly exponential within the selected range.

c/ If the behaviour of diffused strain gauges shows a large scatter, it is very difficult, sometimes even impossible, to attain a quality compensation within a wide temperature range.

The demonstrations of the temperature dependence observed for a temperature difference of $50^\circ C$ are presented in Fig. 2. The Type B circuit is always better for compensation than the Type A circuit.
When applying resistor R_6 with a non-linear temperature dependence of resistance, for example in compliance with the equation

$$R_{6T1} = R_{6TO}.e^{\left(\frac{2600}{T1} - \frac{2600}{T_0}\right)}$$

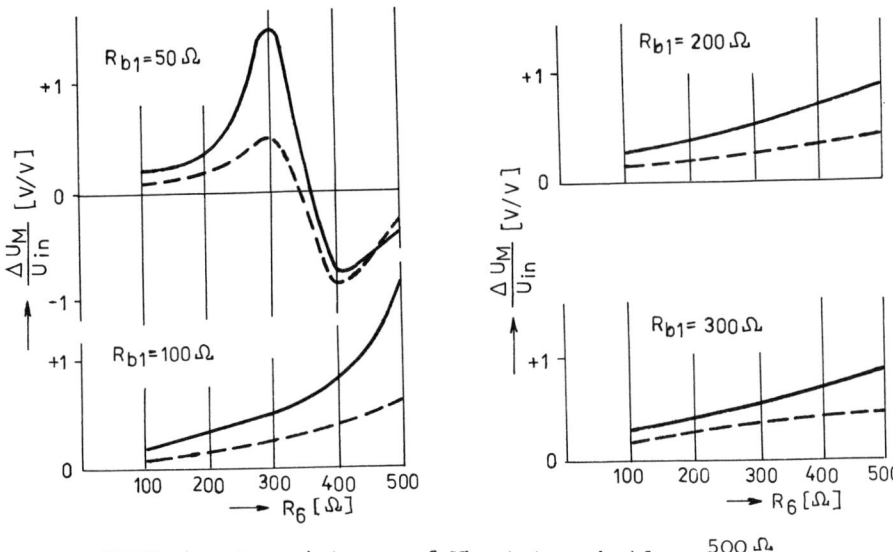

Fig. 2

Dependence of change in bridge supply voltage U_M on
values of compensating resistors

where T_o and T_1 are the initial and the final temperature, respectively
the same type of strain gauges and a temperature difference from -50°C
to $+130^\circ$C, then the following conclusions can be applied to a temperature change of the input bridge voltage:

a/ At a required change in the bridge supply voltage, necessary for compensation of about 15 per cent, a non-linearity in the dependence observed may be expected to be about 1 per cent; for this purpose it is
   more convenient to use a lower value of the input resistance, namely
   the Type B circuit and a higher value of resistance R_6.

b/ The circuit configurations depicted in Fig. 1 allow a temperature
   compensation of the measuring signal even if its change is higher
   than 100 per cent, however, the temperature dependence is stronlgy
   non-linear.

c/ The temperature change of the input voltage is by no means a linear
   function of values of adjusting resistor R_{b1}, particularly within the
   range of its higher values.

d/ The symmetry of the temperature dependence near a temperature of 25°C,
   which implies a certain criterion of compensation modalities within
   a range of positive and negative temperatures, is better for higher

values of resistor R_6.

e/ For a lower difference of temperatures than that observed, all problems even for high temperature changes of the measuring signal are substantially easier. This practically implies that, for example, within a temperature range from $+10^{\circ}$C to $+50^{\circ}$C, it is possible to accomplish a very quality compensation by applying a selected circuit configuration.

f/ Decreasing of the input bridge resistance by connection of strain gauge R_5 is advantageous from the point of view of obtaining the higher temperature change in the input bridge resistance; within the range of positive temperatures, this change is about two-fold, and within the range of negative temperatures, this change is higher by one third to one half when compared with the Type A circuit.

The demonstration of the dependence observed for an input bridge resistance of 250 Ohms is depicted in Fig. 3.

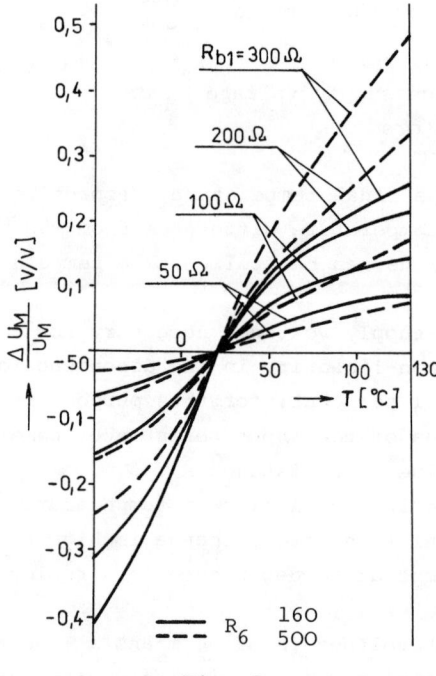

Fig. 3

Dependence of change in bridge supply voltage U_M on temperature

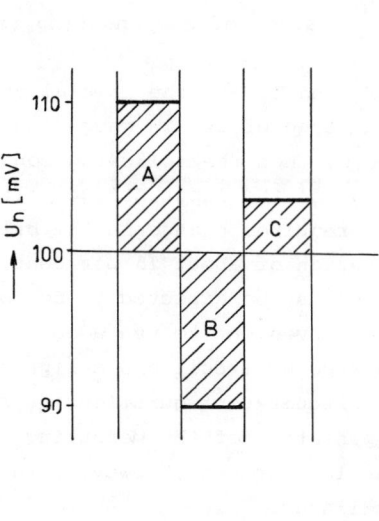

Fig. 5

Advantageous field of application of circuit configurations shown in Figs 4A, B, C

3. Constant-current supply bridge circuit

A high temperature dependence of diffused strain gauges is one of main causes of a high temperature dependence of the measuring signal of these elements when supplied by a constant voltage. This deficiency can partly be eliminated by supplying the bridge circuit from a constant current. It has been proved that some TESLA sensors, Types TM 530 and TM 630, can be used in a number of applications without any compensation 1/. However, in applications of accurate transducers, it is necessary to take into a account a scatter in behaviour of diffused strain gauges in a serial production by applying an additional compensation. For this purpose have been designed the circuits given in Fig. 4. Types A, B and C, where $R_{1,2,3,4,5...}$ are the diffused strain gauges and R_{b2}, R_{b3}, R_{b4}... are the temperature-independent adjusting resistors.

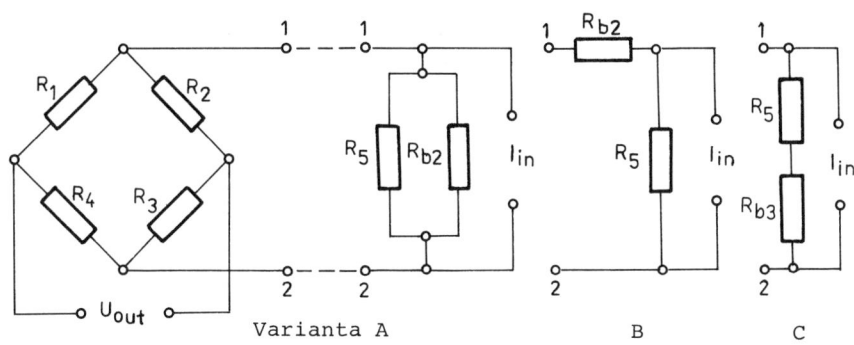

Fig. 4

Constant-current supply bridge circuit

All types of circuits have been mathematically described and the values of adjusting resistors R_{b2}, R_{B3} and R_{b4} have been calculated for the same parameters of diffused strain gauges used in circuits shown in Fig.1, and for the nominal measuring signal of 100 mV which may vary within a range of $\pm$ 10 mV when changing a temperature by $100^\circ C$.

From the analysis of results it follows:

a/ If the temperature change of the measuring signal is positive, the temperature compensation can be realized by the Type A of the Type C circuit; if the temperature change of the measuring signal is negative, the Type B circuit can be applied.

b/ A higher compensation effect can be obtained by applying higher values of resistor R_5, c/ From the point of view of the current-carrying capacity of compensating resistor R_5 it is more advantageous to use higher values of resistance of the compensating element, which may not cause such a decrease of the measuring signal; this fact can be applied particularly in temperature-dependent systems.

Advantageous fields of application of individual circuit configurations are indicated in Fig. 5.

4. Conclusion

The performed analysis of compensating circuits of the system with diffused strain gauges makes easy considering the optimal circuit for particular systems. All types of circuits are advantageous wherever interchangeability of transducers is required.

5. References

1 Lukas J.: Research of pressure gauges with diffused strain gauges, Research Report of Aeronautical Research and Test Institute, V-1473/82

Hybrid techniques of stress analysis

HYBRID TECHNIQUES FOR ANALYZING NONLINEAR PROBLEMS IN SOLID MECHANICS

Prof. Dr.-Ing. Karl-Hans Laermann

FB Bautechnik, BUGH Wuppertal

Pauluskirchstr. 7

D-5600 Wuppertal 2

Regarding the developments in measuring techniques in hardware as well as in software, it becomes obvious that hybrid techniques, i.e. the combination of advanced theories, experiments and the computer-oriented procedures for data acquisition and evaluation, a combination of mathematical and iconic models yield more knowledge and most reliable results in structural analysis. These techniques now are opening the possibility to analyze physical as well as geometrical nonlinear problems in solid mechanics.

Keywords: hybrid technique, nonlinear photoelasticity, integrated photoelasticity, Moiré-technique

1. Introduction

In engineering techniques, measurements are of increasing importance in design, development, testing, production, quality and systems control. Advanced experimental techniques are applied to processes, structures and components of structures. Those techniques ensure that they are more efficient, reliable, economical, safe, and that they achieve their goals. It has been recognized how to utilize the benefits of good measurements in saving energy, material and avoiding waste, in increasing productivity, in improving the quality of products, the safety of structures and the protection of environment. With the increase of technical capabilities, with the tremendous improve of the resolving power of measuring and recording devices, with the combination of mechanical or "iconic" and mathematical models and with the development of computer-oriented on-line evaluation procedures the problems to be solved may become more and more complex. Nowadays, engineers are increasingly interested in multidimensional and multifunctional measurements and their automation.

2. The principle of hybrid techniques

In order to measure mechanical quantities electrically, multiposition measuring devices have been introduced in experimental mechanics not only for data acquisition and processing, but also for the automatic computer-operated control of the experimental as well as the measuring and evaluation process. Such a system not only includes hardware, but with increasing importance software as well. The operational capability of those systems depends more and more on proper software. Recent progress

IMEKO Int. Conf. MEASUREMENT OF STATIC AND DYNAMIC PARAMETERS OF STRUCTURES AND MATERIALS, Plzeň/ČSSR, May 25-28, 1987

in micro-computers and in the diminution of electronic devices has led to a drastic decrease of hardware costs. For any kind of optical methods in experimental mechanics complex systems have been developed for digital-image converting, data storage and on-line processing of the optical quantities, Fig. 1. Such complex systems require more exact and advanced computer-oriented numerical methods for data processing, evaluation and the transmission of results. It must be pointed out that due to the complexity of hardware as well as of software, the danger may arise that the investigator loses contact and control of what happens on the way from the observed phenomena to the output-information, and exact knowledge may be replaced by believing in the infallible correctness of highly sophisticated and complex systems. Therefore it is quite necessary tracking the transmission of energy through the whole system (Fig. 2, 3).

To describe any kind of process, e.g. biological, physical, social or economical processes, models must be developed /1/. Generally this means that such a process or "event" must be transformed into an operational one by verbal description, a mathematical algorithm or "heuristical" mapping or even by formulating an analogy, mapping one process onto another one. Then the reliability and accuracy of results of any analysis strongly depend on the comprehensiveness of such "modelling" of the event. This model must describe the real event as exact as possible and in such a way that the numerous parameters of considerable influence are still controllable.

The physical process in a real structure must be described in order to predict the response of this structure under random loading conditions to determine, for example, the stresses, strains, deformations, or the safety against failure, considering the environmental conditions as well as the material behavior. Generally it is impossible to formulate a "mathematical model", i.e. to derive "true" constitutive equations according to the real event or, if this should nevertheless be possible, to solve these constitutive equations. Assumptions and simplifications must be introduced on very different levels to find an operational model. This leads to an approach of the reality only. Very often, the certainty and admissibility of such approaches are unknown or can hardly be estimated. Therefore the results obtained through mathematical analysis are uncertain as well, despite introducing advanced numerical methods and computers with high capacity.

However, it seems to be possible to use more realistic "iconic" models of the regarded processes for experimental analysis. As such iconic models, the real structure, parts of the structure, a scaled-down replica or an analogous physical process may be considered. The different reactions under the given load conditions may then be observed. But in stress analysis, the observed phenomena are in most cases not identical with the wanted final information. Therefore it is also necessary to introduce mathematical models for evaluation and transmission of experimental results.

As in many cases the material of the iconic model is not the same as the material

194

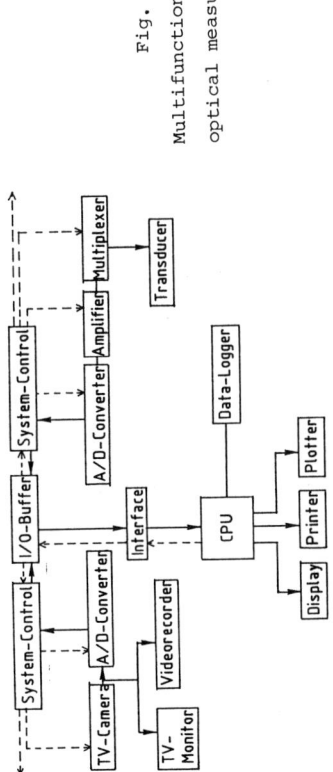

Fig. 1
Multifunctional electrical/
optical measuring system

of the real structure, the different response of material must also be taken into consideration. For the transmission of experimental data from the iconic model to the real event, proper operational models of material behavior must be developed. Therefore it should be pointed out that even the analysis of physical processes by means of iconic models leads to an approximate solution only. However, the approach is much better. More complex and more realistic mathematical models based on advanced theories can be derived and introduced into the analyzing process in order to evaluate the measured data. It becomes possible now to combine theory and experiment in such a way that the constitutive equations must not be solved any more, or they are to be solved experimentally instead of numerically (Fig. 4).

These considerations consequently lead towards the principle of "hybrid techniques" /2/, /3/. It becomes obvious that only a combination of theory and experiment, a combination of mathematical and iconic models yields a better knowledge of structure reactions and most reliable results in stress analysis. In the following, the principle of hybrid technique will be explained for instance by some nonlinear problems.

3. Physical nonlinearity in photoelasticity

As yet, in photoelasticity it has always been supposed Hooke's theory of elasticity to be valid and consequently linear relations between birefringence effects and stresses to be existing. However in areas of high stress concentration, e.g. in the vicinity of crack tips, notches and inclusions and with respect to some of the mainly used photoelastic model materials, considerable uncertainties may result. Therefore the influence of nonlinear-elastic stress - strain relations shall be considered /4/, /5/. On the supposition that the strains are still small, such relations may be formulated according to Kauderer /6/ for an isothermal state:

$$\varepsilon_{ij} = \left[\frac{1}{3K} \varkappa(s) - \frac{1}{2G} g(\tau_0^2) \right] s \delta_{ij} + \frac{1}{2G} g(\tau_0^2) \sigma_{ij} , \tag{1}$$

where $\varkappa(s)$ denotes a compression function, formulated as a potential series

$$\varkappa(s) = 1 + \sum_{\nu=1}^{n} \frac{\varkappa_\nu}{(3K)^\nu} s^\nu \tag{2}$$

and similarly, $g(\tau_0^2)$ denotes a shear function

$$g(\tau_0^2) = 1 + \sum_{\nu=1}^{n} \frac{g_{2\nu}}{(2G)^{2\nu}} \tau_0^{2\nu} . \tag{3}$$

With K and G, the initial values of the compression modulus and the shear modulus respectively, the coefficients $\varkappa_\nu$ and $g_{2\nu}$ describe the nonlinear material response. These values are to be determined by material testing procedures, such as tensile tests and shear tests. Furthermore, in eq.s (2) and (3), s denotes the mean tension, and τ_0 the reference stress. According to Neumann /7/ (see also Coker/Filon /8/, Mindlin /9/), in amorphous material the birefringence effect is assumed to be linear,

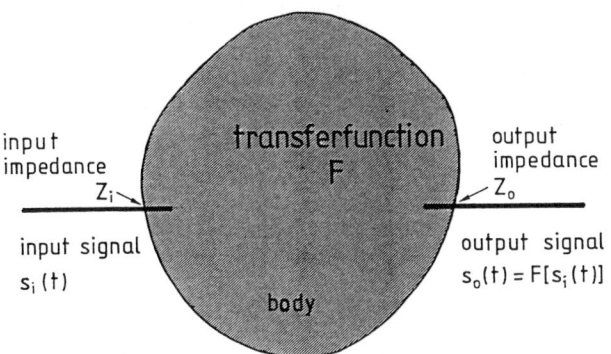

input impedance Z_i

transferfunction F

output impedance Z_o

input signal $s_i(t)$

output signal $s_o(t) = F[s_i(t)]$

body

operational transfer function $F = \dfrac{s_o(t)}{s_i(t)} = \dfrac{b_m D^m + \cdots + b_1 D + b_0}{a_n D^n + \cdots + a_1 D + a_0}$

operator $D^\mu := \dfrac{d^\mu s}{dt^\mu}$

Fig. 2

Transfer function of a single element

System Transfer Function

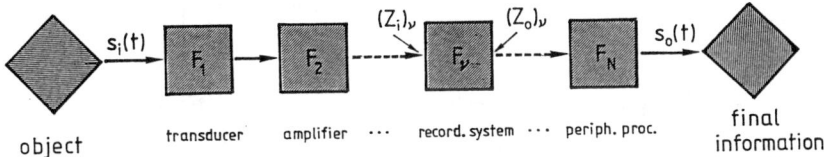

$s_i(t)$ F_1 F_2 $(Z_i)_\nu$ F_ν $(Z_o)_\nu$ F_N $s_o(t)$

object

transducer   amplifier $\cdots$ record. system $\cdots$ periph. proc.

final information

If impedances such that system responds linearly:

$\dfrac{(Z_o)_{\nu-1}}{(Z_i)_\nu} \ll 1 \Rightarrow$ system transfer function: $F_a = F_1 F_2 \cdots F_\nu \cdots F_N$

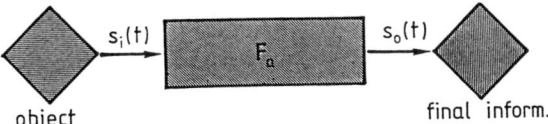

$s_i(t)$ F_a $s_o(t)$

object

final inform.

Fig. 3

System - Transfer - function

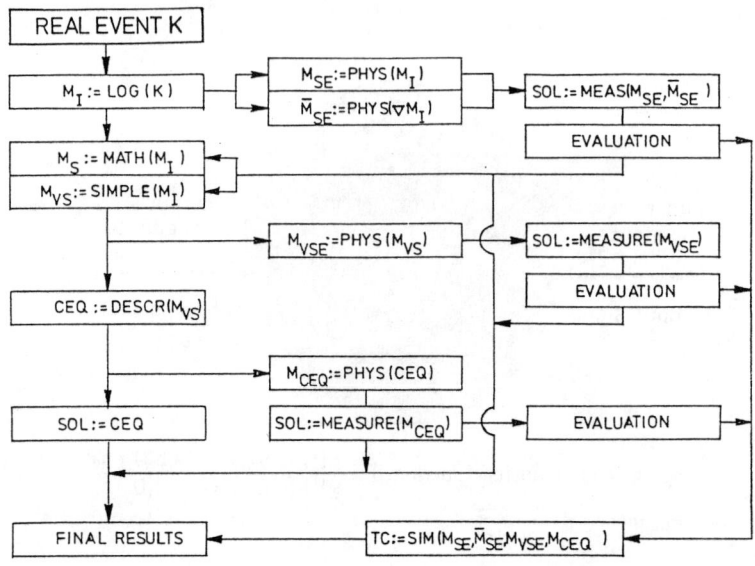

Fig. 4

Bloc-diagram of the principle of hybrid technique

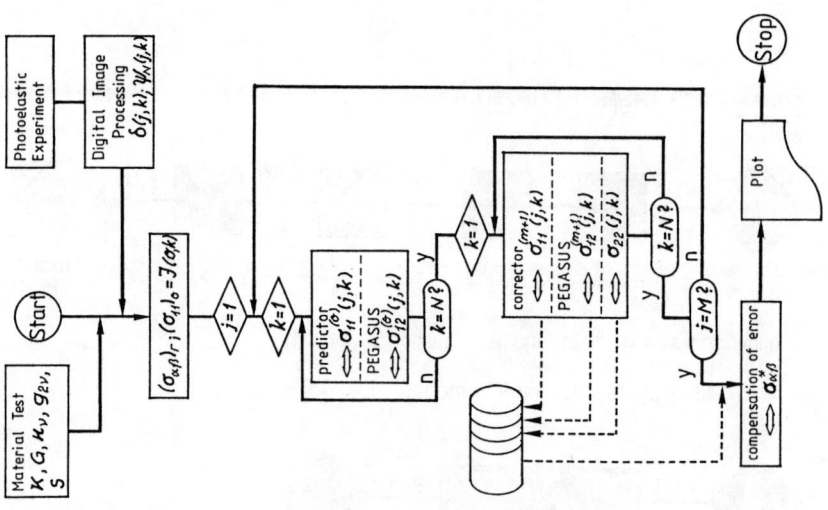

Fig. 5

Bloc-diagram of nonlinear photoelasticity

depending on the mechanically induced strains. And of course there are some materials for which the validity of this assumption has been proved experimentally.

With n_o, the refraction index of the unstrained material, n_{ij}, the refraction tensor, and the strain-optical coefficients d_1 and d_2, the relation between birefringence and strain holds

$$n_{ij}^{-2} - n_o^{-2} \delta_{ij} = \frac{1}{3} d_1 e \, \delta_{ij} + d_2 \, e_{ij} \, . \tag{4}$$

In eq. (4), e denotes the volume change and e_{ij} the strain deviation. Together with eq. (1), a nonlinear relation between the stress tensor and the refraction tensor will be obtained:

$$n_{ij}^{-2} - n_o^{-2} \delta_{ij} = A_1 s \delta_{ij} + A_2 \, \sigma_{ij} \, . \tag{5}$$

For abbreviation it has been introduced:

$$A_1 = d_1 \frac{\varkappa(s)}{3K} - d_2 \frac{g(\tau_o^2)}{2G} \; ; \quad A_2 = d_2 \frac{g(\tau_o^2)}{2G} \, .$$

Considering a plane stress state in the plane (x_1, x_2) and assuming the incident polarized light ray to pass through the model rectilinearly and parallel to the x_3-axis, the stress-optical relation holds

$$\left[1 + \sum_{\nu=1}^{n} g_{2\nu} \frac{1}{(2G)^{2\nu}} \tau_o^{2\nu} \right] \sigma_{12} = \hat{S} \frac{\delta}{d} \sin 2\psi_s \, , \tag{6}$$

where $\hat{S}$ denotes the initial value of the stress-optical module

$$\hat{S} = - \frac{2G}{d_2} \cdot \frac{\lambda}{n_o^3} \, . \tag{7}$$

Eliminating σ_{22}, the reference stress reads

$$\tau_o = \frac{\sqrt{2}}{3} \left[\sigma_{11}^2 - 2 \sigma_{11} \tan^{-1} 2\psi_s \cdot \sigma_{12} + (3 + 4 \tan^{-2} 2\psi_s) \sigma_{12}^2 \right]^{1/2} \, . \tag{8}$$

Furthermore the component σ_{11} will be given by integrating the equilibrium condition

$$\sigma_{11,1} + \sigma_{12,2} = 0 \tag{9}$$

by the discrete predictor – corrector-procedure /10/, the predictor yielding a first approach $\sigma_{11}^{(o)}$, which is to be introduced into eq. (8). The reference stress τ_o now depends on σ_{12} only and will be inserted in the nonlinear equation (6) of grade $(2n+1)$ in σ_{12} to be solved by the PEGASUS-Method /11/, thus yielding a first approach $\sigma_{12}^{(o)}$. In an iterative procedure this approach will be improved by repeated application of the corrector (Fig. 5).

Instead of the reference stress τ_o, the principal shear stress

$$\tau_H = \frac{1}{2} (\sigma_{11} - \sigma_{22}) / \cos 2\psi_s = \sigma_{12} / \sin 2\psi_s \tag{10}$$

may be introduced. As has been proved for different polymers, the nonlinear stress – strain response can be described with sufficient accuracy by considering the first two terms of the shear function only. Then the stress-optical relations hold

199

$$[1 + a \cdot \cos^{-2}(2\psi_s)(\sigma_{11} - \sigma_{22})^2](\sigma_{11} - \sigma_{22}) = 2\,\hat{S}\,\frac{\delta}{d}\,\cos^2(2\psi_s)$$
$$[1 + 4a \cdot \sin^{-2}(2\psi_s)\,\sigma_{12}^2\,]\,\sigma_{12} = \hat{S}\,\frac{\delta}{d}\,\sin^2(2\psi_s),$$

(11)

where

$$a = \frac{1}{4}\,\frac{g_2}{(2G)^2}$$

is taken from material testing. To determine the components of the stress tensor, the cubic eq. (11.2) must be solved, then the well-known shear-stress difference method can be applied. As shown in Fig. 6, the effect of physical nonlinearity indeed is remarkable in areas of high stress concentration.

4. Physical nonlinear stress analysis by strain gages

The surface strains of a workpiece may be measured by means of resistive strain gages. If the material to which the strain gages are applied shows physical nonlinear response, the relations between the measured quantities of strain and the stress tensor hold in an isothermal state /6/

$$\sigma_{ij} = [\mathcal{K} \cdot h(e) - \tfrac{2}{3}\,G \cdot \gamma(\varepsilon_o^2)]\,e\,\delta_{ij} + 2G \cdot \gamma(\varepsilon_o^2) \cdot \varepsilon_{ij}\,,$$

(12)

with the strain function

$$h(e) = 1 + \sum_{\nu=1}^{n} h_\nu \cdot e^\nu\,,$$

(13)

and the shearing function

$$\gamma(\varepsilon_o^2) = 1 + \sum_{\nu=1}^{n} \gamma_{2\nu} \cdot \varepsilon_o^{2\nu}\,,$$

(14)

where ε_o denotes the reference shearing strain. As on the load-free surface the strain state is a plane one, ε_o may be substituted by the principal shearing strain ε_H. Assuming the strains to have been measured considering effects of nonlinear response of the strain gages itself and their cross-sensitivity, the principal shearing strain reads with reference, for instance, i) to a 45°-strain-gage-rosette

$$\varepsilon_H = \frac{1}{\sqrt{2}}\,[(\varepsilon_{11} - \varepsilon_{45})^2 + (\varepsilon_{22} - \varepsilon_{45})^2]^{1/2};$$

(15)

ii) to a 60°-strain-gage-rosette

$$\varepsilon_H = \frac{\sqrt{2}}{3}\,[(\varepsilon_{11} - \varepsilon_{60})^2 + (\varepsilon_{60} - \varepsilon_{120})^2 + (\varepsilon_{120} - \varepsilon_{11})^2]^{1/2}.$$

(16)

The components of the stress tensor on the surface of the workpiece then hold

$$\sigma_{\alpha\beta} = [\mathcal{K} \cdot h(e) - \tfrac{2}{3}\,G \cdot \gamma(\varepsilon_H^2)]\,e \cdot \delta_{\alpha\beta} + 2G \cdot \gamma(\varepsilon_H^2) \cdot \varepsilon_{\alpha\beta}\,,$$

(17)

i.e. a nonlinear equation of grade 2n+1.

Obviously, acquisition and evaluation of the input signals demand much calculation, which can be done economically only in on-line procedures connecting a multi-position measuring system with a computer (see Fig. 1).

200

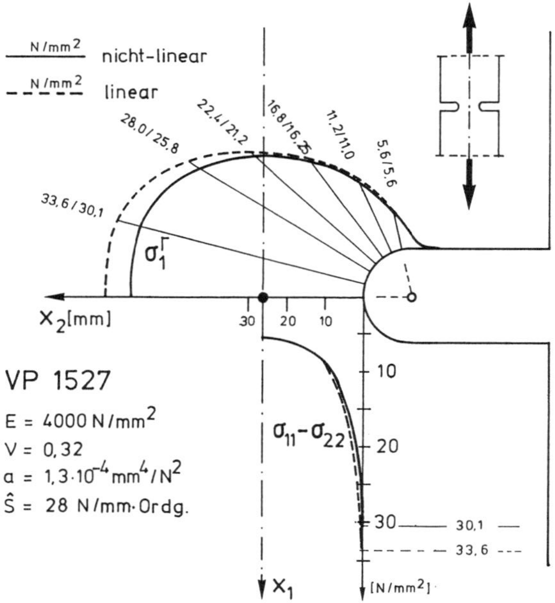

Fig. 6b

Stresses in the notched rod considering physical nonlinear response of material

Fig. 6a

Isochromatic pattern of a notched rod under tension

201

5. Geometrical nonlinearity; Large deflection of plates in bending

Considering thin plates in bending with large deflection, the classical Kirchhoff-Love-theory is not valid any longer. Formulating the equilibrium conditions regarding the deformed state, two simultaneous, inhomogeneous differential equations of the fourth order are obtained, which describe the in-plane as well as the bending stress state /12/

$$B\nabla^2\nabla^2 w = h \cdot L(w,F) + p_a \ ;$$
$$\nabla^2\nabla^2 F = \frac{1}{2} E \cdot L(w,w) .$$

(18)

with B, the bending stiffness, and F, a stress function, and the operator

$$L(r,s) = r_{,11} \cdot s_{,22} + r_{,22} \cdot s_{,11} - 2 r_{,12} \cdot s_{,12} .$$

(19)

To analyze the stress state experimentally, a combination of Ligtenberg's moiré method and the photoelastic reflection technique will be suggested requiring a model with one mirrored surface produced by galvano technique. The moiré technique (Fig. 7) /13/ yields the gradient of the deflection surface $w(x_1, x_2)$

$$grad \ w = \begin{bmatrix} w_{,1} \\ w_{,2} \end{bmatrix} = \begin{bmatrix} k_1 \\ k_2 \end{bmatrix} \frac{d}{2a} .$$

(20)

To obtain higher accuracy, the curvature as the second derivative of w is derived from the experimental data, grad w by cubic spline approximation including smoothing and balancing. Then the bending moments and the stresses $\sigma_{\alpha\beta}^M$ respectively can be computed as well as the principal directions ψ^M of the bending stress state.

To determine the in-plane or membrane stress state, the same plate model in the same experimental set-up will be observed by a reflection polariscope. As the principal directions of both the stress states do not coincide, the principal stresses as well as the directions of the superimposed stress state are functions of x_3, i.e. the direction of the light propagation. The photoelastic experiment then yields characteristic parameters only: The relative phase retardation δ^{M+N} and the so-called "characteristic directions" φ^{M+N} /14/, depending on the superimposed stress state.

Based on Maxwell's electro-magnetic equations, the wave propagation can be described by

$$(E_n) = [U] \cdot (E_0) ,$$

(21)

where (E_0) denotes the impinging, (E_n) the emerging light vector. The transmission matrix $[U]$ includes all informations of the stress state as a function of x_3. As the in-plane stresses are constant over the plate thickness, the bending stresses are linear functions of x_3 and information on the bending stresses is given by evaluating the moiré-experiments, the relation between δ^{M+N} and the superimposed stress state is given by the integral Wertheim's law /15/

$$\delta^{M+N} = \frac{h}{25}(\sigma_1^M - \sigma_2^M)\{(Q+1)^2[1 + \frac{(Q\eta)^2}{(Q+1)^2}]^{1/2}(Q-1)^2[1 + \frac{(Q\eta)^2}{(Q-1)^2}]^{1/2} +$$

$$+(Q\eta)^2 \ln \frac{(Q+1)\{1+[1+(Q\eta)^2/(Q+1)^2]^{1/2}\}}{(Q-1)\{1+[1+(Q\eta)^2/(Q-1)^2]^{1/2}\}} \}$$

(22)

with

$$Q = \frac{\sigma_{11}^N - \sigma_{22}^N}{\sigma_1^M - \sigma_2^M} \quad ; \quad \eta = \tan 2\bar{\psi}^N$$

and σ_1^M, σ_2^M, the boundary values of the principal bending stresses. It must be pointed out that $\bar{\psi}^N$ is related to the principal axes of the bending stress state.

As the bending stresses generally are essentially larger than the in-plane stresses and as the plate thickness is supposed to be small, the "characteristic direction" φ is given with reference to the principal direction of the bending stress state according to

$$\varphi^{M+N} = \frac{1}{4} \arc \tan 2Q^2 \cdot \eta \cdot [(Q^2-1) - (Q\eta)^2]^{-1}.$$

(23)

Obviously the eq.s (22) and (23) for the experimental data δ^{M+N} and φ^{M+N} are to be solved intelligently by computer only. The results Q and η and furthermore the difference of the normal stresses $\sigma_{11}^N - \sigma_{22}^N$ of the in-plane stress state must be transformed onto the coordinate system (x_1, x_2). Further evaluation is possible, e.g. according to the well-known shear-stress-difference method.

If the principal directions of both stress states coincide ($\bar{\psi}^N = 0$), the observed isochromatic fringe order is proportional to the difference of the principal stresses, and the characteristic direction φ is identical with the principal direction of the membrane stress state. This is always true for an axially symmetrical loaded circular plate. Fig. 8a shows the results of the experimental analysis compared to those of the numerical analysis, and Fig. 8b demonstrates the effects of the geometric nonlinearity.

Considering a thin circular plate supported on a yielding subgrade, the constitutive equations hold for a given external loading p_a

$$B\nabla^2\nabla^2 w = h \cdot L(w,F) - \phi \cdot w_{,rr} - \frac{h}{2}\phi_{,rr} + p_a - p \; ;$$

$$\nabla^2\nabla^2 F = \frac{1}{2}E \cdot L(w,w) + \frac{1}{h}(\frac{1}{r}\phi_{,r} - \nu \cdot \phi_{,rr}).$$

(24)

The function ϕ depends on the friction stress in the interface between plate and subgrade; p denotes the normal contact stresses

$$\phi(r) = -\int_o^r t(\bar{r}) \, d\bar{r} .$$

(25)

For numerical evaluation, eq.s (24) are transformed into a finite formulation introducing the sum of the bending moments and the resultants of the membrane stresses:

$$[D](m) = -\lceil n_r \rfloor (w'') - \lceil n_\varphi \rfloor (w') + \frac{h}{2}[A_2](t) - (p_a) + (p)$$

$$[D]((n_r) + (n_\varphi) + [A](t)) = -Eh\lceil w' \rfloor (w'') + ([A_1] - \nu[A_2])(t),$$

(26)

203

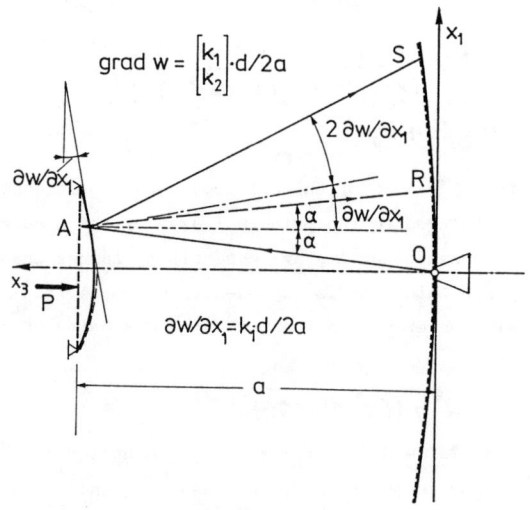

$$\text{grad } w = \begin{bmatrix} k_1 \\ k_2 \end{bmatrix} \cdot d/2a$$

$\partial w/\partial x_1$

A

x_3

P

$\partial w/\partial x_1 = k_i d/2a$

$2\,\partial w/\partial x_1$

$\partial w/\partial x_1$

R

α
α

S

x_1

O

a

Fig. 7

Principle of Ligtenberg's moiré technique

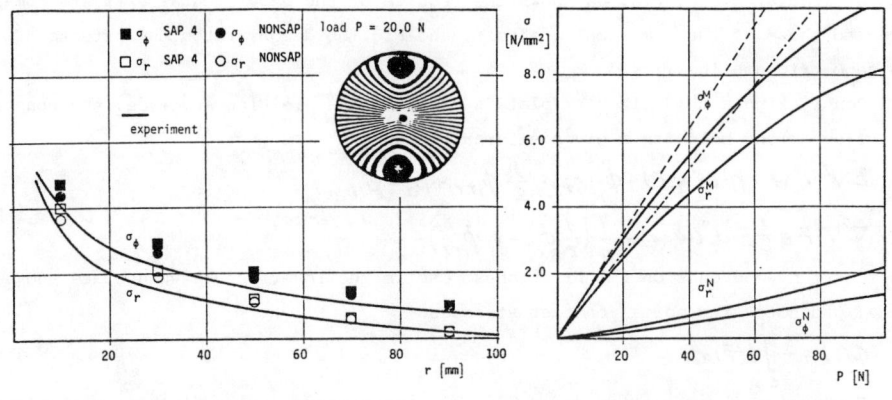

Fig. 8a

Circular plate, centrally loaded, P = 20 N, numerical and experimental results

Fig. 8b

In-plane and bending stresses in a circular plate as function of P

$r = 0,25\ r_a$

where e.g.

$$\lceil n_r \rfloor (e) = (n_r) .$$

By means of Ligtenberg's moiré method the data grad w are obtained, from which the second derivatives of the deflection surface and finally the sum of the bending moment are derived: (w'), (w"), (m).

In a reflection polariscope the isochromatic fringe order along the radius is measured. Then the resultants of the membrane stresses read

$$N_r(r) = -\frac{S}{2} \int_{(r)} \delta(\bar{r}) \frac{1}{\bar{r}} \, d\bar{r} - \phi(r) = \bar{N}_r(r) - \phi(r)$$

$$N_\varphi(r) = -\frac{S}{2} \left[\delta(r) + \int_{(r)} \delta(\bar{r}) \frac{1}{\bar{r}} \, d\bar{r} \right] - \phi(r) = \bar{N}_\varphi(r) - \phi(r) .$$

(27)

$N_r(t)$ and $N_\varphi(r)$ are evaluated from the photoelastic data. With the matrix $[T] = [D][A] + [A_1] - \nu[A_2]$, eq. (26.2) yields the vector (t) of the friction stress

$$(t) = [T]^{-1} \{ Eh[w'] (w") + [D]((\bar{n}_r) + (\bar{n}_\varphi)) \} ,$$

(28)

and furthermore N_r and N_φ can be calculated. Introducing this solution into eq.(26.1) yields the normal contact stress (p)

$$(p) = +(p_a) + [D](m) + \lceil n_r \rfloor (w") + \lceil n_\varphi \rfloor (w') - \frac{h}{2}[A_2](t) .$$

(29)

Thus the contact stresses as well as the internal stress state in the plate has been determined. An example of application is given in Fig.s 9 and 10.

6. Conclusion

Obviously with the principle of hybrid technique modelling of the regarded mechanical event is possible much more realistically. On the other hand it is evident that the numerical calculations can be handled by using computer facilities only.

Considering the afore-mentioned arguments, the principles and methods of experimental analysis in solid mechanics can be applied to a much wider field of problems than ever before. Beyond the "classical" field of static, dynamic and stability problems those problems are investigated now where even very intelligent numerical methods do not lead towards reliable results, regarding geometric nonlinearity, impact loads, wave propagation, system identification, problems of fracture and fatigue and physical nonlinearity, plastic, viscoelastic, viscoplastic as well as non-homogeneous and non-isotropic response of material. Thus, experimental analysis opens the way to correlate the structural design with the technical process, the response of material with the functional purpose of the structure, in order to improve reliability, safety and economy of design, production and operation of structures.

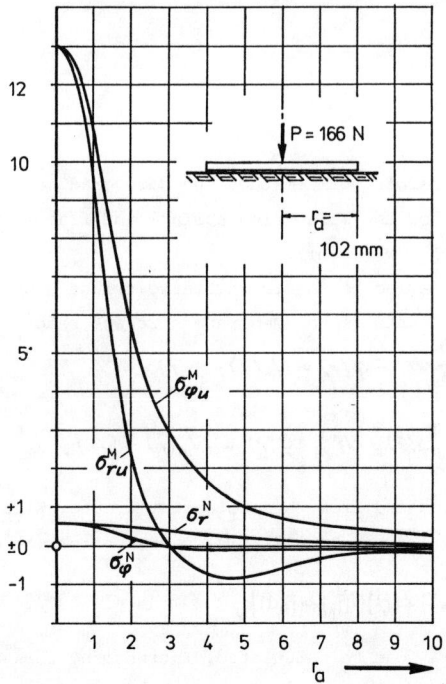

Fig. 9

Circular plate on a yielding subgrade; in-plane and bending stresses

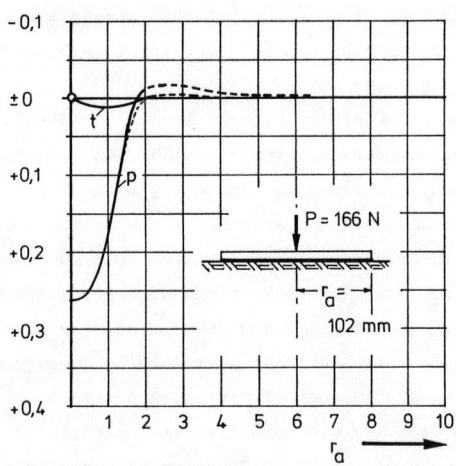

Fig. 10

Circular plate on a yielding subgrade; normal contact and friction stresses in the interface

7. References

1 Laermann, K.-H.: Recent Developments and further Aspects of Exper.Stress Analysis in the Federal Republic of Germany and Western Europe. Exp.Mech., Vol.21/2, 1981.

2 Laermann, K.-H.: On the Principles of Hybrid Techniques in Exper. Stress Analysis. Proc. EAN '82 "Exp. Stress Analysis", Vol. I, Karlovy Vary 1982.

3 Laermann, K.-H.: Recent Trends in Exper. Analysis. Mechanika Teoretyczna i Stosowana, 2/3, 21, Warszawa 1983.

4 Laermann, K.-H.: Über eine Theorie der nichtlinearen Photoelastizität. ZAMM 66, 4, 1986.

5 Laermann, K.-H.: On a Nonlinear Theory of Photoelasticity. Exp. Stress Analysis, Wieringa, H. (ed.), Martinus Nijhoff Publishers, Dordrecht 1986.

6 Kauderer, H.: Nichtlineare Mechanik. Springer-Verlag, Berlin/Göttingen/Heidelberg, 1958.

7 Neumann, F.E.: Die Gesetze der Doppelbrechung des Lichtes in comprimierten oder ungleichförmig erwärmten unkristallinen Körpern. Abh. Königl. Akad. d. Wissenschaften zu Berlin, 3. Nov. 1841.

8 Coker, E.G., and Filon, L.N.G.: A Treatise on Photoelasticity. 2nd Edition, H.T. Jessop (ed.), University Press, Cambridge 1957.

9 Mindlin, R.D.: A Mathematical Theory of Photo-Viscoelasticity. J. Appl. Physics, Vol. 20, 1949.

10 Engeln-Müllges, G., and Reutter, F.: Numerische Mathematik für Ingenieure, 4.Aufl. Wissensch.-Verlag, Bibliograph. Inst., Mannheim/Wien/Zürich, 1985.

11 Dowell, M., and Jarratt, P.: The "Pegasus"-Method for Computing the Root of an Equation. BIT 11, 1971.

12 Laermann, K.-H.: Hybrid Analysis of Plate Problems. Exp. Mech., Vol. 21, No. 10, 1981.

13 Laermann, K.-H.: Über die experim. Untersuchung geometrisch nichtlinearer Probleme der Plattenbiegung. Domke-Festschrift, Lehrstuhl f. Konstruktive Gestaltung, RWTH Aachen, 1982.

14 Aben, H.: Integrated Photoelasticity. McGraw-Hill Int. Book Co., New York, 1979.

15 Laermann, K.-H.: Das Prinzip der integrierten Photoelastizität, angewandt auf die experim. Analyse von Platten mit nichtlinearen Formänderungen. Exp. Stress Analysis, Proc. 7th Int. Conference, Betser, A. (ed.), Haifa 1982

EXPERIMENTAL STRESS AND STRAIN ANALYSIS OF MACHINE PARTS BY PHOTOELASTICITY, MOIRE METHOD AND SPECKLE PHOTOGRAPHY

Prof. Dr.-Ing. habil. Joachim Heymann, Dr.-Ing. Roland Meyer,
Dr.-Ing. Jochen Naumann, Dipl.-Ing. Klaus Uhlig
Technische Universitaet Karl-Marx-Stadt, DDR

The paper describes the application of the photoelastic-coating technique and the phase-grating moire method to solve design problems in the case of three-dimensional disk struc-tures. These methods served to get information from models made of epoxy resin and also from welding-seam regions of original presses. A relatively new field method, the speckle photography, was chosen to investigate machine components. With a computer-aided evaluation set-up YOUNG's fringes are analyzed on-line, and after a smoothing procedure the compo-nents of the displacement vector are determined.

Keywords: photoelastic coating technique, phase-grating moire method, hydraulic open-side press, speckle photo-graphy, computer-aided evaluation of speckle data

1. Introduction

Optical field methods have been applied successfully to solve complicated design problems. Often it will be necessary to adapt the chosen method to the component to be analysed. And as a reciprocal effect, the realization of the demands made by the industrial custo-mers leads to the further development of the applied methods.

2. Photoelasticity and moire method

2.1. Investigation of models

The investigation of variants of three-dimensional disk struc-tures evaluated on the base of strength criteria may be carried out effectively and precisely by means of the photoelastic-coating techni-que and the phase-grating moire method [1] [2]. As models for the

IMEKO 1st TC 15 Conference, Plzen, CSSR, May 25-28 1987

investigations not cast, but cemented models made of epoxy resin are used.

The advantages of this model method are
- the casting of a three-dimensional model and thus the building of a mould does not exist
- the expensive photoelastic freezing method will be avoided
- Poisson's model law will be realized because the investigation is carried out at room temperature
- the geometry of the model may be modified in regions of special interest in order to analyse several variants

An embedded measuring layer which is part of the model provides an average stress value over its thickness after the equation

$$\sigma_1 - \sigma_2 = (n*S)/(2d) \qquad (1)$$

with σ_1, σ_2 principal stresses, n fringe order, S stress-optical coefficient, d thickness of the photoelastic layer.

If there exists a greater stress gradient over the thickness of the measuring layer, the maximum stress value got by photoelasticity may be made more precise by means of the phase-grating moire method since the measurement is carried out directly on the surface of the model.

The measuring principle is "superposition of reference grating and deformed object grating". The result of the measurement is moire fringes representing lines of equal Cartesian displacement perpendicular to the reference grating.

The determination of the strain value is carried out with the following approximative expression

$$\varepsilon \approx p/a \qquad (2)$$

with ε strain perpendicular to the reference grating, p pitch of the grating, a distance between two moire fringes.

The preparation of the phase gratings is made by means of the so-called replica technique [3] [4]. An example of such a three-dimensional disk structure is the welded hydraulic open-side press the model of which is shown in fig. 1. Critical regions of the press frame are the transition radii between front disk and table plate; head plate. In these regions photoelastic measuring layers are embedded and phase gratings prepared. The necessary thin reflective film is evaporated in high vacuum.

Fig. 2 shows the isochromatic-fringe pattern of a chosen transition

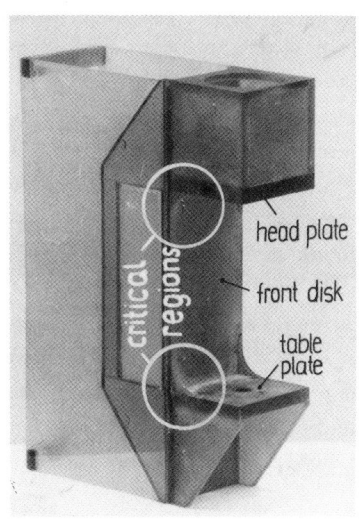

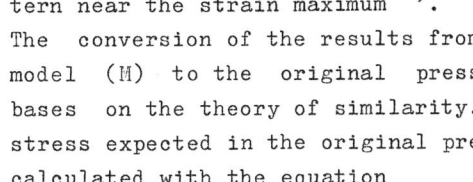

Fig. 1 Model of the frame
of an hydraulic open-side
press made of epoxy resin

radius and fig. 3 shows the moire pattern near the strain maximum [*].

The conversion of the results from the
model (M) to the original press (H)
bases on the theory of similarity. The
stress expected in the original press is
calculated with the equation

$$\sigma_H = (\sigma_M * l_V^2 * F_H)/F_M \qquad (3)$$

with σ_M stress in M, F_M load acting on
the model, F_H load acting on the original press, $l_V = l_M/l_H$ scale of lengths.
The maximum stress of $\sigma_H = 134$ N/mm^2
calculated in the radius between front
disk and head plate after the model
investigation faces a stress value of
$\sigma_H = 143$ N/mm^2 found later in a measurement at the original press. A great
advantage of these model
methods is that a combination with numerical me-
thods, e.g. finite-element
method, may be made in the
design process. If photo-
elastic data are known,
the mesh for the FEM
calculation may be found
easier and more effective.

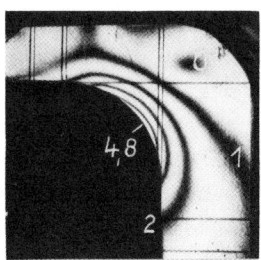

Fig. 2 Isochromatic-
pattern of an embed-
ded measuring layer

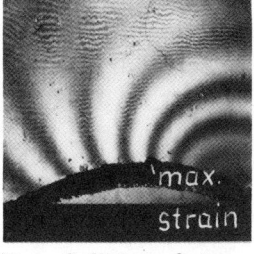

Fig. 3 Moire-fringe
pattern near the
strain maximum

2.2. Measurements at the original press

2.2.1. The photoelastic-coating technique and the phase-grating
moire method in incident-light arrangement are well suited for the
investigation of critical regions of three-dimensional disk structures. Fig. 4 shows the isochromatic-fringe pattern of the region
between front disk and head plate of an open-side press made of steel.

[*] For the solution of special measuring problems moire gratings are
availble from Technische Universitaet Karl-Marx-Stadt, Sektion Maschi-
nen-Bauelemente, DDR

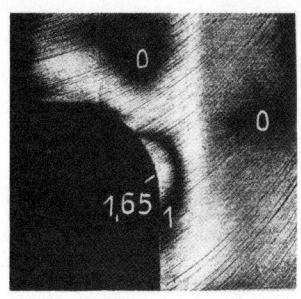

Fig. 4 Fringe pattern of a photoelastic coating on a press made of steel

In order to compare results got by photoelasticity or to get longitudinal strain along lines parallel to the margin tangent, the phase-grating moire method may be applied. For the measurement of strains between $\varepsilon = 0.0001$ and $\varepsilon = 0.001$ the mismatch method must be applied to ensure a high accuracy. The line density of the gratings is 250 lines/mm. The mismatch gratings are made by the help of bars subjected to tension.

The reflective film necessary for incident illumination consists of indium evaporated onto the reference grating in high vacuum. This reflective film is transferred from the reference grating to the object grating during the process of grating relief formation.

2.2.2. The photoelastic-coating technique combined with the phase-grating moire method may also be applied successfully to the investigation of welding-seams.

Because of the small deformations it is necessary to use the mismatch method. It is well-suited to judge the longitudinal-strain propagation on disks or plates connected by a welding-seam. Since the measuring area must be flat the welding-seam often is to be ground. A further application is the investigation of great welding-seams consisting of several welding-layers.

Fig. 5 shows the position of such a welding-seam bet-

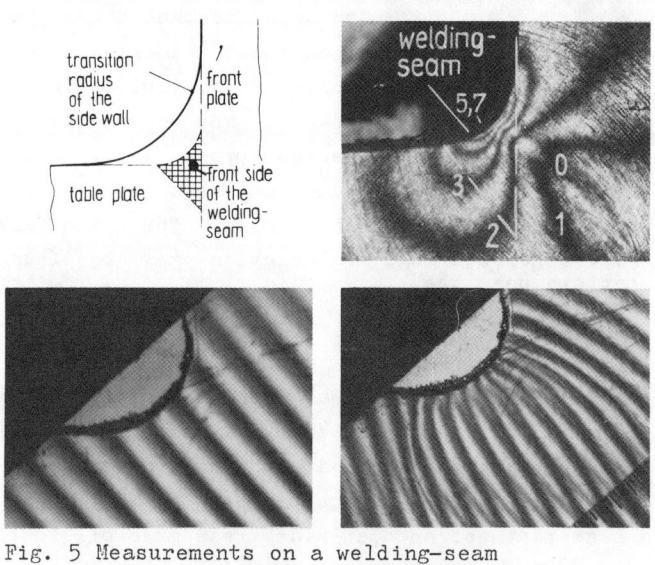

Fig. 5 Measurements on a welding-seam
  a) information on the problem
  b) photoelastic-fringe pattern
  c) moire-pattern, unloaded state
  d) moire-pattern, loaded state

ween front disk and table plate of a press. The isochromatic-fringe pattern, fig. 5b, served to find out the locus of the stress maximum. After removing of the measuring layer a mismatch phase grating of a line density of 250 lines/mm is prepared tangentially to the locus of the stress maximum. The moire-fringe patterns are shown in fig. 5c (unloaded state) and fig. 5d (press under load).

3. Rationalization of data registering in speckle photography

Speckle photography is a coherent-optical field measurement method for determining in-plane deformations of either mechanically or thermally loaded structural members. In our investigations we use a test arrangement with only one beam in connection with double exposure technique. The specklegram is scanned by a laser beam point by point. The Young's fringes arising from that are perpendicular to the displacement vector $\vec{u}$ in the object plane and can be described by the following equation

$$|\vec{u}| = (N * \lambda * 1)/(\beta' * b_N) \tag{4}$$

with λ wavelength, 1 distance specklegram-screen, β' magnification (image/object size), b_N distance between (N+1) Young's fringes.
Relating to the shrink joint described in [5], a semi-automatical evaluation set-up was developed. In addition to this, it may be universally employed to analyze YOUNG's fringes. For measuring the fringes a digital toolmakers microscope (produced by VEB Carl Zeiss, GDR) was coupled on-line with a computer, fig. 6.

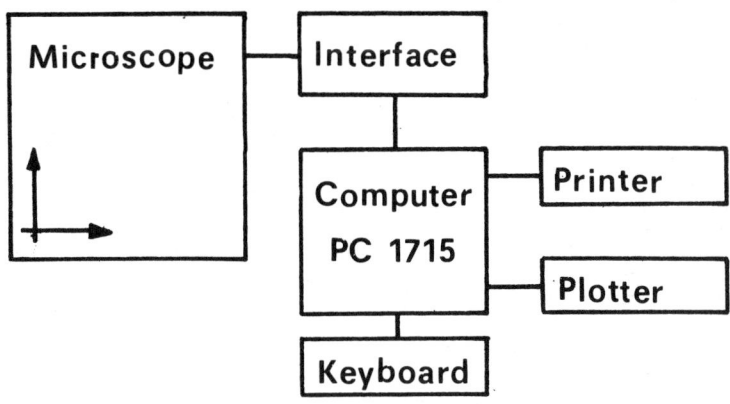

Fig. 6 Block diagram of the evaluation system

In aggreement with the demand to get from a diffraction pattern as much information as possible, the measuring grid shown in fig. 7 was chosen. The measured values may be immediately fed to the computer by a signal given the by operator. Thus an effective, interactive work with the computer is possible. The points of measurement are represented by the intersection points of the measuring tracks with YOUNG's fringes. The measured values are processed in a two-dimensional smoothing procedure by use of a plane of balance

$$n(x_1,x_2) = a_0 + a_1 * x_1 + a_2 * x_2 \tag{5}$$

After computation of the coefficients of equation (5), the components of the displacement vector are determined:

$$u_1 = \lambda * 1 * a_1 / \beta' \qquad u_2 = \lambda * 1 * a_2 / \beta' \tag{6}$$

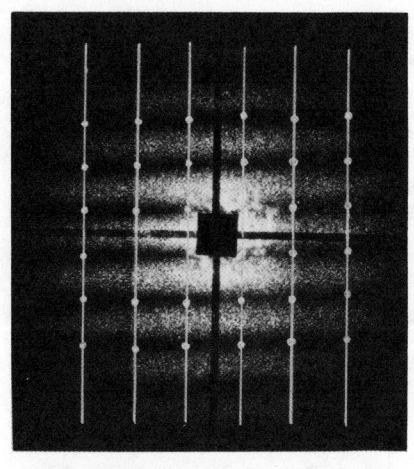

Fig. 7   Measuring grid

References

1 Meyer, R.: Spannungsoptik und Phasengitter-Moireverfahren bei der modellstatischen Analyse von Einstaenderpressen. -1985- Karl-Marx-Stadt, Technische Hochschule, Dissertation A
2 Heymann, J., Meyer, R.: Variantenuntersuchungen von Pressenge-stellen mittels Spannungsoptik und Moireverfahren. Technische Mechanik 3(1982)4, 36-40
3 Heymann, J., Meyer, R.: Beitrag zur Herstellung von Phasenra-stern fuer Dehnungsmessungen mit dem Moireverfahren. Wiss. Z. d. Techn. Hochsch. Karl-Marx-Stadt 14(1972)5, 647-658
4 Naumann, J., Jantschke, B.: Eine Theorie zur Moirestreifenmul-tiplikation bei zwei ueberlagerten Gittern. Wiss. Z. d. Techn. Hochsch. Karl-Marx-Stadt 19(1977)3, 305-316
5 Heymann, J.; Vogel, J.; Naumann, J.: Deformation Analysis of a Shaft-Boss-Connection by Means of Speckle Photography. Proc. 9 th Congress on Material Testing, Budapest 1986, 382-386

CALCULATED EXPERIMENTAL METHODS FOR RESIDUAL STRESS   ESTIMATION

Dr. M.N.Dveres, Dr. A.V.Fomin, Dr. I.A.Razumovsky,
Dr. V.M.Synaisky
Mechanical Engineering Research Institute of the
Academy of Science
Griboedov Str. 4, Moscow Centre, USSR

The paper concerns estimation of a continuous distribution
of residual stresses in plane parts of unspecified shape
in terms of observed data on stresses-strained condition
near slots being made therein. The distribution of residual stresses in a surfaced bimetal part was investigated.
The observed data are compared with the data obtained by
use of x-ray diffraction method.

Keywords: residual stresses, inverse problem, calculated-
          experimental methods

1.Introduction

A problem of residual stress determination in terms of observed
data on stresses condition in the vicinity of slots being made in
parts is of inverse problem class. A stresses-strained condition
picture being observed is a consequence (or response) to residual
stresses released along a cutting line. The residual stresses are
to be determined in terms of the response. The inverse problems are
incorrect ones, their incorrectness being manifested itself in a
fact that to stresses-condition parameters may correspond significantly different distributions of residual stresses along a cutting
line. The problem occurs to be correct and have an analytical solution in a particular case when a small inner area of a large plate
is investigated and a sufficiently thin slot may be considered as
a crack [1].

Correct procedures of residual stress determination may be based on regularizing algorithms for various integral equations [2,3].

IMEKO 5th TC7 Symposium, Plzen, Czechoslovakia, May 25-28, 1987

215

Let us consider two modifications of their derivation for different experiment approaches. For the first case (Volterra equations) a slot is being successfully built up. During this process the measurements are carried out for a parameter of stresses condition and a characteristic point following a slot tip extension. In this manner a distribution of observed parameter along a cutting line in a process of change of part geomentry is carried out. For the second case (Fredgolm equation) the slot corresponds to its final length; the summary effect along some line in the vicinity of the slot is being determined.

2. Volterra integral equation method

In a process of a slot building-up residual stresses are released along its edges; these stresses acting as an edge loading for a new area of a part boundary (Fig. 1,a). Because of a superposition principle a relation of unknown residual stresses σ_y (ξ) to a stresses-condition parameter along a cutting line μ (x) (Fig. 1,b) may be described by use of integral Volterra equation of the first kind

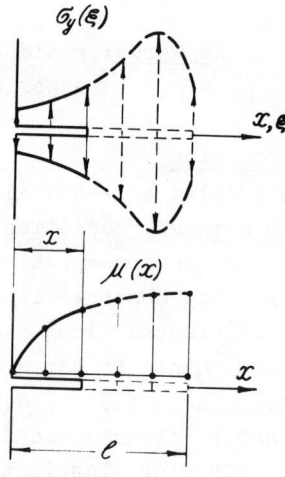

$$\int_0^x B(x,\xi)\sigma_y(\xi)d\xi = \mu(x),$$

$$0 \leq \xi \leq x \leq \ell, \qquad (1)$$

where ℓ is a maximum length of slot. A core $B(x,\xi)$ corresponds to the values μ (x) at a point x due to δ-action taking place at a point ξ ; a range of its definition is triangle, i.e. for $\xi > x$ we obtain $B(x,\xi) = 0$.

Fig. 1

Residual stresses and observed parameters during a slot building-up

Along with the equation (1) we may assume its transformation to be performed on a basis of an additional condition of smoothness of the function being determined μ (x). Then by use of differentiation we can reduce this equation to Volterra integral equation of the second kind

216

$$B(x,x)\mathcal{G}_y(x)+\int_0^x B'_x(x,\mathcal{E})\mathcal{G}_y(\mathcal{E})d\mathcal{E}=\mu'(x),$$

from where we can obtain:

$$\mathcal{G}_y(x)=B(x,x)^{-1}\left[\mu'(x)-\int_0^x B'_x(x,\mathcal{E})\mathcal{G}_y(\mathcal{E})d\mathcal{E}\right].\qquad(2)$$

When employing the Volterra integral equation method the following important factor must be taken into account. During preparations for the experiments it is necessary to consider the experimental condition and sensitivity of measuring equipment for choicing a pitch for a slot building-up which can not be any small.

The data obtained therewith is of a discrete type but it is not connected with a nature of the distribution being determined $\mu(x)$, from which arises a problem of preliminary processing of observed data and constructing some continuous and differentiable analogs within the limit of errors being comparable with the experimental ones. The concrete diagrams of initial data processing may be constructed by some different ways including regulation methods.

3. Method of Fredholm integral equation of the first kind

Let us choice a line S passing in the vicinity of a slot top (Fig 2). A relation of unknown residual stresses $\mathcal{G}_y(x)$ along a slot of length L to values of determined parameter $\mu(S)$ on this line is of a form

$$\int_0^L H(S,x)\mathcal{G}_y(x)dx=\mu(S),\qquad(3)$$
$$S\in S,\ x\in L.$$

A core of an integral operator $H(S,x)$ is determined in a range $S\times L$; it corresponds to determined parameters $\mu(S)$ on a line S due to δ-action at points $x\in L$. In the methodological respect a reduction of a problem solution of Fredholm integral

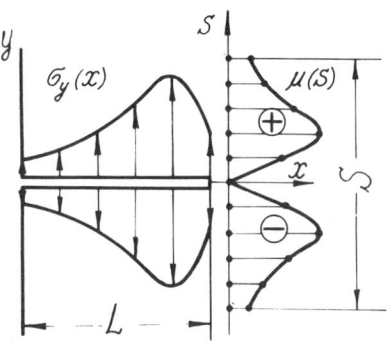

Fig. 2

Residual stresses and observed parameters at constant length of a slot

equation of the first kind is of great importance. Firstly the preparation for experiment is less labour-consuming in comparison with
an approach using Volterre equation since all the experimental data
are taken for the only value of a slot length. Secondly for numerical solution of the Equation (3) a construction of matrix analog of
integral operator demands solution of less edge problems than the
first approach (for the same discretizations).

    Solutions of equations of the first kind (1) and (3) are un-
stable due to the right side. The values μ (x) and μ (S) are
known from an experiment when additive random noises take place of
which value depends on an experimental methods being used and concrete experiment condition. Their accuracy can not be increased
to the extent that the numerical methods of determination of integral operator nuclei permit.

4. Determination of an integral operator cores

    The matrix analogs of operators cores B (x, ξ), B'_x (x, ξ)
and H (S, x) were determined on the ground of numerical solution
of a two-dimensional problem of an elasticity theory by use of a variation – difference method securing high accuracy for complex-sha-
ped parts having high stress gradients. For determination of residual stresses in plane parts a thin rectangular slot was made. For optimization of mesh division of a range at a zone of slot angular
points we used generation of sequence of quazi-uniform meshes which
permits to carry out a controlled evaluation of solution and its ac-

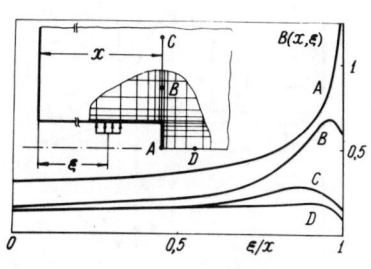

Fig. 3

Numerical determination of
a core B (x, ξ)

curacy as well as to use extrapolation formulas for refinement of solution. As characteristic points of
determination at using the Equations
(1), (2) we took regular points A,
B, C, D near angular points and
about an axis of symmetry (Fig. 3).
With taking as a guide the data obtained by use of photoelastic coating method we used as observed value
a quantity τ_{max} about an axis of
symmetry and a quantity τ_{xy} in
the vicinity of angular points. Obtained values B (x, ξ) in the

vicinity of a slot tip do not practically depend on a slot length, i.e.:

$$B(x, \xi) \approx B(x - \xi),$$

and derivatives with respect to x and ξ coincide, i.e.
$B_x' (x, \xi) \approx B_\xi' (x, \xi)$. Conditionality of linear algebraic equations (1) depends on cores structure due to a choice of check point. For the better conditionality it is expedient to choice a cores with predominant elements of the main diagonal which corresponds to a small vicinity of a slot top, for example of a point A or B (see Fig. 3).

5. Investigation of a specimen of surfaced hull steel

The Volterra integral equation method was used for determination of residual stresses in 9mm-thick overlaying and a basic metal zone bodering on it (Fig. 4). The measurements were carried out by use of photoelastic coating method. A 1,8 mm-thick coating of optically sensitive ED-20 material was pasted to a specimen. A consequtive building-up of 1mm-pitched slot in a surfaced specimen was made by cutting with use of a disc cutter of 0,5 mm thickness. Optical measurements of order of bands m and parameters of isoclinic lines φ at characteristic points A and B (Fig.3) following a slot tip extension were carried out with "Photolastic" reflection polariscope. Fig. 4 shows relations $\tau_{max}^A (x)$ obtained in terms of

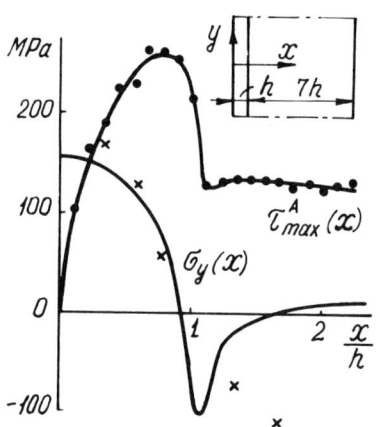

Fig. 4

optically observed data as well as a residual stress diagram $\sigma_y (x)$ constructed in terms of the Equation (1).

Also an analysis of residual stresses condition was performed for an analogous specimen by use of x-ray method. The stresses were determined by used method of multiple incline viewing [4] with "Siemens" automatic diffractometer in cobalt anode radiation. For overlaying the reflection (113) was used; for basic metal the reflection (310) was used which corresponds to diffraction, peaks of

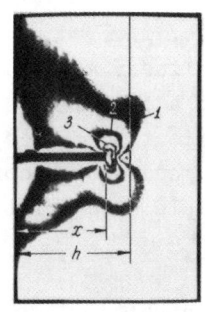

the greatest angles of reflection being
optimal for reveal of elastic strains
due to residual stresses. The filings
of the two indicated metals were used as
a standard. The observed data are represented in Fig. 4 by crosses; the error
being not over than 20 per cents of
maximum observed value.

Fig. 5
Photography of a band
picture of a slot tip
zone for $\ell = 7$ mm

6. References

1 Vaidyanathan, S. Finnie, I.: Determination of Residual Stresses
  From Stress Intensity Factor Measurements. Trans. ASME, Ser. D,
  No. 2, pp. 131-135, 1971.
2 Тихонов А.Н., Арсенин В.Я.: Методы решения некорректных задач.
  М., Наука, 1979, 286 с.
3 Верлань А.Ф., Сизиков В.С.: Интегральные уравнения: Методы, алго-
  ритмы, программы. Справочное пособие. – Киев: Наукова Думка,
  1986, 544 с.
4 Комяк Н.И., Мясников Ю.Г.: Рентгеновские методы и аппаратура для
  определения напряжений. Л., Машиностроение, 1972, 88 с

EXPERIMENTAL INVESTIGATIONS INTO THE PLASTIFICATION OF THIN WEBS

Ing.Miloš Drdácký,CSc - ×Ing.Petr Jaroš,CSc - ⚥Ing.Otakar Weinberg
Institute of Theoretical and Applied Mechanics, Dept. of Experimental
Mechanics, Vyšehradská 49, 128 49 Prague 2, Czechoslovakia
×National Research Institute of Machine Design, Dept. of Experimental
Mechanics, 250 97 Prague 9 - Běchovice, Czechoslovakia
⚥Central Research Institute k.p. Škoda, 316 00 Pilsen, Czechoslovakia

The paper deals with experimental investigations of webs of
welded plate girders subjected to patch loads with regard
to the origin and development of their plastification. The
measurements were completed using resistive strain gages,
acoustic emission method and shadow moiré. A comparison of
these methods is presented.

Keywords: plate girder web, patch load, plastification,
acoustic emission, shadow moiré

1. Introduction

A revision of the Czechoslovak standard for the design of steel
bridges ČSN 73 6205 necessitated an improvement of our knowledge of
the origin a development of plastification of thin webs subjected to
partial-edge loads. Specification of such a behaviour is a require
for the reliable definition of the limit state of such webs.

2. Experimental Investigations

Two series of test girders were investigated, Fig.1. In the first
serie there was studied i) an influence of the applied loading lengths
on the load carrying capacity of thin webs, ii) the onset and the
development of plastic deformations in the web under static loads and
iii) the correlation between the resistance strain gage measurements
and the acoustic emission method. In the second case, the plasticity
in webs with lower slenderness ratio was investigated at different
loading cases. There was further followed a possibility of direct
visualization of web buckling shapes using the shadow moiré technique.
Due to a limited extent of the presented paper let us shortly
describe the tests and comment only the main results.

The 37 grade steel test girders or panels with the web thickness of about 2 mm were loaded in an electromechanical loading frame Testatron. During the loading the surface strains were measured by means of resistance strain gages with the length of 20 mm, in the second serie with the length of 10 mm. The loading length of the upper flange varied from about zero to 200 mm, i.e. to one a half of the distance of vertical stiffeners. The load was distributed through a set of short simply supported distribution beams, e.g. Fig.7.

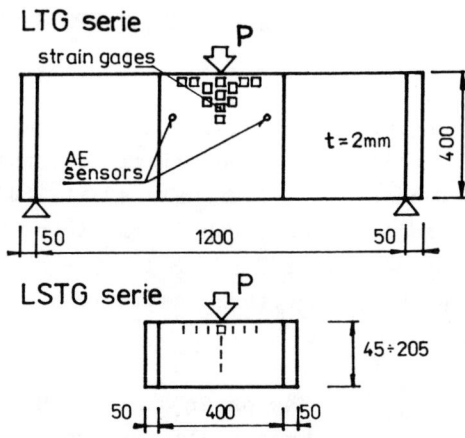

Fig. 1
Test girders

Principal surface strains and principal bending and membrane strains all over the cross-section were evaluated from the measured values. The beginning of plastification was determined by the attainment of the plasticity surface according to Huber-Mises-Hencky hypothesis. In the first serie the plastification was checked also by means of acoustic emission method, where the acoustic signals were recorded by two microphones the position of which is apparent on the Fig.1. Further the web deflections in the cross-section under the applied load as well as the maximum deflections of the both flanges were measured by a set of dial gages. In the second serie the overall web deflections were taken using a shadow moiré technique.

3. Main Results of the Strain-Gage Measurements

The main results are summarized in Figs.2 and 3, where the former shows the different behaviour of the web according to different loading lengths. In the first case the load was applied to the flange by a semicircular edge and the first surface yielding of material occured at 35.6% of the load-carrying capacity and the membrane yielding at 53.8%. In the second case the load was distributed uniformly over a length of s_o = 100 mm and the relevant features occured at 51.3% and 73% of the maximum load.

Fig.3 presents a comparison of these and other experimentally ascertained beginnings of plastification for webs of various

222

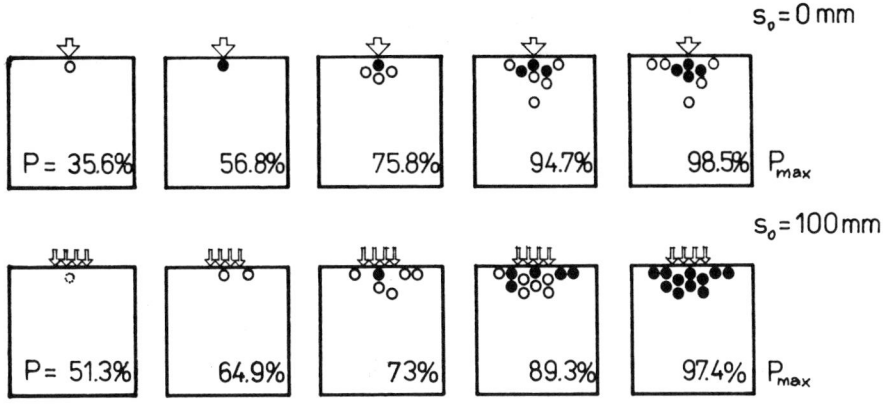

Fig. 2
Web plastification development for two loading cases

slenderness ratios. They involve measurements without acoustic
emission, and with regard to strain gage behaviour in the course of
loading it is possible to assume in a number of cases the origin of
plastic deformation in the web before it has been determined directly
by measurements, (which is indicated by smaller dots in the figure).

The comparison shows that a part of the web can be in a plastic
state already under loads corresponding with 25% of the maximum
load, particularly in the case of webs of lower slenderness.

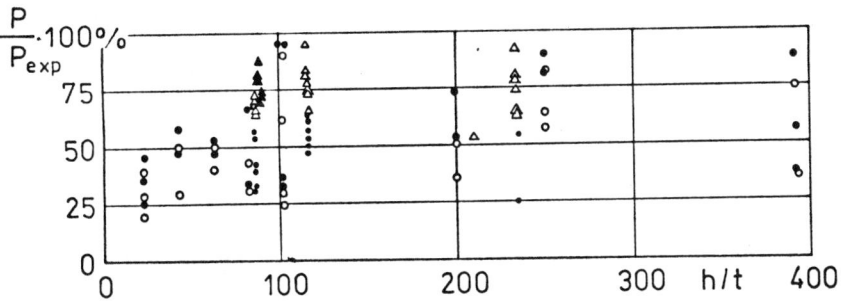

o Onset of yielding on the surface of unstiffened web.
△ Onset of yielding on the surface of stiffened web.
• Onset of membrane yielding in the unstiffened web.
▲ Onset of membrane yielding in the stiffened web.
· Undirectly signalized plastification of the structure.

Fig. 3 Beginnings of plastification - experimental results

223

4. Acoustic Emission Measurements

An overpassing of the yield limit and a subsequent plastic deformation of mild steels in the so called Lueders´s zone generates a significant acoustic emission. This phenomenon can be utilized for the indication of plastification in cases with absence of other disturbing acoustic emission sources. At the presented tests a two channel apparatus Dunegan-Endevco was employed with digital and analog outputs of usual acoustic emission characteristics - number of counts N and number of emission events E_w. The meaning of these terms is illustrated by the explanation note in the Fig.4 where the both parameters are presented in a dependence on the behaviour of the web during the test. For the sake of clarity a load-maximum flange deflection diagram is attached where the corresponding loading steps are marked by capital letters. From this typical picture it follows that the both AE parameters are suitable for applications, even though the ratio N/E_w is changing during the loading, and that an unloading stage brings about only a slight increment of acoustic emission.

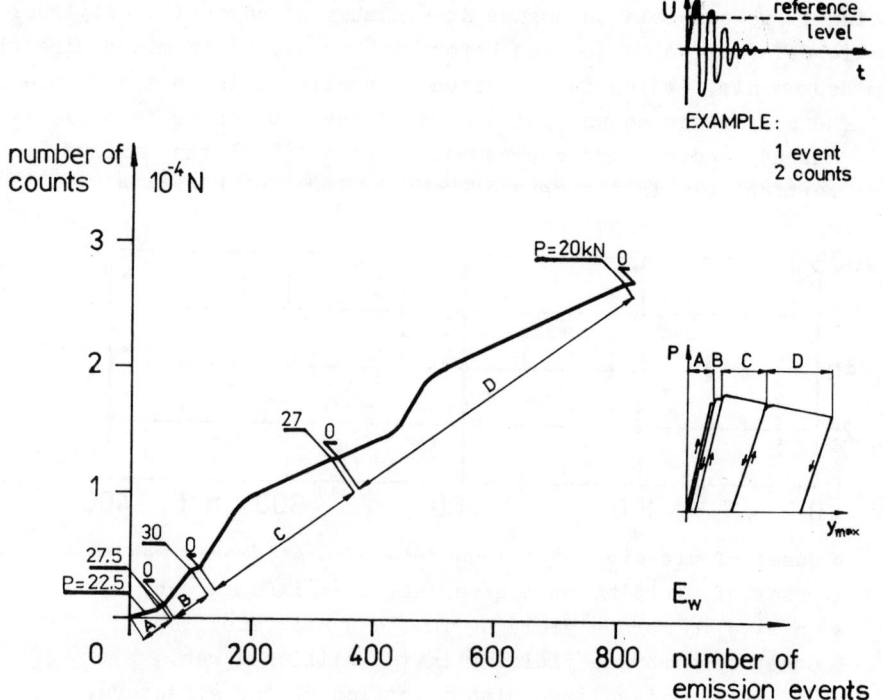

Fig. 4

Acoustic emission characteristics diagram

In the next Fig.5 the number of counts N is plotted against the applied load P and against the maximum flange deflection y_{max}. The Table 1 compares the loads at a plastification beginning determined by strain gages to those given by acoustic emission. From the results it follows that acoustic emission method is sufficiently sensitive and suitable method for investigations into the onset of yielding and plastic deformations indicated by the shift of strain-gage initial readings of the order $\varepsilon \lesssim 100$ microstrains. Moreover this method checks a whole web panel utilizing only one or two sensors without elaborate instalations and evaluations.

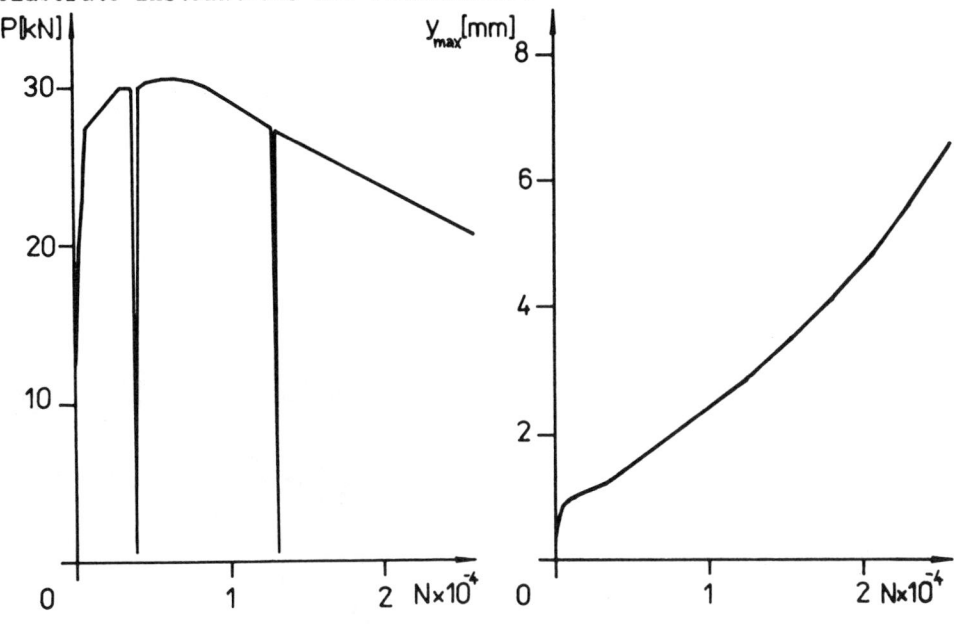

Fig. 5
Load and flange deflection versus number of counts diagram

Test girder	Acoustic emission	Strain gages	Permanent strain at the unloading
LTG 2	12.50 kN	9.4 kN (surface) 14.3 kN (membrane)	---
LTG 3	22.50 kN	15.8 kN (surface) 22.5 kN (membrane)	70 μ-strains at 20.0 kN 100 μ-strains at 22.5 kN

Table 1
Loads at the onset of yielding - comparison of strain-gage measurements with the acoustic emission results

5. Web Deflection Measurements

Web deflections were recorded by means of the shadow moiré method and checked in one section by dial-gauges. The shape of initial web deflections was measured all over the panel utilizing the both methods. Comparing the results for one girder we can see on the Fig.6 that the shadow moiré better gives a true picture of local imperfections as well as of an overall deformation of test specimens. Moreover, the data evaluation is very simple. Fig.7 illustrates web buckling shapes taken in the course of loading. Obviously a rather dense strain-gage installation does not obstruct the topographic measurements.

moiré

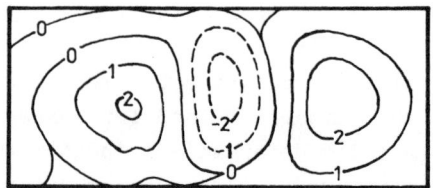

dial-gauges

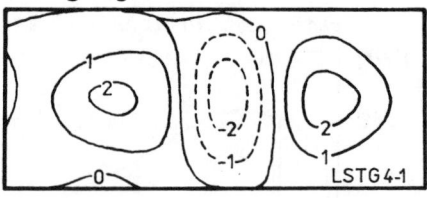

Fig. 6
Contour lines of initial imperfections

6. References

1 Drdácký,M.-Jaroš,P.-Škaloud,M.: Plastification of webs under a partial-edge loading, (in Czech), Proc.of the 21st Conference on Exp.Stress Analysis, Luhačovice, Czechoslovakia, May 23-26, 1983, pp.220-224.

2 Kratěna,J.-Kratěnová,M.: Shadow-moiré at investigations of stability problems of thin-walled girders, (in Czech), Proc.of the 20th Conf.on Exp.Stress Analysis, Karlovy Vary, Czechoslovakia, June 1-3, 1982, pp.51-56

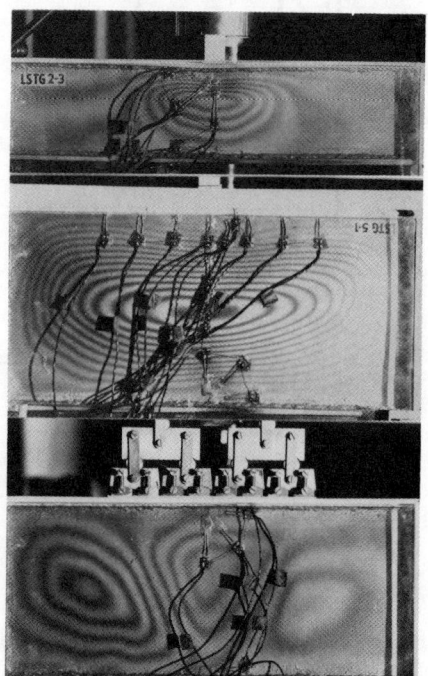

Fig. 7

STRESS RELAXATION BEHAVIOUR IN POLYMERS FOR COMBINED BENDING AND TORSION

M.Sc. Grzegorz Milewski, Dr Bogdan Targosz
Inst. of Mechanics and Machine Design
Technical University of Cracow
Warszawska 24, 31-155 Kraków, POLAND

In the paper the machine for the stress relaxation tests
in polymers as well as the methodology of its description
for combined bending and torsion have been presented.
As an example the experimental data for polyamide has
been given. The X-ray diffractogram analysis of structure
changes of outer surface of examined polymer has been
done, too.

Keywords: stress relaxation,bending,torsion,polyamide

1. The machine for stress relaxation tests in polymers for combined bending and torsion

A testing machine is designed and developed in such a way as
to be capable of exerting bending, torsion and simultaneous bending
and torsion. The ratio of bending and twisting can be adjusted
through the angle Ψ associated with the specimen setting towards
the loading yoke. In the extreme positions one can obtain bending
with constant moment / $\Psi = 90^\circ$ / and pure torsion / $\Psi = 0^\circ$ /.

The machine especially constructed for fatigue and relaxation
tests in polymers has a kinematic control. The required magnitude of
loading is adjusted through the certain set-up of the eccentric what
results in the certain angle of yoke turn /θ/. It corresponds to
the total moment / M / applied to the yoke. The maximal angle of
yoke turn is 45°.

The total moment changes in time / M=M(t)/ are measured by
means of strain gauges cemented on the surface of the measuring
transformer being a claw clutch between a holder of the yoke and
a loading shaft. The gauges form a Wheatstone bridge and their output

IMEKO 5th Symposium, Plzeň, Czechoslovakia, May 25-28, 1987

feeds into a set of voltmeters. Calibration curve / the voltmeter
display dependence on the total loading moment / has linear characteristics. The holders were designed for cylindrical samples.
A schematic arrangement of the testing machine is shown in Fig.1 and
its general layout in Fig.2.

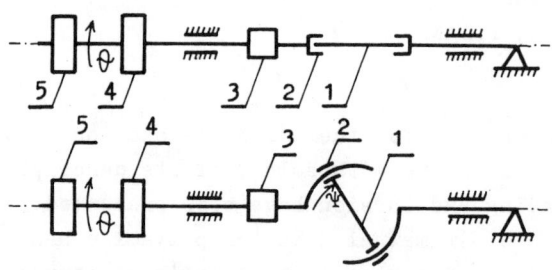

Fig.1 Schematic arran-
gement of the apparatus.

1-specimen,2-specimen's
holder,3-strain gauge
measuring transformer,
4-locking pawl,5-adjus-
table eccentric of a
yoke turn angle, -ad-
justable angle descri-
bing the ratio of ben-
ding moment to twisting
moment

Fig.2 General view of
the testing machine.

1-specimen,2-specimen's
holder,3-strain gauge
measuring transformer,
4-locking pawl,5-set of
Wheatstone bridges,
6-controlling panel,
7-time recorder,8-set
of voltmeters,9-perfo-
rator

2. Stress relaxation description for combined bending and torsion

The terms of kinematic control results in the following displacements

$$\alpha = \vartheta \cdot \sin \Psi \quad , \quad \varphi = \vartheta \cdot \cos \Psi \quad , \qquad (1)$$

where α denotes deflection angle, φ - torsional angle.
The stress state components vary in time according to eq. 2

$$\sigma(t) = M_g(t) / W_g \quad , \quad \tau(t) = M_s(t) / W_o \quad , \qquad (2)$$

where

$\sigma(t)$ - maximal normal stress from bending / in the paper the simplification has been done concerning the linear distribution of normal stresses from bending; only the maximal normal stress is taking into account /,

$\tau(t)$ - shearing stress from torsion,

$M_g(t)$, $M_s(t)$ - bending and twisting moments,

W_g, W_o - cross section characteristics for bending and torsion.

The total loading moment is given as following

$$M(t) = \sqrt{M_g^2(t) + M_s^2(t)} \qquad . \qquad (3)$$

From eq.3 one can calculate

$$M_g(t) = \frac{M(t)}{\sqrt{1 + \left(\frac{M_s(t)}{M_g(t)}\right)^2}} \quad , \quad M_s(t) = \frac{M(t)}{\sqrt{\left(\frac{M_g(t)}{M_s(t)}\right)^2 + 1}} \qquad . \qquad (4)$$

On the present stage of the work the basic assumption concerning the constancy during the test of the ratio of benging to twisting moments has been done, i.e.

$$\frac{M_g(t)}{M_s(t)} = \frac{M_g(0)}{M_s(0)} \qquad . \qquad (5)$$

Taking into account the basic relations for bending with constant moment and pure torsion, i.e.

$$\alpha = \frac{1}{2} \cdot \frac{M_g \cdot l}{E \cdot J_g} \quad , \quad \varphi = \frac{1}{2} \cdot \frac{M_s \cdot l}{G \cdot J_o} \qquad , \qquad (6)$$

where E , G - subsequently Young and Kirchhoff constants / $G = \frac{E}{2 \cdot (1+\nu)}$, ν - Poisson constant /,

one can derive

$$\frac{M_{so}}{M_{go}} = \frac{\alpha(1 + \nu)}{\varphi} \qquad , \qquad (7)$$

what leads to the following relations

$$\sigma(t) = \frac{M(t)}{W_g\sqrt{1 + \left(\frac{\varphi}{\alpha(1+\nu)}\right)^2}} = \frac{M(t)}{W_g\sqrt{1 + \left(\frac{1}{1+\nu} \cdot ctg\,\Psi\right)^2}}$$

$$(8)$$

$$\tau(t) = \frac{M(t)}{2W_g\sqrt{\left(\frac{\alpha(1+\nu)}{\varphi}\right)^2 + 1}} = \frac{M(t)}{2W_g\sqrt{(tg\,\Psi(1+\nu))^2 + 1}} \qquad .$$

3. Experimental data and conclusions

The experiments were carried out with polyamide cylindrical samples at room temperature. The stress relaxation tests were done

for 4 different ratio of bending and torsion:
- bending with constant moment / $\Psi =90°$ /,
- pure torsion / $\Psi= 0°$ /,
- bending with torsion / $\Psi = 30°$ and $60°$ /.

All the experimental results have been worked out according to the following schema [1,2] suggested for stress relaxation description in tensile uniaxial state:
- stress relaxation curves for both components of stress state
    $\sigma =\sigma(t)$, $\tau=\tau(t)$,
- plots for Li method [3] as to determine equilibrium stresses
    $\sigma_\infty = \lim_{t\to\infty}\sigma(t)$, $\tau_\infty = \lim_{t\to\infty}\tau(t)$,
- stress relaxation curves for both components of stress state in a normalized semilogarythmic system of coordinates
    σ^*/σ_0^* vs $\lg t$, τ^*/τ_0^* vs $\lg t$,
where subsequently $\sigma^* = \sigma - \sigma_\infty$, $\tau^* = \tau - \tau_\infty$, $\sigma_0^* = \sigma_0 - \sigma_\infty$, $\tau_0^* = \tau_0 - \tau_\infty$; σ_0 and τ_0 - the initial stresses.

The final stress relaxation curves for different ratio of bending and torsion are illustrated in Fig.3,4,5,6.

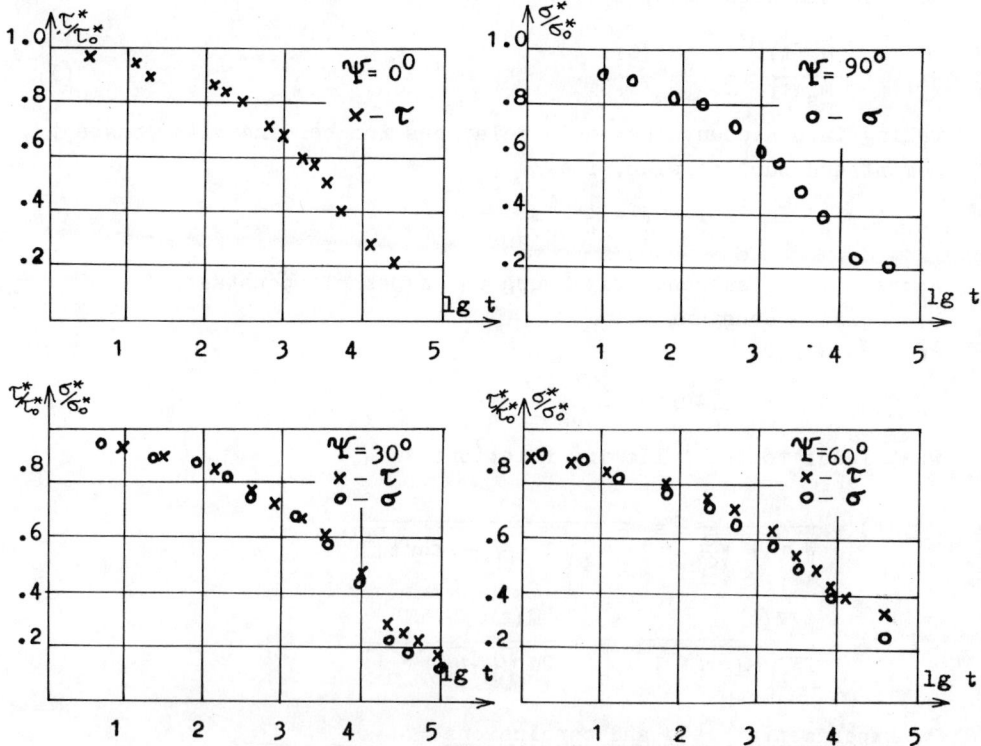

Fig.3,4,5,6 Stress relaxation curves for different ratio of bending and torsion

The following characteristic parameters of stress relaxation have been analysed:
- normalized stress relaxation velocity of both components / describes generally the intensity of relaxation of normal and shearing stresses $F_{\sigma p} = \frac{d\sigma^*}{d \ln t} \cdot \sigma_o^*$, $F_{\tau p} = \frac{d\tau^*}{d \ln t} \cdot \tau_o^*$,
- basic relaxation times of both components / $\tau_{\sigma rel}$, $\tau_{\tau rel}$ - connected with polymer viscoelastic properties ,
- equilibrium stresses σ_∞, τ_∞ / connected with polymer stiffness or flexibility /.
The experiments showed the following regularities:
- maximal normalized relaxation velocity / in the inflection point of the curves / are higher when comparing with relaxation at tensile state / in this case that parameter is treated as a material constant [1]/,
- basic relaxation times /$\tau_{\sigma rel}$, $\tau_{\tau rel}$ / have higher quantities than in uniaxial state / 1 ÷ 2 decades / as well as the equilibrium stresses reach higher magnitudes.
At the present state of the work no rheological models have been applied to stress relaxation in combined stress state.

4. Analysis of polyamide structure changes in relaxation process for combined stress state

Structure investigation was conducted by X-ray diffractogram method on machine DRON-2. The analysis was done according to Warren method [4] , where the changes of structure parameters are compared on the same specimens in the same places before and after stress relaxation tests. As to get the explicit information about the polyamide structure changes due to relaxation the X-ray diffractograms were also done in case of quasistatic loadings in the same stress states and for the same initial stresses / σ_o and τ_o / as in relaxation tests. Two structure parameters changes have been analysed: distance variation between planes (200) and (020) and crystallite dimension change in direction of those planes. The curves illustratind the changes of the last parameter are shown in Fig.7.

Summing up the structure investigation at the present stage one can notice the explicite changes of crystallite dimension of polyamide due to relaxation when comparing with virgin samples. No distance changes between the planes (200) and (020) have been observed.

231

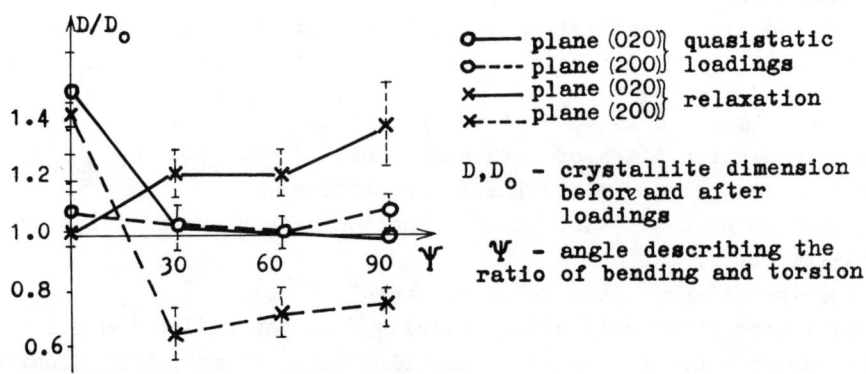

Fig.7 Crystallite dimension changes in polyamide due to quasistatic
loadings and relaxation for combined stress states

5. References

1 Kubát J., A similarity of stress relaxation behaviour of high polymers and metals, Dr's diss., Stockholm, 1965

2 Milewski G., Rychwalski W., Stress relaxation models in polymers,
Inż. Mater. N° 1/1983 / in Polish /

3 Li J.C.M., A method for describing internal stresses in materials,
Can.J.Phys.45,493, 1967

4 Urbańczyk G.W., Physics of fibre, WNT,Warszawa, 1976 /in Polish/

PHOTOELASTIC AND STRAIN GAUGE ANALYSIS OF THICK-WALLED SHELL OF REVOLUTION

Željko Goja, M.Sc.; Mirko Husnjak, M.Sc.; Stjepan Jecić, dr prof.
Faculty of Mechanical Engineering and Naval Architecture,
University of Zagreb, Yugoslavia

The paper deals with the stress distribution in
the parts of thick-walled shell of revolution
formed by cylindrical, toroidal and spherical
parts and subjected to an internal pressure.
Stresss distribution analysis has been done
experimentally both by, using the method of
photoelasticity, and by strain gauge measurements.
It has been found that the maximum stresses are on
the toroidal region of the shell. The stress
distribution throush the thickness of the wall is
nonlinear in the small curvature radii region.
Good agreement of the photoelastic and strain
gauge results has been found.

Keywords: photoelastic, stress, thick-walled shell

1. Introduction

Thick-walled rotationally symetric shells are used as parts for
many reservoirs, containers and holders. Analytical stress analysis
for rotationally symetric shells, composed as a combination of
cylindric, toroidal and spheric parts of variable thickness and
complex shape, can be carried out with considerable difficulties
only approximately. Numerical analysis using finite element method
requires, for shells of variable wall-thickness, a net of three-dimensional elements with many degrees of freedom.

Experimental stress analysis for three-dimensional case using
photoelasticity enables achievement of good results even for very
complex shells. This is confirmed by excelent agreements with results obtained by strain gauge measurements on original construction.

2. Object of Investigation and Means of Stress Analysis

The purpose of this investigtion was to determine stress distribution in a newly-shaped container for CO (67,2 1), and in case
of statisfying results, accept it for current production. The bottom
of container is composed of thick-walled spherical, cylindrical and
toroidal shell. Dimensions of the head are given in Fig. 1. The
container is loaded by inner pressure.

According to different standards (JUS, DIN, ISO) such a container must resist to any failure in head region.

Theoretical analysis of this problem is possible by using the
theory of Novozhilov, but only approximately taking the constant
thickness for all parts of construction.

Neglecting all quantities of order h/R, it is necessary to
solve differential equation:

$$\frac{d^2 \tilde{T}}{d\theta^2} + [(2\frac{R_m}{R_\varphi} - 1)\,ctg\,\theta - \frac{1}{R_m}\frac{dR_m}{d\theta}]\frac{d\tilde{T}}{d\theta} + i\frac{R_m^2}{R_\varphi c}\tilde{T} = \frac{R_m^2}{R_\varphi c}F(\theta) \qquad (1)$$

Here $F(\theta) = T_m^* + T_\varphi^*$ are membrane forces and $\tilde{T} = \tilde{T}_m + \tilde{T}_\varphi$ complex forces as in Fig. 2.

The meaning of expressions is follows:

$$\tilde{T}_m = T_m - \frac{i}{c}\frac{M_\varphi - \nu M_m}{1 - \nu^2} \qquad\qquad \tilde{T}_\varphi = T_\varphi - \frac{1}{c}\frac{M_m - M_\varphi}{1 - \nu^2} \qquad (2)$$

In the above equations subscript denotes circular, and m meridional directions. The constant c is:

$$c = \frac{h}{\sqrt{12(1 - \nu^2)}} \qquad (3)$$

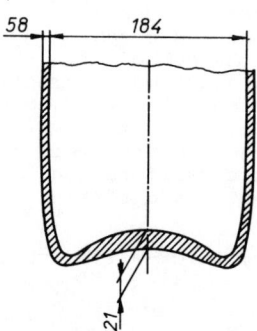

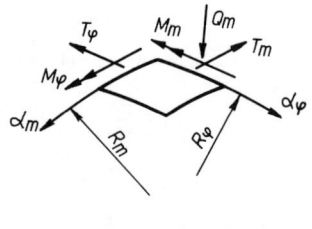

Fig. 1 Object of investigation   Fig. 2 Forces on a shell element

Solutions of these differential equations for separate elements of shell of revolution are possible by application of certain transformations and by method of asymptotic integration. Constants of integration are calculated taking into account boundary conditions. Applying this procedure to the structure shown in Fig. 3, only approximative results can be obtained. However, this was done as a clue for photoelastic analysis and for selection of regions for strain gauge application.

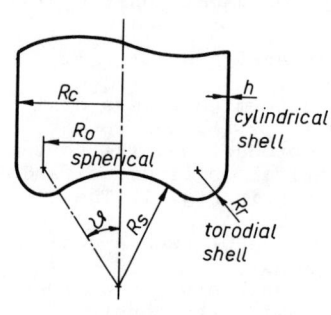

Fig. 3 Scheme of the structure

Numerical procedure in a case of thick-walled shell of revolution can be successfuly carried out for both linear and nonlinear analysis of stress and strain. Due to the variable wall thickness in a given shell, convetional shell elements can't be used. Forms obtained by degradation of isoparametric three-dimensional elements are suitable in such cases.

Photoelastic analysis of three-dimensional model enables determination of the whole stress field both in meridional and circular cross sections. Agreement with results obtained by strain gauge measurements confirum the reliability of the method used.

3. Photoelastic stress analysis

For photoelastic analysis, a rough mould from Araldite B (CIBA-GEIGY) was formed. The model was machined to the final dimension by a specially adapted tool.

Stress freezing was carried out by thermal procedure. The shell was loaded by pressure of p = 19,6 kPa, while the temperature was 423 K. Afterwards the shell was cooled down to the room temperature, at temperature rate of 2,5 K/h. The load was removed and the model was cut in to several meridional and hoop slices. The mode of cutting is shown in Fig. 4.

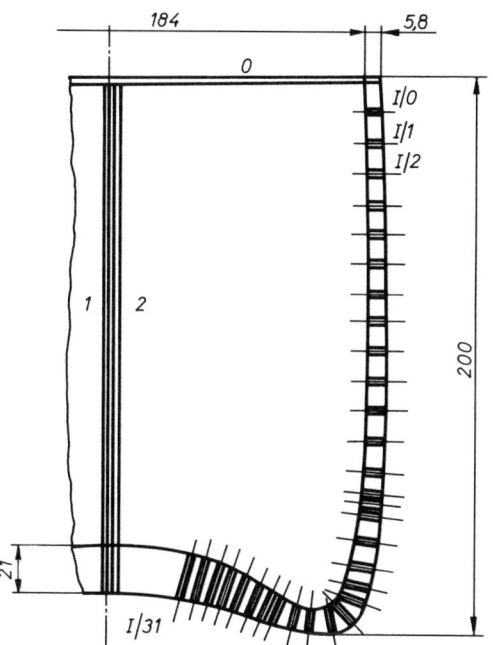

Fig. 4   The mode of slicing

For meridional slice the main photoelastic equation is:

$$\sigma_m - \sigma_r = \frac{f_\sigma N_1}{h_1} \qquad (4)$$

and for the hoop slice:

$$\sigma_\varphi - \sigma_r = \frac{f_\sigma N_2}{h_2} \qquad (5)$$

where f_σ is photoelastic constant, h is a slice thickness, N_1 i N_2 are fringe orders and σ_r, σ_φ, σ_m radial, circumferential and meridional stresses. Taking into account the similarity law (estimated error 5% to 8%) it is possible to calculate stresses in the original using the expression:

$$\sigma_i^o = \sigma_i^M \frac{\lambda}{\lambda_l} \frac{E^o}{E^M} \; ; \; (i = \varphi, m) \qquad (6)$$

Stresses denoted by subscript M reffer to model, and those denoted by subscript o to original.

E^o i E^M are respectively moduli of elasticity while λ and λ_l are the scales for strain and dimensions.

Taking into account the relation:

$$\frac{\sigma_i^o}{p^o} = \frac{\sigma_i^M}{p^M} \qquad (7)$$

stress analysis of original construction was carried out. The results are shown in Fig 5.a) in the form of diagram, where stress distribution and on the inner surface is shown. Diagram in Fig. 5.b) shows the stress distribution on outside surface. Stress analysis through the wall thickness confirms assumption about nonlinear stress distribution in the areas where the bending is predominant.

235

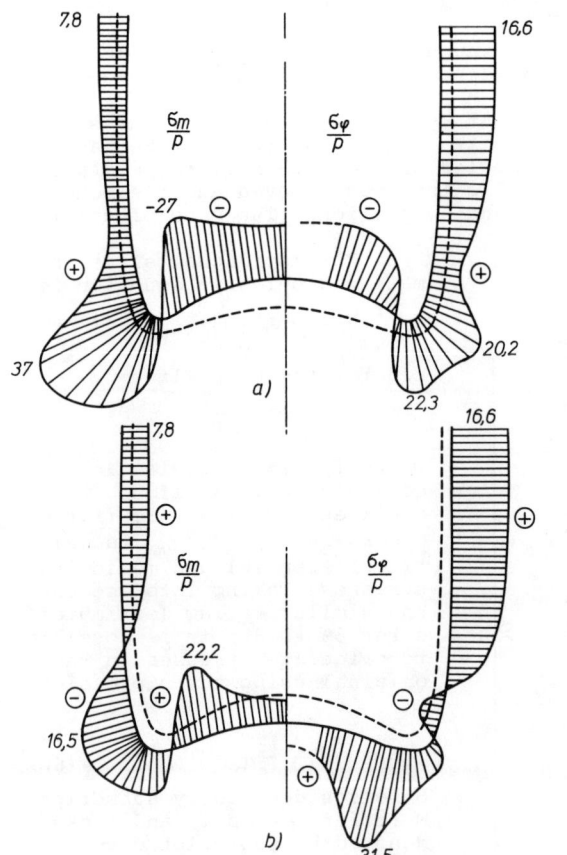

Fig. 5 Stress distribution on the
inner (a) and outer (b)
surface

Fig. 6 Stress distribution
through wall
thickness

4. Strain gauge stress analysis

In order to confirm the results obtained by photoelastic method
and to find out the maximum pressure which causes the failure of the
vessel, strain gauge measurements were performed.

Instrument layout for strain gauge measurement is shown in
Fig. 7, and comparsion of photoelastic and strain gauge measurements
are shown in Fig. 8.

236

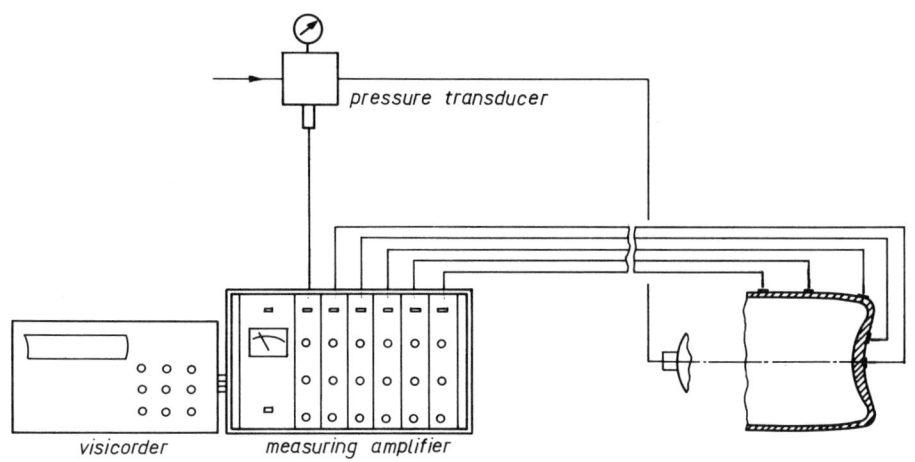

Fig. 7 Experimental setup for strain gauge measurements

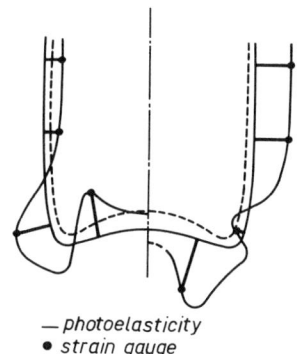

— photoelasticity
• strain gauge

Fig. 8   Comparison of photoelastic
         and strain gauge data

5. Conclusion

The areas of maximum effective streses according to von Mises theory were found to be at the juncture between thoroidal and spherical shell (see diagram Fig. 5). These results are in very good agreement with strain gauge measurements.

The failure of the pressure vessel during pressure test repeatedly confirm this fact.

On the base of these results, a new container head was designed, which was more appropriate.

6. Reference

V. V. Novoshilov: Theory of thin shells (in Russian),
Leningrad, 1962

Computer aided evaluation of measurements

COMPUTER AIDED RECONSTRUCTION OF DISPLACEMENT
FIELDS IN HOLOGRAPHIC AND SPECKLE INTERFEROMETRY

Dr. Wolfgang Osten - Dipl. Phys Roland Höfling

Akademie der Wissenschaften der DDR

Zentralinstitut für Kybernetik und Informationsprozesse

Kursraße 33, 1086 Berlin, DDR

Institut für Mechanik

POB. 408, 9010 Karl-Marx-Stadt, DDR

A procedure is presented for the fast computer aided
evaluation of holographic interferograms and speckle
correlograms. The factors of the image formation process, which affect the quality of the optical signal
are described and methods for the improvement of the
images are discussed. The algorithms are implemented
on a commercial image processing system ROBOTRON A6472.

Keywords: holographic interferometry, speckle interferometry, digital image processing

1. Introduction

In recent years holographic interferometry (HI) and speckle
techniques have been widely used in the field of industrial
inspection and experimental mechanics. In an interferogram the
gray values of the fringes and their loci contain the necessary information about the optical phase difference at each
object point. These values characterize the difference between
the two states of the object which are to be compared and hence the displacement field. Up to now, most interferograms have

had to be evaluated by reading the fringe orders and their positions manually. This procedure, however, is very inaccurate
and time consuming. Increasingly these data have been processed by computers using image analysis methods to enable semi-
automatic or fully automatic analysis of the fringe pattern.
In this paper we describe the application of a commercial image processing system ROBOTRON A6472 for the real time generation of speckle correlograms and for the fast on line evaluation of interferograms.

2. Description and simulation of the image formation process
   in holographic and speckle interferometry

For the development of algorithms in digital image processing it is convenient to start from certain a-priori knowled-
ges about the image formation process. In holographic and
speckle interferometry the interesting information for the calculation of displacements or derived mechanical quantities is
the phase distribution $\delta(\vec{r})$ encoded in the observed fringe
pattern. For an accurate digital acquisition, recognition, enhancement and interpolation of these fringes it is necessary
to consider their special characteristics. With the well-known
relation for the two-beam interference we can describe the intensity distribution in the reconstructed image of an holographic interferogram /1/:

$$I(\vec{r}) = \langle I_o(\vec{r}) \rangle \left[(1-\alpha)^2 + 4\alpha\cos^2\delta(\vec{r})/2 \right] R(\vec{r}) \qquad (1)$$

with

$\langle I_c(\vec{r}) \rangle$ – basic intensity
α^2 – proportion between the intensities of the first
and second exposure
$R(\vec{r})$ – coherent noise (speckle).

242

Neglecting systematic errors of the electronic acquisition system (for instance spatial nonuniformity of the TV-target sensitivity) sources of optical and electronical errors, which affect the quality of the image are essentially:

* Speckle noise by reason of coherent laser light

  The speckles are greater the smaller the viewing aperture is. But a small viewing aperture is needed in holographic interferometry to get a high depth of focus which projects the fringes, that are localized in space, on the surface of the object and in speckle interferometry to resolve the speckles on the TV-target.

* Superimposed background intensity by reason of nonuniform illumination (gaussian intensity distribution in the laser beam) and reflexion of the object surface.

* Superimposed "parasitical" fringes (diffraction pattern) by reason of light diffraction at dust particles on the surface of optical elements.

* Variations of the fringe visibility in the interferogram by reason of localisation phenomena of the fringes.

* The random and time dependent noise by reason of the electronical registration, transmission, amplification and digitazing of the optical signal.

An important role with regard to the automatical segmentation (finding the regions of fringe maxima -"ridges"- and minima -"valleys"-) of the fringe pattern plays the normalization of the background intensity and the smoothing of speckle noise. Additionally to this we must take into consideration the electronic noise in digital speckle pattern interferometry (DSPI). In the static case the correlogram results from the subtraction of two stored speckle images and following contrast spreading. Small statistical and time dependent fluctuations of the intensities between the two images cause a visible decrease of the SNR, because the signal amplitudes are nearly identical with the noise amplitudes.

In order to improve the software development process the image formation process has been simulated for HI as well as for DSPI. On this way the simulation of speckle patterns was the first step. To create synthetic speckle patterns we follow the natural process occuring with the formation of subjective speckle and use a simplified model of it /2/. Fig. 1 contains such a synthetic speckle pattern with the corresponding histogram, showing the well-known negative exponential intensity distribution.

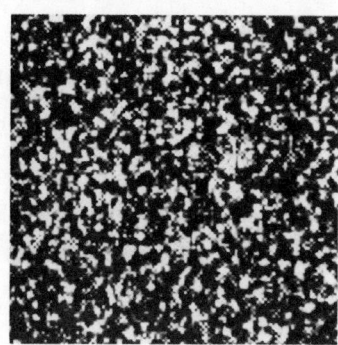 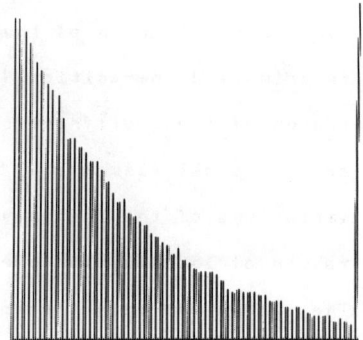

Fig. 1

Synthetic speckle pattern and histogram

Adding a reference wafe we are able to create patterns, which are sensitive to a given displacement field. In this way the full procedure of correlogram construction and displacement analysis from it may be checked.

The background intensity distribution we have approximated taking into consideration the gaussian intensity distribution in the cross-section of a laser beam, its expansion with help of optical elements and its projection on the surface. The completed process of simulated image formation is shown in fig.2.

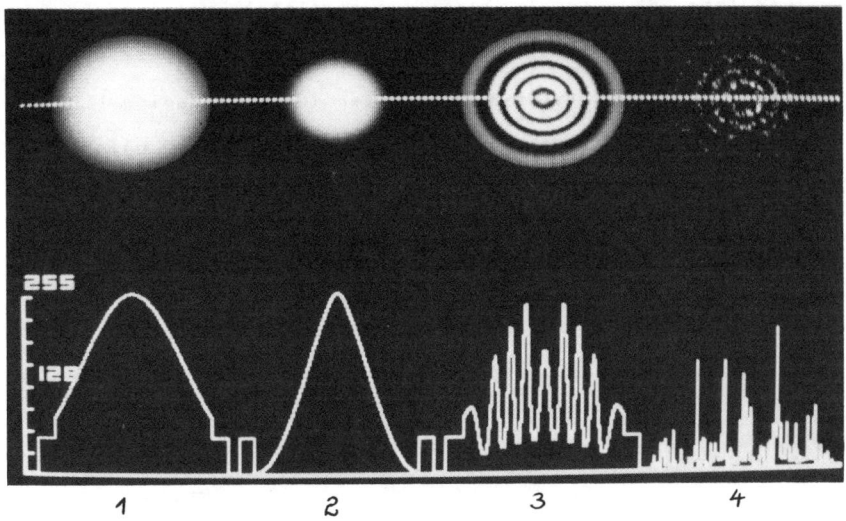

Fig. 2

Steps of creating a synthetic holographic interferogram

for a central loaded circular plate

1 - background intensity
2 - phase distribution
3 - interferogram without noise
4 - interferogram with simulated speckle noise

3. Experimental setup for DSPI

The scheme of our speckle pattern interferometer is drawn in fig. 3. The expanded and collimated light from a 30 mW He-Ne laser is directed to a beam splitter thus illuminating a 50 mm spot on the diffusely scattering reference and object surface. One of the objects is a central loaded circular plate, sprayed white. The reflected light of both surfaces is collected again by the beam splitter and imaged onto the pickup tube of a TV-camera. A numerical aperture (NA=5.6) has been found sufficient

245

for the camera lens, although the resulting speckles (typically less then 15 m in size) are not completely resolved by our vidicon. The camera signal is fed to the image processing system and digitized to be stored after frame averaging (12... 16 frames to suppress the time dependent electronic noise) in one of the four 512X512X8 bit image memories. The squared and contrast enhanced difference between a stored reference frame and the live camera image is presented on the image display at video rates using the fast pipeline processor K2072 of the image processing system.

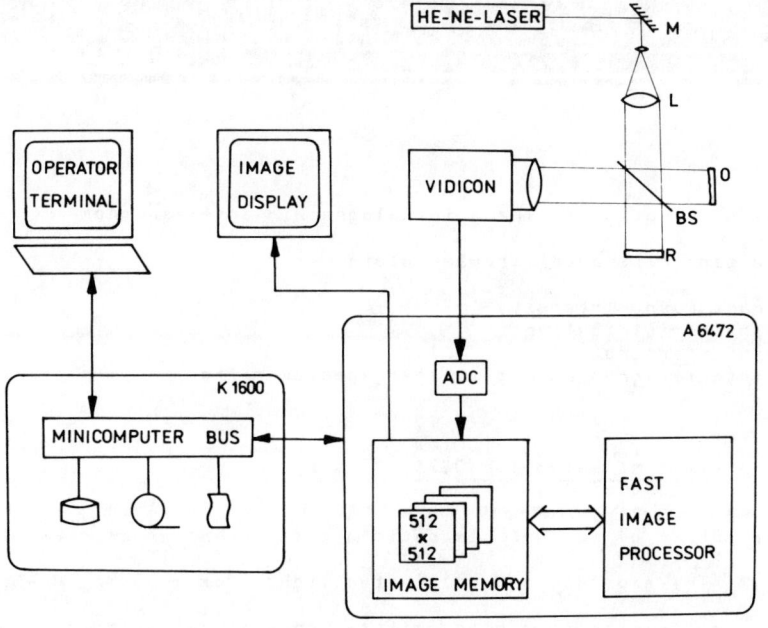

Fig. 3

Schematic diagram of the experimental setup

M – mirror, L – lenses , O – object, R – reference surface, BS – beam splitter

Fig. 4 shows the influence of frame averaging in DSPI on the quality of the correlogram.

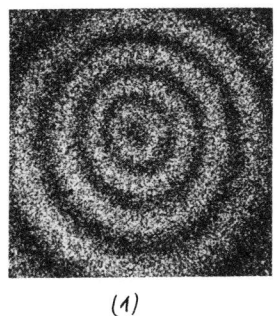
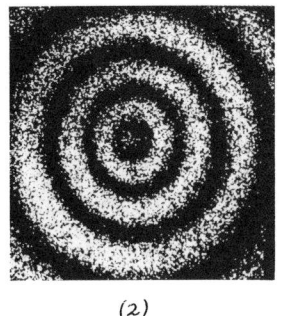

(1)   (2)

Fig. 4

Influence of frame averaging

(1)   Correlogram with high electronic noise (single frame)

(2)   Correlogram from frame averaged speckle patterns

4. Computer aided processing and evaluation of interferograms

The procedure is based on the fast automatic finding of the skeleton of an interference pattern and on its interactive evaluation. The main steps are /3/:

* Preprocessing of the interferogram

   - frame averaging

   - normalization of background intensity by image division

   - speckle smoothing with mean filters and a special geometric filter /4/

* Segmentation

   - finding of "ridges" and "valleys" in the intensity relief

   - fringe skeletonization by line thinning of a binary image

   - improvement of the skeleton

* Interactive quantitative analysis
  - fringe numbering
  - interpolation of the phase values on a mesh (32X32 grid points)
  - calculation of the displacement field
  - graphic presentation as gray value mountain with coloured lines of equal displacement and pseudo-3d-plotting

Numerous experiments with synthetic and natural interferograms have been successfully performed. The processing time for the full procedure is generally less than 6 min. Fig. 5 shows the holographic interferogram of a central loaded circular plate with intensity profil along the dotted line, its skeleton and the calculated displacement field as pseudo-3d-plot.

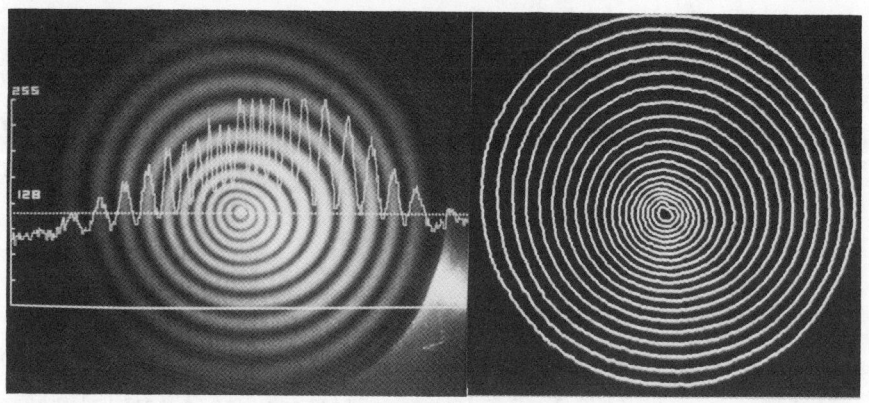

Interferogram   Skeleton

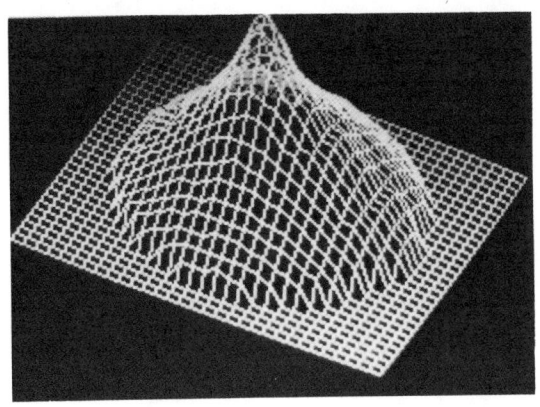

Pseudo-3D-Plot

Fig.5

Steps of quantitative evaluation

5. References

1 Höfling, R.,:   Contributions to computer simulation of in-
   Kaschte, A.,   terferograms using the digital image pro-
   Osten, W.   cessing system A6471.
                    FMC-Series 26(1987)

2 Höfling, R.,:   Displacement measurement by image processed
   Osten, W.   speckle patterns.
                    Optica Acta (to be published)

3 Osten, W.,   :   Schnelle automatische Auswertung von Inter-
   Saedler, J.,   ferogrammen mit Verfahren der digitalen
   Wilhelmi, W.   Bildverarbeitung.
                    Laser Magazin (to be published)

4 Crimmins,T.R.:   Geometric filter for speckle reduction.
                    Appl. Opt. 24(1985)10,1438-1443

APPLICATION OF THE BAND SELECTABLE FOURIER TRANSFORM
FOR STRUCTURAL DYNAMIC TESTING

Dr. Friedrich Wahl
Technical University Magdeburg
Department of Mechanical Engineering
3010 Magdeburg, GDR
PSF 124

A method is presented to measure complex frequency response
functions of structures by means of periodic narrow band
random signals. These signals are generated using the inverse
Band Selectable Fourier Transform (inverse ZOOM).
An algorithm is given readily programmable. The application
of the method gave excellent results for structures with
closely spaced or very lightly damped modes.

Keywords: dynamic testing, frequency response, FFT, ZOOM

1. Introduction

Modern methods of measurements of frequency response functions
of mechanical structures are based on broadband exiting signals
analyzed in a desktop computer or in a microprocessor. In reference
/1/ the advantages and disadvantages of broadband test signals are
discussed. Many practical experiences have shown that periodic random signals are the most useful test technique, especially if curve
fitting is of paramount importance. There are two reasons:

- Periodic random exitation gave excellent results in a small
  amount of testing time. Leakage is not a problem because the signals are periodic in the measurement time window of the Fourier
  analyzer.
- Successive records of frequency domain data may be averaged to
  remove nonlinear effects, noise, and distortion from the measurement.

The only drawback is that peridic random signals are synthesized by
the digital Fourier transform in a frequency range from zero frequency (DC) to Nyquist-frequency. This transform is spread over

a fixed number of frequency lines N (typically N=1024) which limits
the frequency resolution between lines. One way to improve resolution is to increase the block size N. However this is an inefficient
way because of digital processing time and the limited computer
memory size.

There are two different principles to improve the frequency resolution without increasing the block size N /2/,/3/. These methods are
used today in ZOOM-FFT analyzers. The method described in reference
/3/ is based on the Band Selectable Fourier Transform (BSFT). In this
paper the inverse BSFT is re-derived and the application of this
transform for generation of test signals with high frequency resolution is shown. This paper contains the following material.

Paragraph 2 describes an improved method for computing periodic random signals by means of the complex FFT without redundancy. In par-
graph 3 the BSFT of reference /3/ is expanded to complex-valued
time series and in paragraph 4 the inverse problem of the complex
BSFT is re-derived. In paragraph 5 a method is presented to compute
periodic narrow band random signals by means of the inverse BSFT.
Applications of these results are outlined in paragraph 6 for measuring frequency response functions of structures.

2. Generation of periodic random signals

The complex diskrete Fourier Transform of N complex-valued time
samples $z(n)$ is defined by

$$Z(k) = 1/N \sum_{n=0}^{N-1} z(n) \exp(-j\, k2\pi n/N); \quad k=0,1,\ldots (N-1) \qquad (1)$$

where the $Z(k)$ are the N complex-valued frequency lines.
The inverse discrete Fourier transform brings $Z(k)$ back to $z(n)$
defined by

$$z(n) = \sum_{k=0}^{N-1} Z(k) \exp(j\, k2\pi n/N); \qquad n=0,1,\ldots (N-1) \qquad (2)$$

It will be assumed that the transforms (1) and (2) are carried out
in a computer as Fast Fourier Transform (FFT). The scheme for generation of peridic random signals is illustrated in Fig. 1.

The method most commonly used is seen on the left. It is generated
a complex spectrum $U(k)$ with uniform amplitudes and random phase.
To generate real time series this spectrum must be complemented by
the complex conjugate $U^*(k)$ as seen. The inverse FFT (2) gives a
real-valued pseudorandom signal. This method is redundant. The
frequency resolution is

$$\Delta f = 2f_N/N = f_S/N, \qquad f_S: \quad \text{Sampling frequency}$$

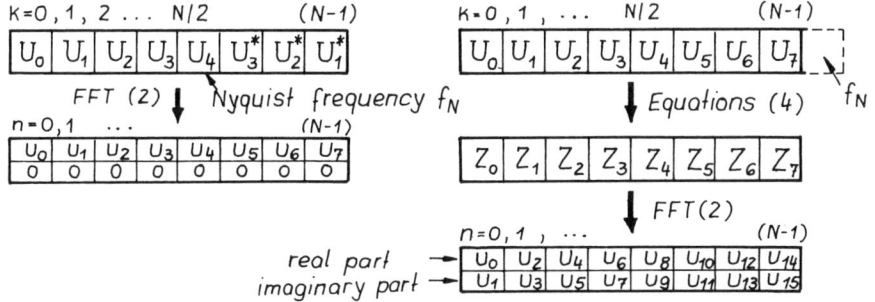

Fig. 1
Calculation of real-valued and complex-valued time series (N=8)

On the right of Fig.1 is seen a method developed in /4/, removing
the redundancy and doubling the frequency resolution. In this method
the spectrum U(k) is predetermined over the whole block size N.
Before carrying out the inverse FFT according to (2) a complex spectrum Z(k) is calculated defined by

$$
\left.
\begin{aligned}
Z(k) &= U(k) + U^*(1) + j(U(k)-U^*(1)) \exp(j \pi k/N); \quad k=0,1,\ldots N/2 \\
Z(1) &= -Z^*(k) + 2U^*(k) + 2U(1) \qquad\qquad\qquad\qquad\quad l=N-k
\end{aligned}
\right\} \quad (3)
$$

The inverse FFT then gives N complex-valued or 2N real-valued time
samples. The frequency resolution is

$$\Delta f = f_N/N = f_S/2N$$

This principle of generation periodic random test signals is integrated in the program package ASAM /5/. The scheme of the test signal
generator and the data acquisition is shown in Fig. 2.
In the practical application of the method described in Fig. 2 there
are two problems:

1. The digital-analog-converter (DAC) decreases the amplitude spectrum near the Nyquist frequency f_N/6/. So the generation process
   must be controlled by the computer. This is not necessary if the
   spectrum only comprises lines to half of the frequency f_N.

2. If the analog-digital-converter (ADC) is controlled by a multi-plexer, the highest accuracy is obtained if the input and output signals are sampled alternately. This means that the sampling frequency of the ADC is the half of the DAC. To avoid aliasing all the terms $U(k)$ for $k \geq N/2$ must be zero (see also Fig. 1).

With $U(k)=0$ for $k \geq N/2$ and $U(0)=0$ from equations (3) follows

$$\left. \begin{array}{l} Z(k) = U(k) + jU(k) \exp(j\pi k/N) \\ Z(N-k)=-Z^*(k)+ 2U^*(k) \qquad\qquad ; k = 1,2,\ldots(N/2-1) \end{array} \right\} \quad (4)$$

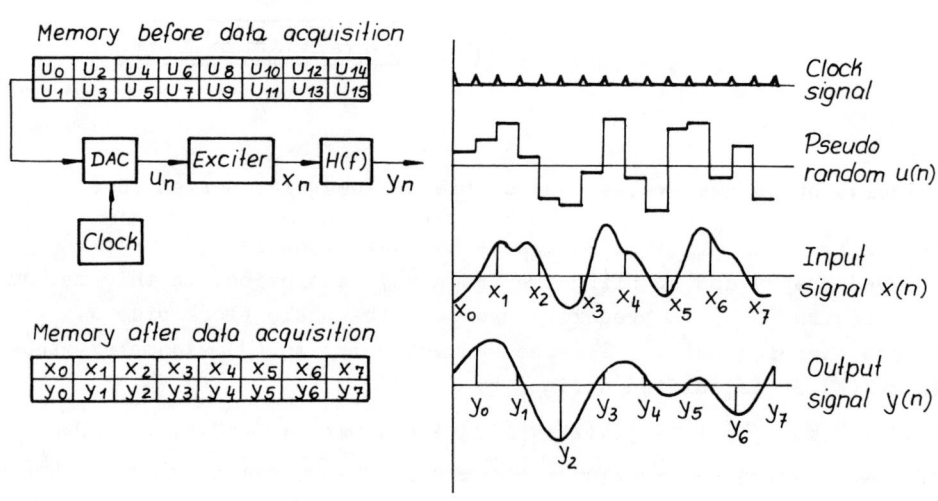

Fig. 2
Scheme of data aquisition

3. The Band Selectable Fourier Transform (BSFT)

The BSFT is a method which increases the frequency resolution with-out increasing the FFT-block size N. In Fig. 3 the scheme of BSFT is shown. The only way to improve the frequency resolution is to trans-form the increased block size of the $2\beta N$ real-valued or βN complex valued samples. β the so called ZOOM-factor is of order $10 \div 100$. The frequency resolution

$$\Delta f = f_S/2\beta N$$

is now improved by the factor β compared to the Baseband-FFT. The method of the BSFT is described in /3/ for real valued time series. In /7/ this method is extended to complex-valued time series.

254

The main steps of this method are:

The $2\beta N$ real-valued samples $u(n)$ are defined as βN complex-valued samples $z(n)$ assuming that the samples with even index are the real parts, with odd index the imaginary parts. Then the complex series $z(n)$ is splited up into subseries $\bar{z}_i(n)$ defined by

$$
\left.
\begin{aligned}
\bar{z}_0(n) &= (z(0),\ z(\beta),\ z(2\beta),\ldots\ z(N\beta - \beta)); \\
\bar{z}_1(n) &= (z(1),\ z(\beta+1),z(2\beta+1),\ldots\ z(N\beta - \beta - 1)); \\
&\ \vdots \\
z(\beta-1)(n) &= (z(\beta-1),z(2\beta-1),\ \ldots\ z(\beta N-1))
\end{aligned}
\right\}
\quad (5)
$$

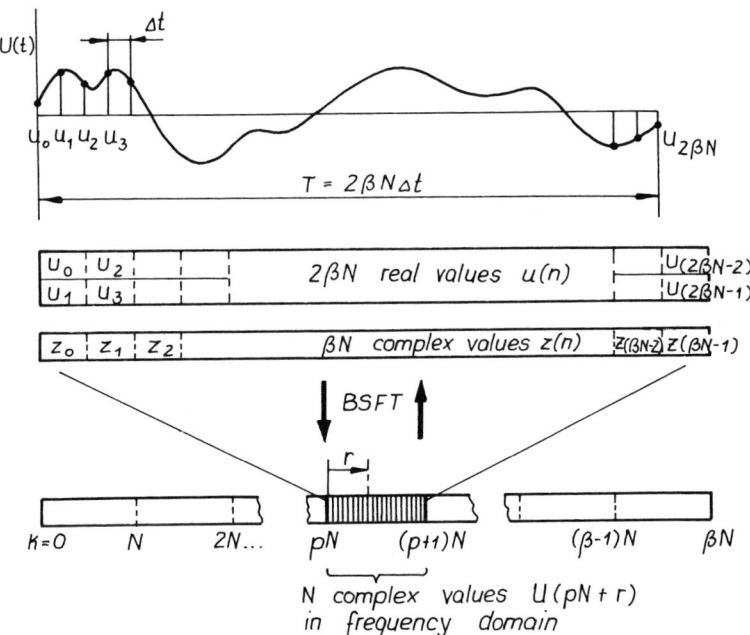

Fig. 3
Complex Band-Selectable Fourier Transform (BSFT)

Each subseries is transformed by the Baseband-FFT (1) of the block size N. One yields β blocks of complex-valued spectral lines

$$
\bar{Z}_i(r); \quad
\begin{aligned}
r &= 0,1,\ldots\ (N-1) \\
i &= 0,1,\ldots\ (\beta-1)
\end{aligned}
$$

Let the spectral lines $Z(k)$, $k=0,1,\ldots(\beta N-1)$ be splited up into partial blocks of size N (see Fig. 3). Then it is possible to compute the $Z(k)$ in any block separatly by combining the individual results of $\bar{Z}_i(r)$. For the p-th block ($p=0,1,\ldots(\beta-1)$) the spectral lines are given by

$$Z(pN+r) = 1/\beta \sum_{i=0}^{\beta-1} w_{pi}\bar{Z}_i(r) \alpha_i(r); \qquad r=0,1,\ldots (N-1) \tag{6}$$

where

$$w_{pi} = \exp(-j\ p2\pi i/\beta); \qquad \alpha_i(r) = \exp(-j\ r2\pi i/\beta N) \tag{7}$$

are the phase correction factors to compensate the time shift of the subseries (5). In terms of the $Z(pN+r)$ the true spectral lines of the p-th block are then given by

$$4U(pN+r) = Z(pN+r) + Z^*((\beta-p)N-r) -$$
$$-j(Z(pN+r) - Z^*((\beta-p)\ N-r))\ \exp(-j(pN+r)\pi/\beta N)$$
$$r=0,1,\ldots (N-1) \tag{8}$$

4. The inverse Band Selectable Fourier Transform

With the equations (6) to (8) it is possible to invert the BSFT. Let $U(pN+r)$ the spectral lines in the p-th block. Then, from the equation (8), it follows that the $Z(pN+r)$ are given by

$$Z(pN+r) = U(pN+r) + U^*((\beta-p)\ N-r) +$$
$$+j(U(pN+r) -U^*((\beta-p)N-r))\ \exp(j(pN+r)\pi/\beta N)$$
$$r=0,1,\ldots (N-1) \tag{9}$$

To compute the $\bar{Z}_i(r)$ of the subseries $\bar{z}(n)$, note from (6) that because of the orthogonality relationship of the phase correction factors

$$\sum_{p=0}^{\beta-1} w_{mp}^*\ w_{pi} = \begin{cases} \beta & \text{for } m = i \\ 0 & \text{for } m \neq i \end{cases} \tag{10}$$

one obtains the result

$$\bar{Z}_i(r) = \alpha_i^*(r) \sum_{p=0}^{\beta-1} w_{ip}^*\ Z(pN+r); \qquad \begin{aligned} r&=0,1,\ldots(N-1) \\ i&=0,1,\ldots(\beta-1) \end{aligned} \tag{11}$$

When transforming each partial spectrum $\bar{Z}_i(r)$ by the inverse FFT the subseries $\bar{z}(n)$ defined in (5) are obtained and the final time series $u(n)$ can be easily arranged.

5. Generation of periodic narrow band random signals

The method outlined in paragraph 4 is available for generation of periodic narrow band random signals. Let the spectrum $U(k)$ be given as

$$U(k) = U_k \exp (j \pi R_k / \beta N); \qquad k=0,1,\ldots (\beta N-1) \qquad (12)$$

in which the U_k are the amounts of $U(k)$ and the R_k are random numbers uniformly distributed in $\underline{/0, 2\beta N/}$.

In the following deliberation is assumed that the $U(k)$ are all zero except in any block p (see Fig. 3).
The algorithm of the inverse BSFT described in paragraph 4 then becomes very simple.
From equations (9) and (11), one then obtains formulae for the partial spectra $\bar{Z}_i(r)$ readily programmable. These formulae are:

$$\bar{Z}_i(0) = 0$$
$$\bar{Z}_i(r) = U_r(\cos\varphi_1 - \sin\varphi_2) + U_1(\cos\varphi_3 + \sin\varphi_4) +$$
$$+j(U_r(\sin\varphi_1 + \cos\varphi_2) + U_1(\sin\varphi_3 - \cos\varphi_4)); \qquad (13)$$
$$\bar{Z}_i(1) = U_r(\cos\varphi_1 + \sin\varphi_2) + U_1(\cos\varphi_3 - \sin\varphi_4) +$$
$$+j(U_r(-\sin\varphi_1 + \cos\varphi_2) + U_1(-\sin\varphi_3 - \cos\varphi_4));$$

with

$$\varphi_1 = 1/2\,\beta((p N+r)2i + R_r)\,2\pi/N$$
$$\varphi_2 = 1/2\,\beta((p N+r)(2i+1) + R_r)\,2\pi/N$$
$$\varphi_3 = 1/2\,\beta(((\beta-p)\,N-1)\,2i - R_r)\,2\pi/N \qquad (14)$$
$$\varphi_4 = 1/2\,\beta(((\beta-p)\,N-1)(2i+1) - R_r)\,2\pi/N;$$

$$r = 1,2,\ldots N/2 \;;\; 1 = (N-r)$$
$$i = 0,1,\ldots (\beta-1)$$

The following steps are necessary in a computer programm:
For $i=0,1,\ldots(\beta-1)$:

1. Computing the partial spectra $\bar{Z}_i(r)$ from equations (13) and (14).

2. Transforming the $\bar{Z}_i(r)$ by the inverse FFT and recording the sub-series $\bar{z}(n)$ to obtain the real-valued time series $u(n)$.

This algorithm becomes especially simple and fast if the ZOOM-factor is a power ot two. Then the φ_μ defined by equations (14) are calculated without multiplications, only shift and add operations are needed. If the φ_μ are written as

257

$$\mathcal{G}_\mu = (K_\mu + \varepsilon_\mu)\, 2\pi/N; \qquad \mu = 1,2,3,4 \qquad (15)$$

with integer $K_\mu < N$ and $|\varepsilon_\mu| < 1$
then

$$\sin \mathcal{G}_\mu \approx \sin(2\pi K_\mu/N) + \varepsilon_\mu \cos(2\pi K_\mu/N)$$

$$\cos \mathcal{G}_\mu \approx \cos(2\pi K_\mu/N) - \varepsilon_\mu \sin(2\pi K_\mu/N)$$

The sin- and cos-functions in these relations are not calculated
because they are always stored in the memory for the Baseband-FFT.
So the computing time for the partial spectra $\bar{Z}_i(r)$ is small in comparison to the β times FFT.

6. Measurements of structural frequency response functions

The method presented in paragraph 5 for measuring structural
frequency response functions with high resolution is integrated in
ASAM /5/ too. The general scheme of the data acquisition is already
shown in Fig. 3.
In general the measurement begins by analyzing the baseband frequency range of interest (method of paragraph 2). If high resolution is
required the input parameters of the inverse BSFT are selected
(block number p, ZOOM-factor β). Then the high resolution measurement begins with generating a narrow band random signal. Via DAC,
amplifier and exiter the structure under investigation is excited.
After the structure is vibrating in a steady state condition, a
measurement is taken and the power spektra are formed. Then further
uncorrelated signals are generated and measurements are taken again.
After averaging the power spectra the frequency response function
is computed.
To transform the measured input and output signals in the frequency
domain the BSFT of section 3 is not used because it is time consuming. The extremely sharp roll-off and out-of-band rejection of the
test signal enables to analyze the incoming data by frequency shift
and resampling. This technique is shown in Fig. 4. In practice the
shift operation and resampling are exchanged, reducing the incoming
data to N samples per signal. The succeeded shift operation is formed in time domain multiplying the samples by

$$\exp(-j2\pi np/\beta); \qquad n = 0,1,\ldots (\beta N-1). \qquad (16)$$

Because of resampling and the scheme of data aquisition shown in
Fig. 2 follows, that

$$n = (\beta/2)\,i\,; \qquad i = 0,1,\dots (\beta N/4 - 1) \qquad\qquad (17)$$

Thus, the shift operation (17) takes on the values 1 and -1 only. Therefore no multiplications are needed and the Fourier transform of the incoming resampled data is accomplished by the ordinary Baseband-FFT independent of the ZOOM-Factor β.

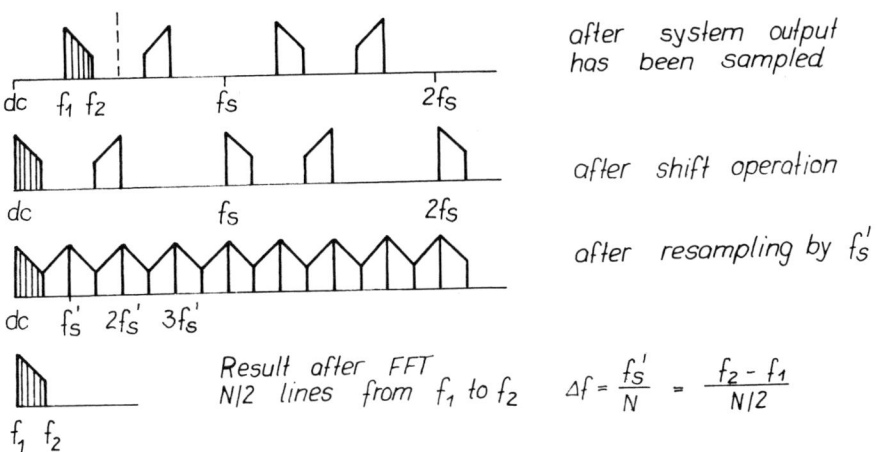

after system output has been sampled

after shift operation

after resampling by f_s'

Result after FFT N/2 lines from f_1 to f_2 $\qquad \Delta f = \dfrac{f_s'}{N} = \dfrac{f_2 - f_1}{N/2}$

Fig. 4

Signal flow of the BSFT after data aquisition

A typical example to illustrate the described methods is shown in Fig. 5. A lightly damped plate type structure was tested using periodic random discussed in paragraph 2. Four averages were taken to average out the noise from the measurement. The result of the inertance is shown in Fig. 5 above. It is seen that the frequency resolution $\Delta f = 2$ Hz is not sufficiently small to describe the closely spaced resonances near 584 Hz. A ZOOM-measurement was applied therefore, using the BSFT-technique. The selected parameters were $\beta = 16$ and $p = 4$. Four averages were taken too. The improved frequency resolution $\Delta f = .25$ Hz yields an excellent result which makes possible an accurate identification of the modes and parameters associated with each mode. The whole test time was 32 seconds or about eight times longer than the baseband measurement.

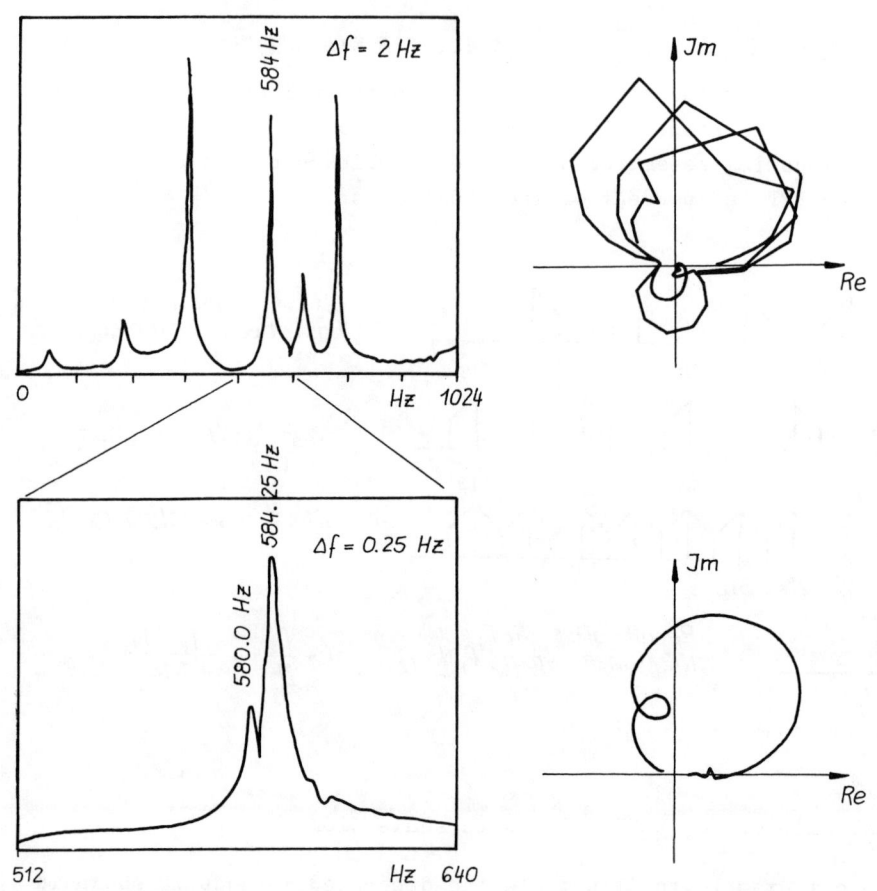

Fig. 5
Baseband- and BSFT-measurement

7. References

1 Lingener, A.: Analysis of Mechanical Systems Excited by Random
   Vibrations. Proc. IUTAM-Symposium Random Vibrations and Reliability, Frankfurt/Oder, 1982, pp. 172-194
2 McKinney, H.W.: Band Selectable Fourier Transform. Hewlett-
   Packard Journal, 26, No. 8, pp. 20-24, April 1975
3 Thrane, N.: ZOOM-FFT. Brüel & Kjaer Technical Review, No. 2,1980
4 Wahl, F.: Ein effektives Verfahren zur experimentellen Ermittlung
   von Systemkennfunktionen mit Hilfe von Pseudozufallssignalen,
   Technische Mechanik, 3, Heft 1, S. 11-17, 1982

5 Schmidt, G.: Kurzbeschreibung eines Programmsystems zur digitalen Verarbeitung analoger Meßsignale. Forschungsberichte 6.Tagung Festkörpermechanik, Dresden, Sept. 3-6, 1985, S.XLIII/1-9

6 Sperling, L.: Zur Simulation zufälliger Belastungen mittels Fourierreihen nach der Spektraldichte. Forschungsberichte 6.Tagung Festkörpermechanik, Dresden, Sept. 3-6, 1985, S. XLIV/1-15

7 Wahl, F.: Die bandselektive Fouriertransformation und ihre Anwendung zur Simulation von Schmalbandprozessen. Technische Mechanik 3, Heft 4, S. 23-28, 1982

ON THE DIGITAL PROCESSING OF FRINGE PATTERNS

Mgr inz. Andrzej Pietrzyk
Instytut Mechaniki Konstrukcji Inzynierskich
Politechnika Warszawska
Al. Armii Ludowej 16, 00-637 Warszawa, Poland

Digital processing of fringe patterns obtained in experimental mechanics is considered. Image coding and processing workstation based on a sequential 16 bit microcomputer is presented. Software used for fringe pattern analysis is briefly described. Possible extensions of the system to enable it to control also testing machine and data aquisition from electric transducers (strain gauges, LVDT, thermocouples) as well as the possible building-in of special hardware to improve image processing is mentioned.

Keywords: microcomputers, image, experiment, mechanics

1. Introduction

In several methods used in experimental mechanics fringe pattern is obtained as a result of an experiment. There are displacement measurement methods based on the phenomenon of light interference, like various moire techniques, holographic interferometry or speckle. In all these methods specimen made of real structural materials are usually tested. Thus, fringe pattern frequently appears on non-homogeneously bright background. A very popular method of (mainly qualitative) stress analysis based on the phenomenon of birefringence is photoelasticity. In this group of techniques models made of special transparent materials are often tested instead of real structures or structural members. Thus, it is possible to assure homogeneous background brightness. There are also several other methods giving visual information, although of different type (e.g. brittle lacquer technique, USG, X-raying or thermovision).

Digital computers are presently commonly used in experimental data analysis. In several applications they are also used for data logging

---

IMEKO 1st TC15 Symposium, Plzen, Czechoslovakia, May 25 - 28 1987

and/or the control of testing machines. The wide dissemination of personal computers with memories ranging at 1MB and the introduction of
integrated, "flash" A/D converters with sampling frequencies exceeding
10 MHz enables the automatization of visual information processing.
With some kinds of computers easily available on the market it is possible to create a versatile worksatation having all of the afore mentioned capabilities. The presence of a microcomputer with high processing power at the test stand enables the application of hybrid experimental - numerical techniques.

2. Workstation

To make the choice of the microcomputer to be used one has to
formulate the requirements it should fullfil. In our case they were
formulated as follows:
- the system should be fairly cheap
- the microcomputer should be a standard, easily available one
- the system should be versatile, not restricted to image
processing
- the computing power should be big enough to run finite element
programs
- image coding should be performed in 512x512 pixels with grey
levels resolution of 6-8 bits per pixel
- image coding should be done in real-time

To fullfil this requirements an IBM-PC compatible computer was
chosen. Initially the AT version as planned, but the XT was finally
decided due to high cost of the image processing cards. This kind of
computer is easily available and fairly cheap. It has 8 expansion slot
which makes it a versatile system, easy to expand by simply adding various available build-in cards. The software available for this computer is very rich. There are ready to use finite element programs and
compilers for various high level languages. The system contains a 20MB
Winchester type hard disk drive, two floppy disk drives of 360kb each,
640kb RAM memory and an arithmetic coprocessor. The clock frequency is
switchable between 4.77 and 8 MHz. The schematic diagram of the system
is shown in figure 1.

To enable image processing a special add-on card was chosen. This
card contains a flash A/D converter, which performs digital coding of
black & white TV camera signal in form of 512x512 pixels with 256 grey
levels per pixel in real-time. It has also it's own memory of 1 MB, so
that four subsequent TV frames can be coded one by one. Alternatively,

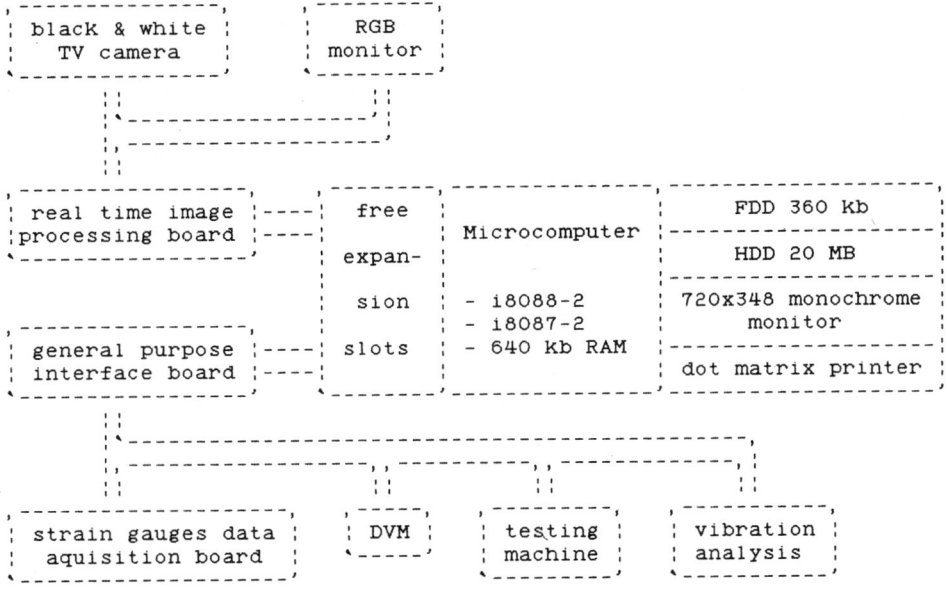

Fig. 1

Block diagram of the system

the continuous frame grabbing mode can be selected. The board has also a set of three D/A converters that are used to display the coded image on the colour TV monitor in the so-called pseudo-colours directly when it is being coded. The data transfer between the image processing card and an IBM-PC/XT compatible computer is possible both through ports and through the DMA channel. The card contains a set of eight 256bytes look-up tables for each of the three output channels. Look-up tables are user defined. The use of a set of three such cards together in one computer is possible. If each card is connected to one of the colour TV camera RGB channels it is possible to digitize and process colour images in real-time with the spatial resolution of 512x512 pixels and with 256 intensity levels for each of the fundamental RGB colours. The image processing card is supplied with the software library. This soft ware can be used in one of the two modes: as an interpreter or as a routines library for high level language. In the first case, a set of commands is used to control image coding, processing, displaying and data transfer, each command being executed immediately when it is entered from the keyboard. The set of commands includes taking a snap, defining the lookup tables, controlling data transfer between the card and an IBM-PC resources. Image processing commands include histogram calculations of the defined window, histogram displaying and convólution of an image with a user defined 3x3 kernel. In case the routines library mode is selected it is possible to incorporate card control commands as subroutines into programs written in the Microsoft Corp. C or Fortran77 programming languages. Thus, it is easy to create one's own set of image processing routines.

3. Image processing software

The software used for fringe pattern processing was prepared earlier and was tested on different computer. The discussion of the problems encountered in this task was presented in [1] and therefore would only briefly be reviewed here. In general, fringe pattern analysis was divided into several steps. The first step consists in image filtration in order to suppress the noise disturbing fringe pattern. A number of alternate algorithms was prepared to perform this task (e.g. image smoothing, features detection and non-isotropic filtration, spatial frequency filtration based on Fourier analysis). The use of the particular algorithm depends on the quality of the coded image and the type of fringe pattern. It is possible to skip this step if it is not necessary. Image smoothing techniques consist in exchanging the actual pixel with the mean or median value of it's neighbourghs. This can be regarded as a low pass filter. Mediane filter better preserves sharp edges but it converts peaks into step-like objects thus decreasing their spatial determination. Spatial frequency filtration is effective in noise reduction for simple patterns of regular geometry i.e. for patterns with narrow spatial frequency spectrum. However, it affects fringes localization. For more complicated patterns it is difficult to fit a proper filter to avoid changing of fringe pattern itself during noise filtration. Non-isotropic filtration requires some knowledge on the pattern. For fringe patterns this should be performed along ranges which represent fringes in the grey level surface. Therefore, the probable directions of fringes in small windows should be known. To find them, special templetes are defined and convolved with the image. They are designed to fit ranges passing in various directions through the window. High responce means good fit of templete to the actual image. The set of directions obtained is inconsistent due to noise. To improve consistency the relaxation method [2] is used. Once a consistent set of directions is determined it is used as a mask for non-isotropic filtration. This method is particularly recommended for this kind of

266

fringe patterns in which fringes are of the approximately the same width everywhere in the image. Unfortunately, this condition does not hold for photoelastic images. The second step consists in the recognition of fringes. This is achieved either by thresholding or by maximum searching range tracing algorithms. The automatic threshold selection is possible by the analysis of histogram. Thus, it is possible to select different threshold value for each window accounting for the slow changes of background brightness. The most important and the most difficult step consists in fringe labeling. By now, human interaction is required at this step. For some kind of pattern it seems quite a problem to distinguish peaks from valleys in the contour map obtained from experiment. The last step consists in the analysis of strains and stresses revealed by the recognized fringe pattern. Least squares surface fitting is used to obtain approximate values between the contour lines Finite element method is used for the separation of principal stresses in case of photoelastic fringe.

4. Further extensions

As it was mentioned before the currently used system is, thanks to it's free expansion slots, ready for further development. Several applications for image processing are planned concerning mainly non-destructive testing methods like X-raying, gamma-raying and thermovision. For example, the stereometric reconstruction of reinforcement bars in reinforced concrete structures investigated by gamma rays is considered. Another development consists in the instalation of the popular IEEE-488 interface board. This interface enables the computer to control several laboratory devices like digital volt meters, strain gauge data aquisition systems, vibration analyzers. This interface can also be used for testing machine control during experiment. Since the system is used at the technical university, it's versatility is particularly important. It can be utilized to illustrate several novel experimental techniques in the course of education saving substantialy theuniversity's spendings. Finally, it has to be noted that the development of hardware dedicated to image processing is rapid and new devices are being introduced on the market. They are based on the CCD matrices of 1024x1024 elements. These units are particularly suited for laboratory applications due to their high spatial resolution and to the fact that the detecting elements positions are fixed. This feature distinguish them from vidicon based cameras. Thanks to the different technology they can be expected to be cheap soon as well.

5. Refences

1 Pietrzyk, A.: Digital image processing in experimental   mechanics.
Proc. VIIIth International Conference on Experimental Stress Analysis,
Amsterdam, The Netherlands, May 12-16, 1986, pp. 281-290.

2 Hummel, R. A. - Zucker, S. W.: On the Foundations of Relaxation Labeling
Processes. IEEE - PAMI, 5, No. 3, pp. 267-287, May 1983.

AUTOMATIC RECORDING OF TRANSIENT PROCESSES IN DIES

Prof. Kamil Vrba, MSc, PhD,
Milan Forejt, MSc, PhD,
Assoc. Prof. Radimír Vrba, MSc, PhD
Technical University of Brno
Obránců míru 65, 662 41 Brno, Czechoslovakia

The principle of a mechanical strain transient recorder and the method applied is described. One of 32 special measuring strain gauge channels is analysed. The transient recorder, originally designed for nut-forming machine tests, is widely applicable for all similar diagnostic purposes.

Keywords: mechanical strain transient recorder, self-calibration

1. Introduction

In the production of joining elements and machine parts modern, highly effective forming methods on progressive automatic presses come increasingly to be applied, with the production speed ranging from 100 to 600 pcs/min.

The basic problem in dimensioning the dynamically loaded tools is the knowledge of true external loads and stress distribution at critical points. Some of the most heavily loaded tools used in these technologies are sleeve bandage dies, such as those used in the forming of bolt nuts. The dynamic investigation of these tools consists in recording of very short transient mechanical stress pulses. Mechanical signals are converted to electrical signals with the aid of strain-gauge resistors. These electrical pulses can be recorded and subsequently processed by the strain-gauge analyser described below.

2. The method of investigating mechanical processes in the die

High production speeds of present-day multistage automatic forming machines give rise to rapid deformations accompanied by considerable mechanical and thermal loading of the bandage die used, with the mechanical stress propagating basically by impact. In the case of TPM 12 automatic machines the pulses of stress are of about 100 ms duration, with type MS automatic machines they are as short as 10 ms.

The method of determining stress by means of attached strain gauges has not lost any of its significance. The main precondition here is to have the possibility of attaching the strain gauges at suitable points of the tool under examination and to connect them to a suitable recording analyser which can simultaneously record the required number of signals, store them in digital memory and subsequently process them.

The usual equation

$$\sigma = \frac{E}{K} \cdot \frac{\Delta R(t)}{R} \tag{1}$$

expresses mechanical stress in dependence on the relative change $\Delta R(t)/R$ of electrical resistance of the strain gauge, on the modulus of elasticity E in tension of the die material, and on the coefficient K of deformation sensitivity of the strain gauge. For time-variant mechanical processes it further holds that

$$\frac{\Delta R(t)}{R} = K \cdot \varepsilon(t) . \tag{2}$$

The dynamic deformation acting on the simple strain gauge is illustrated by the solid line in Fig. 1. The deformation pulse propa-

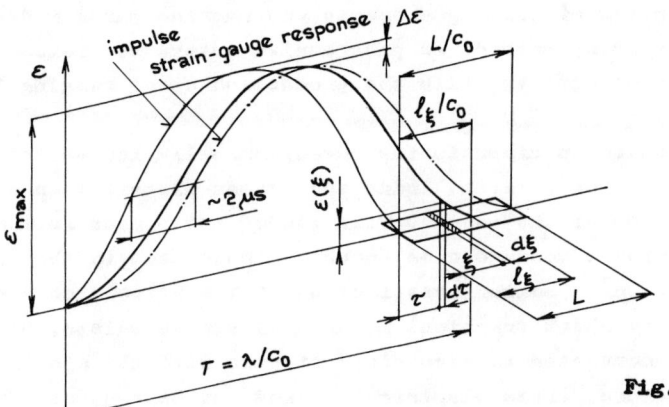

Fig. 1

Time vs dynamic deformation waveform

gates through the material at a finite velocity c_0. If the deformation coefficient K is constant along the strain gauge axis, then in keeping with Fig. 1 it will hold that

$$\frac{\Delta R(t)}{R} = K \frac{c_0}{L} \int_{t}^{t+L/c_0} \varepsilon(\tau) \, d\tau , \tag{3}$$

where $\angle$ is the strain gauge length; in deriving equation (3) we made use of the differential $d\xi = c_0.d\tau$. Equation (3) shows that owing to its non-zero length (e.g. $\angle = 4$ mm) the strain gauge always generates a signal corresponding to the mean value of the relative deformation on an interval of length $\angle$. The deformation course as distorted by this phenomenon is shown by the dashed line in Fig. 1. The distortion concerns the pulse shape, the amplitude and the time delay.

3. Arrangement of the analyser of transient processes

Fig. 2 shows the measuring system arrangement of strain gauges

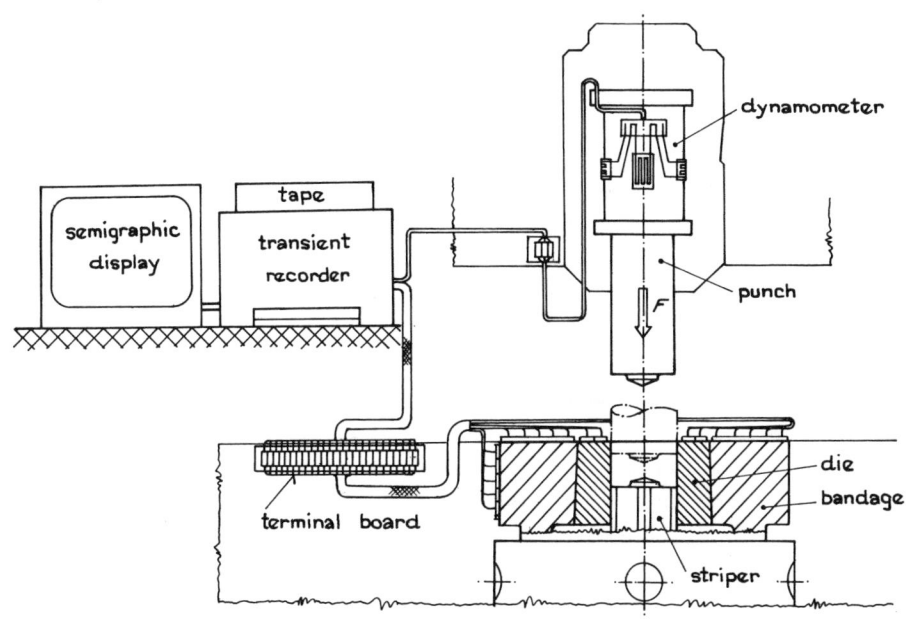

Fig. 2   A sketch of the measurement site

on the die, and the overall survey of the measuring facilities. The strain gauges are attached orientated according to the main stresses, i.e. in both radial and tangential directions, on the face of the bandage and of the die, and also along the outer circumference of the bandage. Attached on the face side is a tenfold combined HBM 2/120 KY41 strain gauge with alternate 90° orientation. The other strain gauges used are of the HBM 3/120 LY11 type. In addition, the punch carries a dynamometer for the determination of forming force F.

Fig. 3 shows the block diagram of the automatic measuring of mechanical stresses. The whole system is controlled by a microcomputer

complemented with fast control logic circuit to provide faster opera-

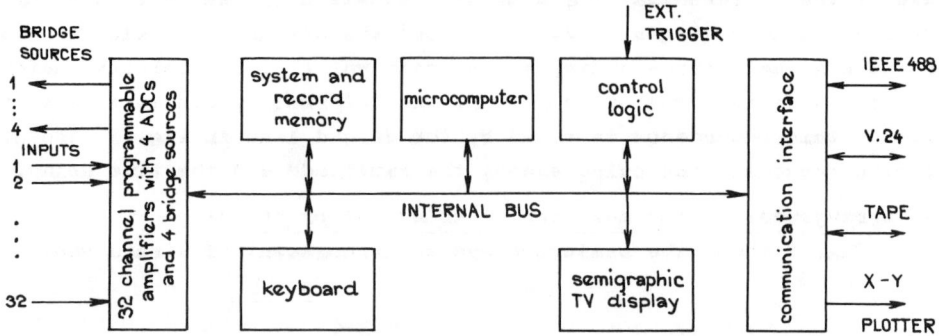

Fig. 3   Block diagram of the analyser

tion during the recording. The system and record memory has a capacity
of 48 Kword. During recording, digitized samples of measuring voltages
from as many as 32 measuring strain gauge bridges are stored in the
record memory. The bridges are fed by a given supply voltage which is
optimally programmed for the maximum possible output signal of the
bridge.

Output signals from the individual strain-gauge bridges undergo
basic preprocessing in channel units. The channel unit compensates the
basic unbalance of the bridge and the input offset voltage of operational amplifiers, and calibrates the whole amplifiing path between the
bridge output and the A/D converter.

Both these operations are controlled by microcomputer and they
automatically take place always prior to every measurement.

Fig. 4 gives a simplified schematic diagram of the channel amplifier with adjustable gain, with nulling circuits, with gain calibration and S/H module at the output. The amplifier itself is formed by a
standard instrument amplifier with three operational amplifiers (OA2,
OA3, and OA4) whose gain can be switched over by means of an electronic multiplexer MX2 controlled by digital signals from the microcomputer. Altogether, four gain values can be set, namely $A_1 \doteq 5, 27.5,$
151, and 834. At specified instants of time the output signal of the
channel amplifier is sampled by a S/H module (in all channels simultaneously). The output voltages v_2 are then successively processed
for an octet of channels in a block amplifier with electronically switched gain ($A_2 \doteq 2, 3, 4.7,$ and 7.2) and in a subsequent 9-bit A/D converter (these circuits are not shown in the figure).

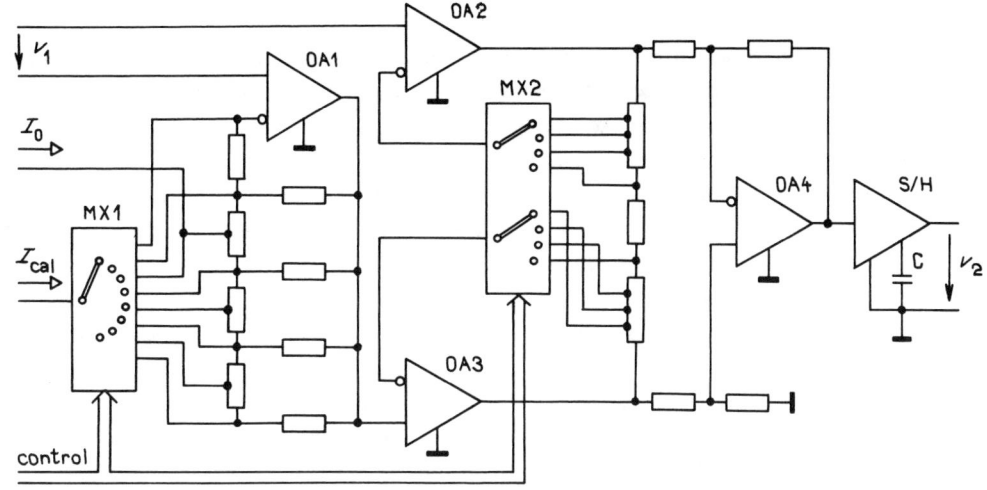

Fig. 4   Simplified schematic diagram of the channel amplifier

Worth noticing is the nulling and calibrating circuit connected ahead of the channel amplifier. It is made up of the operational amplifier OA1 with a connected network of resistors, the multiplexer MX1, and two sources of digitally adjusted nulling and calibration currents I_0 and I_{cal} , which are supplied from computer-controlled 8-bit D/A current output. The nulling and calibration proceeds as follows:

a) The amplifier is connected to the strain-gauge bridge which remains without mechanical load.

b) The current is set to be $I_0 = I_{cal} = 0$ mA, gain is set at $A_1 \cdot A_2 = 84$, and then the amplifier output voltage is measured by the A/D converter.

c) In depedence on the result of measurement the nulling current I_0 is set such that the output voltage of the amplifier is approximately zero. In this way both the basic unbalance of the bridge and the input offset voltages of OA1, OA2, and OA3 are compensated.

d) The required gain $A_1 \cdot A_2$ is set in accordance with the needs of the problem to be solved; simultaneously, the multiplexer MX1 is switched to the corresponding position. From the changes in calibrating current and from the corresponding changes in the output voltage of the whole amplifier the gain A of this link is calculated and stored in microcomputer memory. At the same time the exact value of the current I_{cal} is calculated which corresponds to an exact nulling of the channel, and this value is used to null the channel amplifier.

e) This procedure is repeated for all the other channels used.

The measuring channels calibrated and compensated in the above way make it possible to process signals from strain-gauge bridges for the values of elongations measured $\varepsilon = \pm(100 \text{ to } 100\ 000)\ \mu m/m$.

A/D converters operate with a resolution of 9 bits and a conversion time of not more than 2 μs. The number of recording channels can be determined interactively from the keyboard.

The recording of the signal under investigation can be initiated manually, by an external pulse or, which is the most frequent case, internally from the level and slope of any of the 32 input signals. The entire communication with the analyser takes the form of interactive dialog. Two selected sets of most frequently used record parameter adjustments of the analyser can be stored in the CMOS RWM memory with standby battery supply.

The built-in microcomputer is used above all for communication control between the analyser, operator and peripherals, and it partially solves simple mathematical tasks necessary for calibration and self-compensation.

4. Evaluation of the waveforms measured

The waveforms recorded can be displayed on the semigraphic display or plotted on an XY-plotter or transferred by IEEE 488 (IMS 2) or V.24 interface bus to an external computer, or they can be stored in a magnetic cassete memory.

According to the measurement parameters (specified interactively by the operator) and to the stored gain values of the channel units the built-in microcomputer will compute the scales for the relative elongation axis and for the time axis. When the waveforms are displayed, we can use the cursor to select any arbitrary point and then read straight from the display the numerical values of elongation and time.

5. References

1 Forejt,M.: Determination of radial and tangential stresses in a compound die under operating conditions (in Czech). Research report F-2621-002. Faculty of Mechanical Engineering of the Technical University of Brno, Dept. of Forming, 1985

2 Vrba,K.-Vrba,R.: Automatic analyser of mechanical signals (in Czech). Technical report. Faculty of Electrical Engineering of the Technical University of Brno, Dept. of Microelectronics, 1986

Fracture mechanics

JT_J-CONTROLLED CRACK GROWTH –   IMPROVED J-RESISTANCE TESTING

Dr.rer.nat.Peter Will
Prof.Dr.sc.nat. Bernd Michel
Dr.Ing. Uwe Zerbst
The Institute of Mechanics
Academy of Sciences of GDR
Karl-Marx-Stadt, PSF 408, GDR 9010

Incorporating modern, generalized integral criterion of fracture mechanics into elastic-plastic failure and J-R testing
is presented here. An appropriate modelling of stable crack
growth based on the energy balance at the crack front is
proposed. The engineering analysis of J-resistance curves is
reformulated. A modified failure assessment line is developed
taking into account energy dissipation during stable crack
growth.

Keywords: fracture mechanics, J-R testing, failure

1. Introduction

It has become common practice in the recent years to characterize
the behaviour of ductile cracks by means of the J-integral /1/. It is
widely accepted as a criterion for crack growth initiation. It was
introduced as a characterizing parameter for the crack tip field.
Assuming elastic materials J is not only a singularity parameter but
is indeed also useful as an energy release rate. Confusion can arise
when the symbol J is used to denote the variation with crack length of
areas under load extension curves. Originally J is defined to be the
energy used to create new crack surface. Although energy dissipation
occurs as data are collected experimentally, the energy release rate
does not coincide with J. At actual crack a plastic zone is formed at
the crack tip upon loading. In general, a fracture process region has
been proved to exist around the crack tip, where structural processes
take place and conventional continuum mechanics does not work any
longer.

2. Elastic-plastic stable crack growth

Bearing in mind just mechanical and thermal energies the   energy release rate per unit thickness reads as [1,2]:

$$d\phi/dt = J_k v_k{}^c + I = [-\int_{\Gamma+\Gamma_s} u_{i,k}\sigma_{ij}n_j d + \int_A (\S k_{,k} - f_i u_{i,k} + \sigma_{ij} u_{i,jk}) dA$$

$$- \int_{A+A_\bullet} (u_{i,k}\sigma_{i3})_{,3} dA + \int_{\Gamma_0} (n_k + \delta_{k3}) \S (k+e) d\Gamma] v_k{}^c + \qquad (1)$$

$$\int_{\Gamma_0} (n_k + \delta_{k3}) \S (k+e) \Delta v_k d\Gamma$$

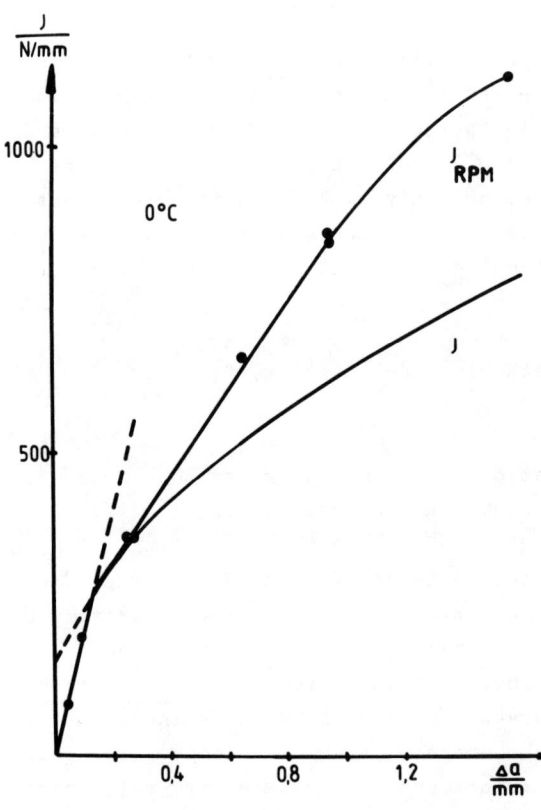

Fig.1 J-R curve (HS60-3Ni)

The path $\Gamma+\Gamma_s$ (unit outward normal n) is an arbitrary closed contour in the (1,2)-plane surrounding the area A and the interior, finite process region A_0 at the crack tip. The last one is assumed to move with the velocity v_c. and **u** are the stress tensor and the displacement vector respectively. Suffixes following a comma denote local differentiation. The vector **f** corresponds to body forces. The symbol e is the specific internal energy. Note that the three components of the local J-integral are related to a fixed plane x_3=constant. Neglecting the specific kinetic energy equation (1) reduces to:

$$\frac{d\phi}{da} = J + \S e_0 dA_0/da \qquad (2)$$

The self-evident generalization of the J-integral represents the net energy rate available to create new material surface. $\S$ eo denotes the materialspecific internal energy defining appropriately the boundary of the finite process region. Then the fourth integral in equation (1)

278

vanishes. In small scale yielding J can be interpreted as the known elastic parameter adjusted by an effective crack length.

The process of crack growth is accompanied by energy dissipation, which can occur in many different forms in addition to those used up in the formation of free surface. The second term in (2) regarding the expansion of the process region can be expressed as:

$$\S e_o dAo/da = c_1 JT_J \qquad , \qquad T_J = n/(n+1)e_o^{-1} dJ/da \qquad (3)$$

where the dimensionless factor c_1 takes into account the real crack configuration geometry. T_J denotes a modified tearing modulus. n is a strain hardening exponent associated with eventual, nonlinear deformations. Assuming that the dissipation rate in the process region is negligible to that of the plastic zone the region A_o may be replaced by the plastic zone surrounding the crack tip.

In contradiction to the generalization of the Griffith-Orowan concept accounting for the total energy used up per unit crack extension the proper resistance of materials controlling stable crack growth is assumed to be a materialspecific absorbed energy per unit crack growth which compensates for the excess energy during crack advance. The latter concept yields an instability criterium:

$$JT_J \geq 0.5EC_R/\S_F^2 \qquad (4)$$

Inequality (4) emphasizes the consequence of $J-T_J$ diagrams [3] regarding design of structures. Equation (4) associated with stable crack growth implies the nonlinearity of the J-resistance curve for a wide range of crack extension.

$$J = [J_c^2 + C_R(a-a_c)]^{1/2} \qquad a \geq a_c \qquad (5)$$

The quantities J_c and a_c are recognized temporarily as integral parameters. A physical interpretation will be given in section 3. Relationships (3,4,5) hold true for power law hardening materials.

3. J-resistance testing

It seems relevant to point out here, that the foregoing discussion has focused on the interpretation of J in terms of the net energy rate to create new crack surface. Measuring techniques for the J-resistance curve from compliance measurements are based upon its meaning as the difference in the potential work done between two crack lengths [4]. The great experimental expense of the multiple specimen method has been prompted several estimation procedures [5]. These methods relating J linear to the area under the load displacement curve are strictly valid only for a nonextending crack. Separating the fracture energy from other forms of energy dissipation during crack growth has often been in error. A more appropriate value of J holding for any path leading to the current values of the crack length a and the deflection u can be determined [6,7] from equation (6):

$$
J = \int_0^u \frac{\eta}{W-a_o} P \, du' - \int_{a_o}^a \gamma \frac{J}{W-a'} \, da' \tag{6}
$$

with η =2, γ =1 for deeply cracked bend specimen. P and W-a are the current load per unit thickness and the remaining ligament respectively. Taking into account pseudo crack advance due to severe local deformation at the crack tip (blunting) the introduction of equation (5) into equation (6) and linearizing yields a self-consistent J-R-curve formula:

$$
\int_0^u \frac{2}{W-a_o} P \, du' = J_{RPM} = D_1 + D_2(a-a_c) \tag{7}
$$

with

$$
D_1 = J_c \left[1+0.5(a_c-a_o)/(w-a_o) \right]
$$

$$
D_2 = J_c/(W-a_c)+0.5C_R/J_c
$$

Relationship (7) holds for small crack extensions $a-a_c \ll W-a_o$. The regression coefficients D_1, D_2 fitting the experimental data J_{RPM} permit the evaluation of the parameters J_c and C_R which control stable crack growth. The most commonly applied standard ASTM E 813 [8] has used best straight line fit of the J-resistance curve to identify an engineering value of J_c at initiation of crack growth. Note that the nonlinear J versus crack growth behaviour (5) can not be approximated unambigously by a straight line. Therefore the engineering critical value at the onset of crack growth should be determined appropriately as the intersection point of the J-R curve (5) with the measured blunting line. In figure 1 the results of an instrumented, precracked charpy testing are presented using relationships (5,7) to account for the effects of crack extension occuring during the test.

4. Elastic-plastic failure

Fracture of a component occurs in several stages. The initiation of fracture associated with the parameter J_c is followed by the period of stable crack growth which bases upon the equation (4). In a final propagation fracture is bounded by the plastic collapse, corresponding with large-scale plastic yielding of the remaining ligament. Interpolating between these limits failure can be expressed in terms of a failure assessment diagram. The energy balance (2) may be used to assess a structure's defect tolerance over the full range from the cleavage fracture to the plastic collapse.

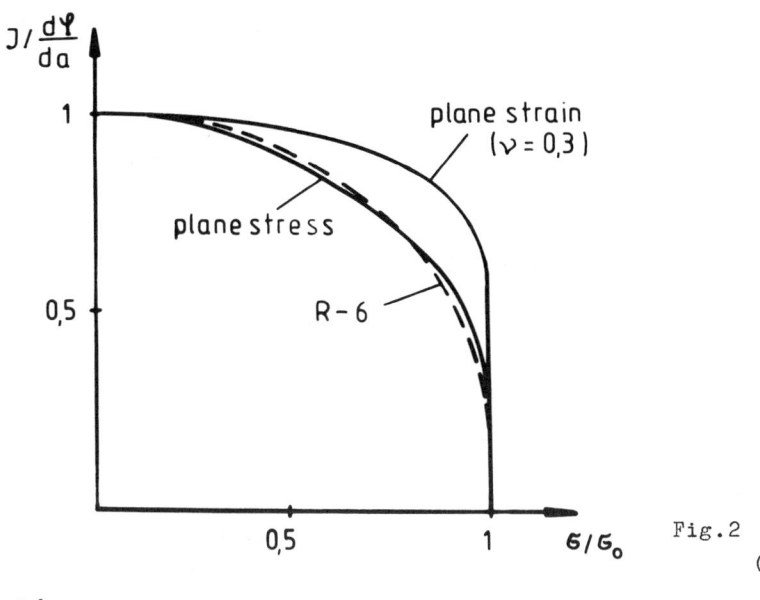

Fig.2 FA-diagram (formulae 8)

$$J / \frac{d\varphi}{da} = (1 + c_2 \ln [\sec \frac{\pi \sigma}{2 \sigma_F}])^{-1} \qquad (8)$$

The modified failure assesment diagram (8) of a dugdale crack differs from the original result of Milne [9], where the ratio of the applied stress intensity factor to the initiation fracture toughness has been related to the ratio of the applied load to the collapse load. The energy based approach (8) denoting the ratio of the cleavage fracture energy to the total energy used up per unit crack growth extends the failure concept to include stable crack growth under load control. In order to accomodate arbitrary flawed configurations the yield stress 6_F should be replaced by the plastic collapse stress. Aiming at to be in accordance with the small-scale yielding prediction the parameter c_2 has been chosen as:

$$c_2 = \frac{4}{\pi^2} \left[\begin{array}{ll} 1 & \text{plane stress} \\ [1+1.5\nu(7\nu +4\nu^3 - 8\nu^2 - 3)]/(1-\nu^2) & \text{plane strain} \end{array} \right. \qquad (9)$$

The parameter c_2 correlates with the classical effect of the apparent high toughness surface layers combined with the lower toughness layers at the midthickness. Relationship (8) is presented graphically in figure 2. For comparison the assessment diagram of Milne (R-6) is also shown.

5. References

1   Michel,B.-Will,P.:Integralkonzepte für die Rissbewertung – eine aktuelle Trendanalyse von Risskriterien der modernen Bruchmechanik. Z.angew.Math.Mech., 66, No.2, pp.65-72, 1986

2   Will,P.:Energy release rate for 3-dimensional ductile crack configuration. Z,angew.Math.Mech., 64, No.8, pp.367-368, 1984

3   Ernst,H.A.-McCabe,D.E.-Landes,J.D.:Perspective on J-T plots. Application of fracture mechanics to materials and structures, Martinus Nijhoff Publishers, pp.415-427, 1984

4   Begley,J.A.-Landes,J.D.:The J-integral as a fracture criterion. ASTM STP 514, pp.1-20, 1972

5   Keller,H.P.:Experimentelle Methoden zur Bestimmung des J-Integrals. Fortschr. Ber. VDI-Z., 18, No.10, pp.4.1-4.32, 1981

6   Ernst,H.-Paris,P.C.-Rossow,M.-Hutchinson.J.W.:Analysis of load-displacement relationship to determine J-R curve and tearing instability material properties. ASTM STP 677, pp.581-599, 1979

7   Steenkamp,P.A.J.M.-Latzko,D.G.H.-Bakker,A.:Engineering application of elastic-plastic fracture assessment methods: An exploratory study. Nucl.Eng.Des., 87, pp.51-65, 1985

8   ASTM:E 813, J_{IC}, a measure of fracture toughness, American Society for Testing and Materials, 1981

9   Milne,I.:Failure analysis in the presence of ductile crack growth. Mat.Sci.Engng., 39, pp.65-79, 1979

Dynamic Measurements of Initiation Toughness
at High Loading Rates

James W. Dally and Donald B. Barker
Mechanical Engineering Department
University of Maryland
College Park, MD 20742
USA

An experimental method is described for measuring the dynamic initiation toughness of a sharp stationary crack. A plane specimen is utilized which consists of a central region 50mm wide and 200mm long with
integral dog bone ends. The loading is accomplished by the detonation of
four small explosive charges which produce two tensile stress waves upon
reflection from the dog-bone ends. The stress waves meet at the midpoint
of the specimen and reinforce to produce a relatively large uniformly
stressed region with a very high loading rate. The crack is positioned
at the midpoint of the specimen at the location where the reinforcing
tensile stress waves meet.

A series of photoelastic experiments were conducted using Homalite
100 as the model material to observe, in a full field view, the arrival
of the dilatational waves, the subsequent development of the stress field
at the tip of the stationary crack and the initiation of the crack. The
isochromatic fringe pattern was also used to determine the instantaneous
value of the stress intensity factor $K(t)$ after the characteristic fringe
loops developed in the region near the crack tip.

Finally, $K(t)$ was measured using a single strain gage positioned and
oriented so that its signal output was proportional to $K(t)$ and independent of the next two higher order terms in the series representation of
the strain field. A method was developed to determine the instant of
initiation from the strain-time trace. Results obtained from the photoelastic and strain measurements of the dynamic initiation toughness K_{ID}

were consistently higher than the static value of K_{IC}.

INTRODUCTION

For most materials the dynamic initiation or impact fracture toughness, K_{ID}, is lower than the static fracture toughness K_{IC}. Due to
strain rate effects, the yield stress increases, plastic flow and/or
creep is inhibited and fracture toughness decreases as loading rates
increase. This behavior is well recognized for intermediate loading
rates which are obtained in experiments involving impact due to free
falling weights. However, at higher loading rates, which can be achieved
with an airgun, flyerplate and explosive loading, the functional relationship between K_{ID} and the loading rate dK_I/dt has yet to be
established. There are several studies [1,2] which indicate that the K_{ID}
associated with very high loading rates is significantly higher than the
static fracture toughness.

Relatively few high loading rate experiments have been conducted due
to the complexity of the experiment and the difficulties associated with
extracting the stress intensity factor from data that can be acquired.
The few high loading rate experiments that have been conducted include
loading methods such as flyerplates, e.g. Shockey and Curren [3], split
Hopkinson bars, Costin et al [4] and Klepaczko [5], electromagnetic
loading, Ravi-Chandar [6], and projectile impact loading, Kalthoff [7].
The conclusions regarding the influence of loading rate on toughness is
mixed. Materials were different from one study to the next and methods
of analyzing data and reporting results were different. There appears to
be clear evidence indicating that $K_{ID} > K_{IC}$ for very high loading rates
but the data is too sparse to provide a functional relationship for any
material.

The purpose of this paper is to present a new experimental method for
determining the dynamic stress intensity factor at initiation that does
not require elaborate laboratory procedures or equipment. The new method
for measuring K_{ID} should offer several advantages when compared to the
modified Hopkinson-Kolsky bar used by Costin et al [4]. Costin used a

circumferential notched and a cracked round bar as the specimen. The crack was loaded with an explosively induced stress waves generated from one end of the bar. The load $P(t)$ and the crack opening displacement $COD(t)$ were recorded simultaneously with respect to time to give the data necessary for an indirect determination of K at initiation.

This new method for determining K_{ID} utilizes a different loading system, a plane specimen instead of a round bar and a more direct approach to the measurement of $K(t)$. The specimen is dog-bone in shape as illustrated in Fig. 1. The central region is 50mm wide and 200mm long and the dog-bone ends serve to support four small explosive charges. These ends also serve to reflect the stress waves into the central region of the short bar specimen. Symmetry of the loading, which is essential in producing two plane fronted stress waves, is accomplished by detonating the small explosive charges with four closely matched bridge wire detonators that are connected in parallel to a single high voltage (2000 KV) firing circuit.

This approach to the specimen design and the dynamic loading has four advantages when compared to the modified Hopkinson-Kolsky bar method for measurement of K_{ID}.

1. With a plane specimen the crack tip is accessible and near field measurements are possible using several different experimental methods.

2. The short bar, 200mm long, reduces attenuation permitting higher loading rates to be achieved with stress wave loading.

3. Double ended loading produces two tensile stress waves which reinforce to double the loading rate.

4. Plate materials with texture introduced during the rolling process can be investigated in the as rolled thickness.

The dynamic behavior of the explosively loaded dog-bone specimen was characterized by full field photoelastic experiments. These experiments verified the symmetry of the loading, showed that a plane fronted dilatational wave propagated in the test section of the specimen and indicated that the crack initiated before the arrival of the distortion stress wave. The dynamic initiation toughness K_{ID} was measured for Homalite 100

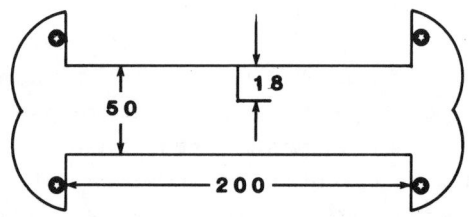

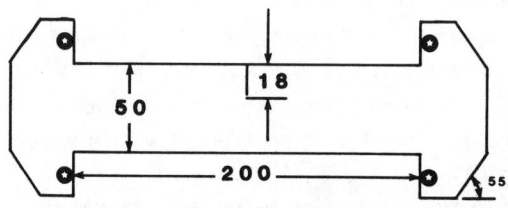

Fig. 1 Geometry of dog bone specimens

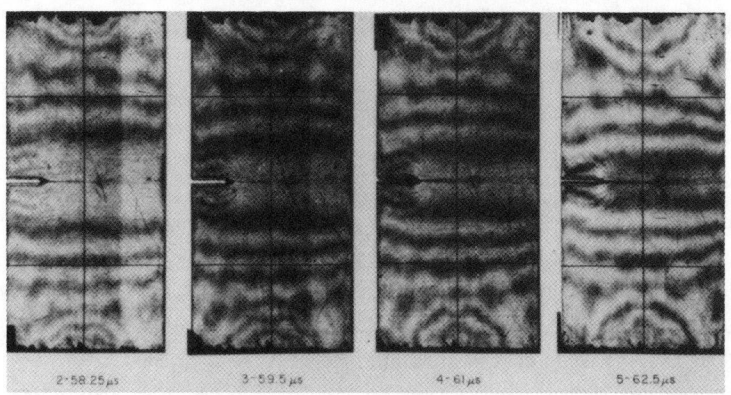

Fig. 2 Stress wave propagation toward the center region of the specimen

286

using both photoelastic and strain gage measurements in the near field region adjacent to the crack tip. These experiments and methods of data analysis are described in subsequent sections of this paper.

PHOTOELASTIC CHARACTERIZATION OF THE DOG-BONE SPECIMEN

A sequence of 16 dynamic isochromatic fringe patterns were recorded with a Cranz-Schardin camera operating at a framing rate of 667,000 fps. The photoelastic model was fabricated from a sheet of Homalite 100, 9.4 mm thick to the dimensions shown in Fig. 1. Selected frames showing the key features of the dynamic behavior are presented in Fig. 2-6.

The fringe patterns presented in Fig. 2 show the two dilatation stress waves propagating toward the center region of the dog-bone specimen. It should be noted that the waves are essentially plane, that symmetry has been achieved and that the distortional waves are absent during this time period.

Frames 6 to 9 presented in Fig. 3 illustrate the reinforcing period where the two stress waves overlap to produce a region of nearly uniform stress in the central region. Also evident is the reflection from the crack face which unloads the corner regions near the crack as can be seen in the left central portion of the photos. Fringes associated with the distortional wave can be observed near either end of the test section but they are well removed from the central region where the crack is located. A stress singularity exists at the crack tip during this time period, but it cannot be observed. Apparently the singularity is so localized that the optical system is not sufficient to resolve the dense fringe pattern at the tip of the crack. Note, the absence of distortions in the fringe pattern in the central region of the specimen which indicates that the stress waves scattered from the crack tip in the radial direction are very low in magnitude.

Initiation of the crack is illustrated in Fig. 4, frames 10-12, where a radial fringe pattern centered at the crack tip is evident. This pattern is quite local to the tip of the crack and appears embedded in the relatively uniform stress field over the central region of the specimen.

287

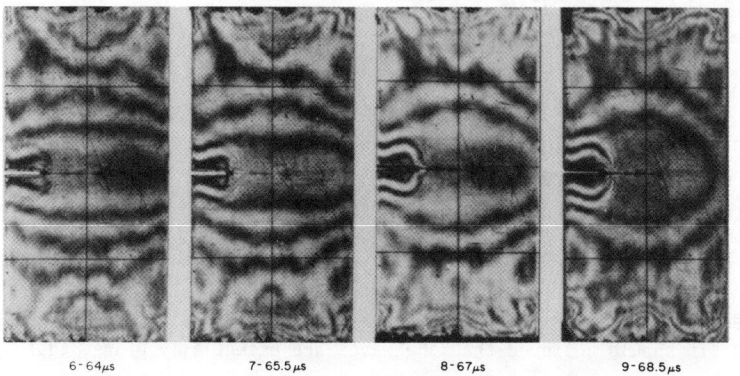

Fig. 3   Stress wave reinforcing at the central region

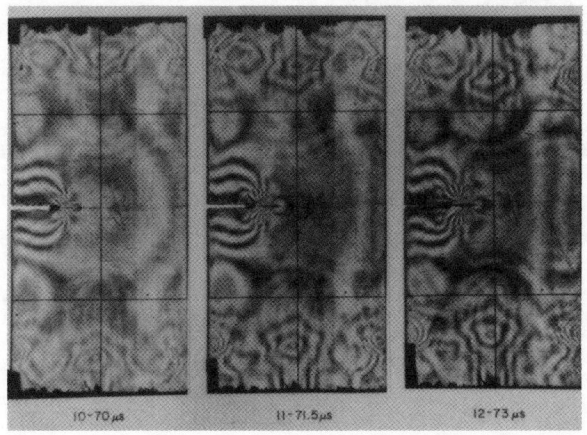

Fig. 4   Initiation of the crack before frame 10

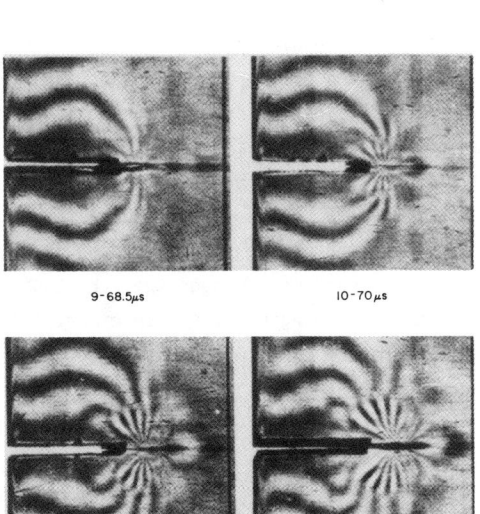

9-68.5μs   10-70μs

11- 71.5μs   12-73μs

Fig. 5   Enlargements showing shear wave generated at initiation

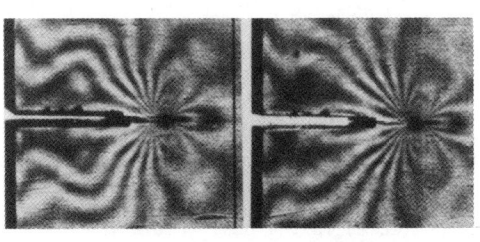

13-74.5μs   14-76.5μs

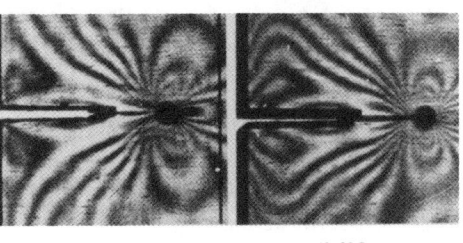

15-79μs   16-82.5μs

Fig. 6   Classical fringe loops during propagation phase

289

The distortional wave from the loading source can be observed entering the field of view at the top and bottom of the photos, but the loading of the central region and crack is due entirely to the dilatational wave, and waves reflected and scattered from the crack.

Enlargements of the near field region over the time period from 68.5 to 75 μs are presented in Fig. 5, frames 9-12. These fringe patterns clearly show that the initiation of the crack has produced a disturbance which propagates radially outward and changes the stress distribution in the uniform field in the central region of the specimen. Velocity measurements of this disturbance confirm that it is due to a shear wave generated when the crack initiates.

This shear wave attenuates as it propagates radially outward, and the fringe pattern behind the front changes dramatically as indicated in Fig. 6, frames 13-16. At frame 13 the radial fringes have changed to the classical isochromatic fringe loops which can be utilized to determine the stress intensity factor at the tip of the crack. For the remainder of the event, the classical fringe loops expand and grow in number as the crack propagates into an increasing K field.

PHOTOELASTIC DETERMINATION OF K

Five different photoelastic experiments were conducted in order to determine K at initiation. The photographs were analyzed with a digitizing tablet that gave a resolution of less than 0.01 mm as measured on the model. The data collected in the post initiation event was processed using a six parameter dynamic representation of the stress field and an overdeterministic method of relating K to the fringe position as described in reference [8]. Typical results showing crack tip position and K as a function of time after arrival of the stress wave at the crack tip are shown in Fig. 7. The results indicate that K increases approximately linearly with time, and it is possible to extrapolate the K-t curve back to the origin which corresponds to the time of arrival of the dilatational wave at the crack location.

The time of initiation was determined three ways. First an estimate of initiation time was determined from a direct inspection of the fringe

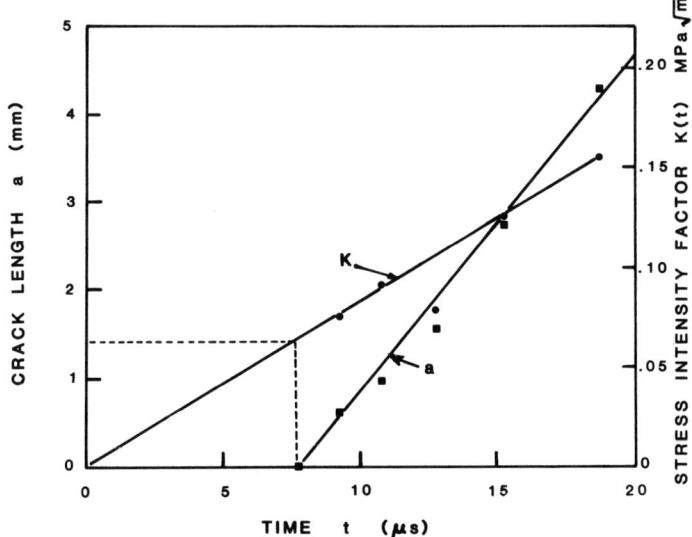

Fig. 7   Crack tip position and stress intensity factor as a function of
         time

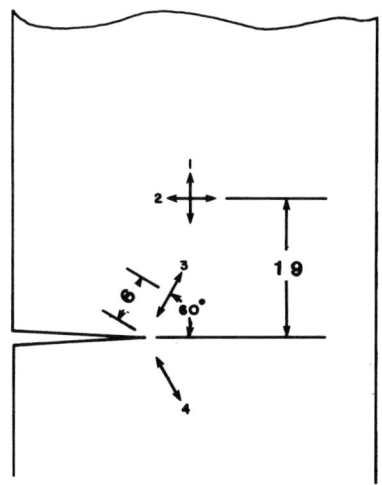

Fig. 8   Strain gage locations on the specimen

patterns. Initiation time was also calculated by extrapolating the line representing crack tip position versus time back to the initial crack tip location. Finally, crack initiation time was taken as identical to the time of initiation of the shear wave emitted from the crack tip as described in the previous section. All the initiation time determinations agreed to within ±0.5 μs. This result corresponds to the time resolution of the digital oscilloscope that provided the time for each frame in the high speed recordings.

The value of K at the initiation was determined from the K-t curve to give K_{ID} as illustrated in Fig. 7. The crack tip velocity was established from the slope of the a-t curve at the instant of initiation. For the example presented here, the crack velocity was 391 m/s and the value of K_{ID} was 0.646 MPa√m. Results for a series of five experiments are presented in Table I.

TABLE I

PHOTOELASTIC RESULTS FOR HOMALITE 100

Exp No.	Crack Velocity	K_{ID}	dK/dt	Initiation time*	Wave Arrival time	Loading time
	m/s	MPa√m	MPa√m/μs	μs	μs	μs
1	378	0.501	0.081	68.8	61.8	7.0
2	427	0.646	0.066	67.5	60.2	7.3
3	368	0.711	0.090	69.3	62.3	7.0
4	391	0.646	0.083	70.8	63.0	7.8
5	355	0.771	0.062	68.5	61.0	7.5
AVERAGE	380	0.655	0.076	69.0	61.7	7.3
STD DEV	33.5	0.101	0.011	1.21	1.09	0.34

*Time measured from the instant of detonation of the explosive charges

These results show that the average loading time is 7.3 μs from the arrival of the stress waves until crack initiation. This corresponds to a loading rate of $dK/dt = 0.076$ MPa√m/μs. The dynamic crack initiation toughness $K_{ID} = 0.655$ MPa√m showed considerable variation. This variation is due to several different factors which affected the measurements. These factors include differences in the sharpness and straightness of the crack tip, variations in the material toughness from one area to another in the sheet and experimental error in the determination of K_{ID}. The largest difference from the average value of K_{ID} was 23 percent and the coefficient of variation was 0.154.

The static value of the crack initiation toughness K_{ID} for Homalite 100 determined with a compliance experiment was 0.445 MPa√m. The difference in the static and dynamic values of the initiation toughness is 47 percent with the dynamic response being higher than the static.

STRAIN GAGE DETERMINATION OF INITIATION TOUGHNESS

Dally and Sanford [9] have recently shown that a single strain gage can be used to determine the stress intensity factor K for the opening mode. The placement and orientation of the gage is critical if reasonable accuracies are to be achieved. First, the gage must be placed at a position where $r_g > 0.5$ h, where h is the thickness of the specimen. Next, the gage must be placed with an orientation angle $\alpha = 60$ degrees and positioned along a radial line with $\theta = 60$ degrees, to eliminate the errors due to the second and third order terms in the series expansion. Note with $\nu = 1/3$ then $\alpha = \theta = 60°$. With these constraints on the placement and orientation of the gage, the relation for K in terms of the strain e_g is given by:

$$K = [(8/3)\pi r_g]^{1/2} Ee_g \qquad (1)$$

It should be noted that this relation was derived for the static loading conditions. Its use in the dynamic case is speculative and subject to confirmation when suitable dynamic solutions become available. Its use here is justified by a favorable comparison to dynamic photoelastic results.

Strain gages with a 0.75 mm gage length were position on the specimen
at locations shown in Figure 8. Gages 1 and 2 were employed to determine
the strain associated with the incoming stress wave, and gages 3 and 4
were used to sense K in accordance with Eq. (1). The strain gage signals
were recorded on a digital oscilloscope using conventional bridge and
amplifiers for preconditioning the signal. The response from gage 2
which recorded the transverse strain at the center of the specimen indicated that $e_t = 0$. This result shows that the specimen is sufficiently
wide to produce plane strain conditions in the central region. The
response from gage 1, presented in Fig. 9, shows that the strain is
nearly linear with time and a peak strain of about 2400 μm/m is achieved
in 23 μs. The small deviations from the linear ramp are due to signal
noise and the effects of waves scattered from the boundaries of the specimen and the crack tip.

The strain gage signal from the crack tip gages, presented in Fig.
10, show the symmetry and reproducibility of the loading from the two
ends of the specimen. Again the strain time trace is essentially linear
with a maximum strain of 1500 μm/m achieved in about 10 μs. This trace
is proportional to the K-t relation and gives the data necessary to
determine K_{ID} providing that the initiation time can be established.

Attempts to measure initiation time with continuity gages and optical
sensors were not successful because of poor reproductibility and poor
time resolution. To obtain the initiation time, the strain-time trace
from gages 2 and 3 were compared. It was noted that the strain recorded
by the gage near the crack tip began to decrease after a peak of about
1500 μm/m which is much lower than the anticipated peak strain of about
3600 μm/m expected at the gage due to the two superimposed strain waves.
The decrease in the strain peak is due to the unloading wave generated at
crack initiation.

The unloading begins when the shear wave shown in Fig. 5 reaches the
strain gage 3 or 4. At this instant, defined as t_d, the trace begins to
deviate from linearity as shown in Fig. 11. The time increment δt for
the arrival of the shear wave is given by:

$$\delta t = r_g/c_2 \qquad (2)$$

294

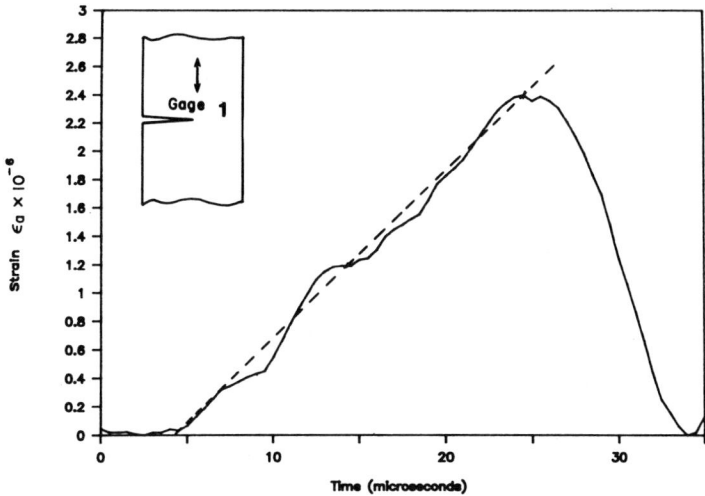

Fig. 9   Response from gage no. 1

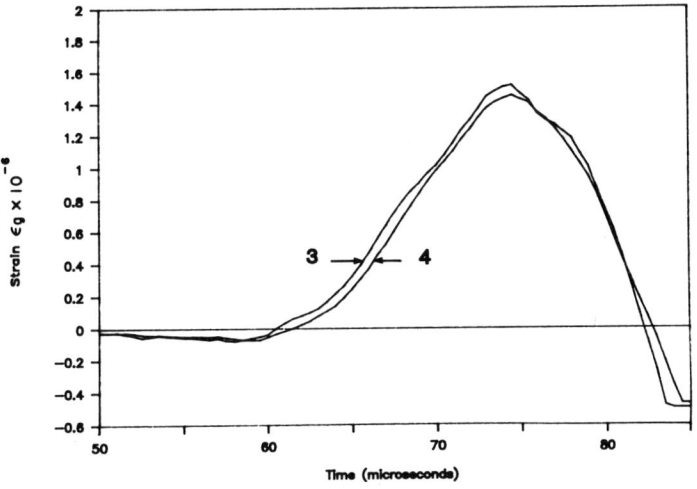

Fig. 10   Strain as a function of time for gages no. 3 and 4

295

and the initiation time t_i is given by:

$$t_i = t_d - \delta t \qquad\qquad\qquad (3)$$

The location of t_i and the determination of the strain at crack initiation e_g is illustrated in Fig. 11.

The results from five different strain gage experiments are presented in Table II.

TABLE II

Dynamic Initiation Toughness K_{ID} Determined from Strain Gage Measurements

Exp No.	K_{ID} (MPa√m)
1	0.564
2	0.702
3	0.573
4	0.589
5	0.563
AVERAGE	0.598
STD DEV	0.059

The strain gage determination of K_{ID} compares favorably with the photoelastic measurements providing the comparisons are made based on average values. The difference is 8.8 percent with the strain gages giving slightly lower values. Errors in the strain gage determinations include time resolution on the strain time records, which is limited by the sampling rate of the digital oscilloscope, in addition to those listed in the previous section. Regardless of these errors the coefficient of variation corresponding to the strain gage results was only 10 percent. These comparisons appear to indicate that the static relation provided in Eq. (1) can be used with reasonable accuracy in dynamic initiation experiments.

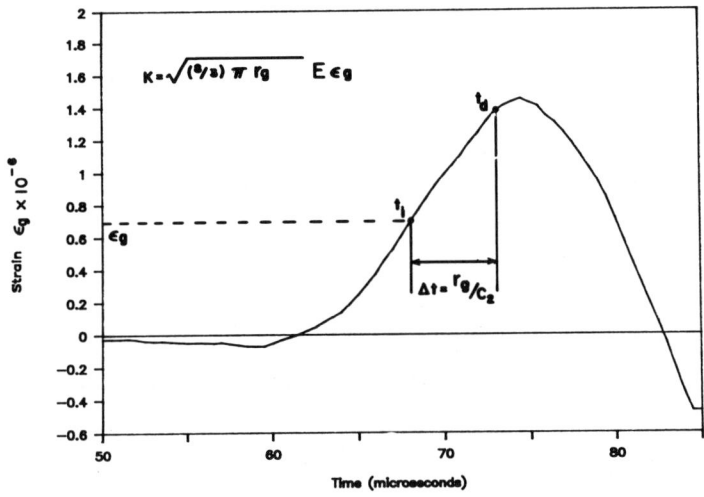

Fig. 11   Method of determining initiation time and e_g

CONCLUSIONS

Several conclusions can be drawn from the results of this study which include:

1. The dog-bone specimen with the four explosive charges exhibits symmetrical loading with two stress waves providing very high loading rates. The incoming waves are plane fronted and rein- force to give a large relatively uniform stress field in the central region of the specimen. The specimen is sufficiently long to delay the distortional waves until after crack ini- tiation.

2. For Homalite 100 crack initiation occurred with an average loading time of 7.3 μs after the arrival of the stress waves at the center line of the specimen. The corresponding loading rate is dK/dt = 0.076 MPa$\sqrt{m}/\mu$s.

3. The results obtained for K_{ID} were 0.655 and 0.598 MPa$\sqrt{m}$ for pho- toelastic and strain gage determinations respectively. These values are about 40 percent higher than the static initiation toughness K_{ID} = 0.445 MPa$\sqrt{m}$, but not as high as those reported in reference 6.

4. The strain gage method of measurement of dynamic K shows con- siderable promise even though it is based on a static theory. The method is relatively simple to employ and can be used by industrial laboratories equipped with commonly available dynamic instrumentation.

5. Experimental error on individual determinations are relatively large (about 25 percent); however, these errors can be reduced to a more acceptable value of about 10 percent if five measur- ments are made and the results of the individual values are averaged.

298

ACKNOWLEDGEMENT

The authors appreciate the support and encouragement of ORNL and NSF in conducting this research program.

References

1   Knauss, W.G. and K. Ravi - Chandar, "Some Basic Problems in Stress Wave Dominated Fracture", Special Issue Dynamic Fracture, Intl. Jrnl. of Fracture, 27, 3-4, 1985, pp. 127-143.

2   Shockey, D.A.; Kalthoff, J.F.; Homma, H. and D.C. Erlich, "Response of Cracks to Short Pulse Loads", Workshop on Dynamic Fracture, California Institute of Technology, Pasadena, CA, 1983, pp. 57-71.

3   Shockey, D.A. and D.R. Curren, "A Method for Measuring K_{IC} at Very High Strain Rates," Progress in Flow Growth and Fracture Toughness Testing, ASTM STP 536, 1973, pp. 297-311.

4   Costin, L.S., J. Duffy and L.B. Freund, "Fracture Initiation in Metals Under Stress Wave Loading Conditions", ASTM STP 627, 1976, pp. 301-318.

5   Kelapckzko, J.R., "Application of the Split-Hopkinson Pressure Bar to Fracture Dynamics, Proc. 2nd Conf. Mech. Prop. High Rates of Strain (J Harding, ed.), Oxford, The Institute of Physics, 45, 1979, pp. 201-204.

6   Ravi - Chanar, K. and W.G. Knauss, "Investigation Into Dynamic Fracture: I. Crack Initiation and Arrest", International Journal of Fracture, vol 25, pp. 247-262, 1984.

7   J.F. Kalthoff, "Fracture Behavior Under High Rates of Loading", Engineering Fracture Mechanics, Vol. 23, no. 1, pp. 289-298, 1986.

8   Sanford, R.J. and Dally, J.W. "A General Method for Determining Mixed Mode Stress Intensity Factors from Isochromatic Fringe Patterns", Engrg Fract. Mech. Vol. 11, pp. 621-633, 1979.

9   Dally, J.W. and R.J. Sanford, "Strain Gage Methods for Measuring the Opening Mode Stress Intensity Factor K_I", Proceedings SEM Spring Conference on Experimental Mechanics, 1985, pp. 851-860

TEST METHODS IN THE MECHANICS OF THE BRITTLE
CEMENT BASED COMPOSITES

Prof.Dr. Andrzej M. Brandt
Institute of Fundamental Technological Research /IFTR/, Polish
Academy of Sciences
Swiętokrzyska 21, 00-049 Warszawa, Poland

Three main trends of development of the testing and test
methods in the mechanics of cement based composites are considered. These methods concern: local effects, multiple measurements made in one test and non-elastic phenomena of various kinds. The examples of successful test techniques are
given and discussed.

Keywords: Cement based composites, mechanics, measurements

1. Introduction

The rapid progress of test methods and measuring equipment in research of mechanical properties of structures and materials is significant for a few last years. Also in the investigations concerning
concretes and other cement based brittle materials that progress is
observed. It is perhaps worth-while to consider main directions of development and the principal achievements in the domain of measurement
methods.

The scope of the paper is limited to the investigations of the
material properties on specimens and elements. Test methods for structures and structural elements are not considered here. Similarly, the
construction of equipment is not analysed but only its possibilities
and performance for material testing.

The aim of the paper is to analyse the impact of new methods and
equipment on the test programs and to discuss the new trends.

2. Stimulation for progress in research

Intensive research programs in the field of the mechanical properties of cement based materials are carried on for various reasons
related to the new material compositions, particular loading conditions, required durability in corrosive environment, etc.

301

The variety of concrete-like materials is increasing with introduction of polymers, fibres, different admixtures and additives. The
mechanical properties of these materials should be determined in view
of their applications. The structural and non-structural elements are
now often produced of specially designed materials and exposed on particular actions and conditions, like increased abrasion, chemical at-
tacks, mechanical impacts or thermal shocks. The material design to
satisfy all imposed requirements is possible only after thorough and
complex testing in which qualitative but also quantitative relations
between material components, their composition, internal structure
and material properties are determined.

3. Main directions of development and the measuring techniques

Starting from the classical testing methods of mechanical properties in which mainly mean values of critical stress or strain and deformation were determined, the following new directions may be mentioned:
- local effects are observed and measured with good results,
- several effects at a large number of points of a tested specimen
  are measured simultaneously,
- phenomena beyond the elastic deformations and continuity of materials are investigated.

In all these three directions not only special measuring devices
are necessary but also continuous recording of their indications as
well as computerised storage and analysis of obtained data.

The above listed trends are illustrated by a few examples taken
from recent research programs. There are many other experimental methods available, like SEM observations or use of infrared microscopy
which are not mentioned here because of the volume limitation of the
paper.

4. Examples

4.1. The position and dimensions of the damaged zone ahead of the moving crack tip in a large concrete plate 0.12x3.00x3.50 m were determined by three methods, [1]:
- A system of plane rosettes, each composed of three electric resistance strain gauges was glued on the plate and their indications
shown where the crack crossed a gauge or passed in the vicinity. The
accuracy of strain measurements, including the gauge drift, was better than 30.10^{-6} (Fig. 1).

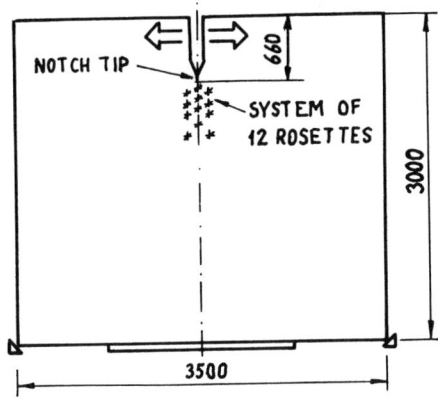

Fig.1 Concrete plate with a notch and system of rosettes [1]

- A system of emitters and receivers was fixed on both faces of the plate to measure the time of travel of the sound between the devices with an accuracy of 1.10^{-9} s. The increase of the time indicated certain material damage.
- Four acoustic emission sensors were used and their indications allowed to record and locate all events in the material. The treshold of detection was 6 dB above ground noise.

The results obtained by all 3 methods were in a good agreement.

4.2. The compact tension specimens were tested and the critical values of basic fracture parameters K_{Ic} and J_{Ic} were determined on, [2]:
- polymer concrete specimen and calculation of K_{Ic} and J_{Ic} by FEM,
- photoelastic model,
- concrete specimen with photoelastic coasting and reflection surface (Fig.2).

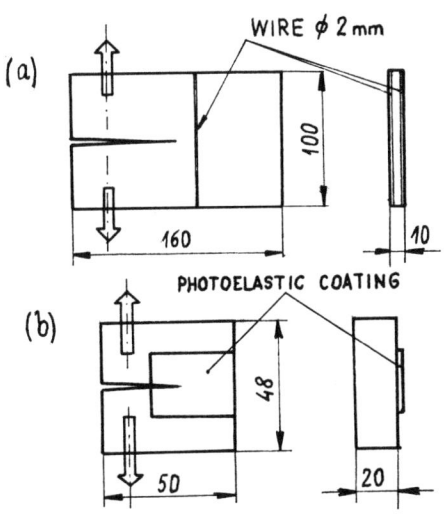

Fig.2 Compact tension specimens (a) photoelastic model (b) concrete specimen [2]

The results of both experimental methods were similar and confirmed by calculations.

4.3. The interface between steel fibre and cement matrix was observed during pull-out test and relative displacement of fibre with respect to the matrix was measured by two methods:
- optical microscope observation with the accuracy of measurement of 5 - 10 μm and photographs of the interface related to $\sigma-\varepsilon$ curve (Fig. 3),
- speckle photography with the accuracy of 0.5 μm (Fig. 4).

Both methods furnished complementary results, allowing to determine quantitatively the situation in the interface and to follow the crack propagation along the fibre under increasing pull-out force, [3]. It was concluded from the tests concerned among others the distribution of bond stress of fibre along the embedment.

303

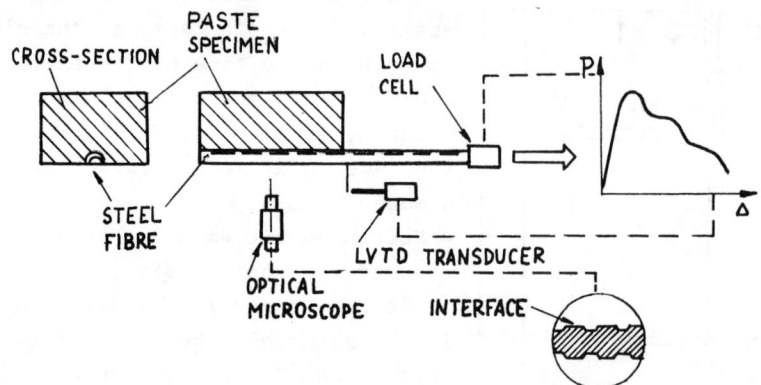

Fig. 3 Microscope observation during a pull-out test [3]

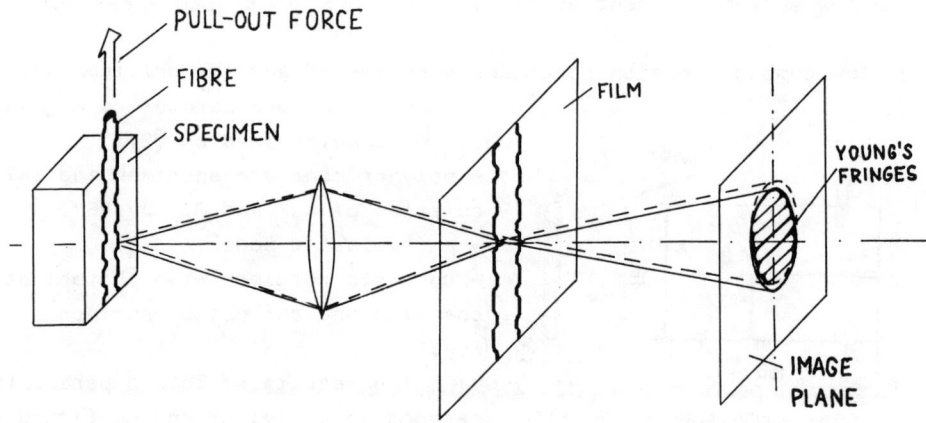

Fig. 4 Speckle photography of a specimen under load [3]

4.4. The energy changes in brittle cement matrix wich are related to deformation processes and crack opening may be detected and measured by thermistors (Fig.5). It has been shown in [4] that the pearl thermistors of very reduced dimensions may monitor local temperature variations with high sensitivity of about $0.0002^{\circ}K$. Using that kind of local measurement interesting results have been obtained and the conclusions proposed in [4] concerned the relation between temperature variations and energy changes connected to the crack propagation.

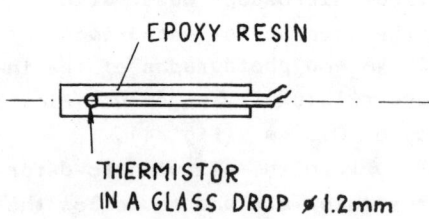

Fig. 5 Pearl termistor [4]

4.5. Complex testing of crack propagation and fracture toughness is carried on at present at the Delft University of Technology [+] to determine various mechanical properties of mortar, concrete and concrete reinforced with short steel fibres and to discover the relations between these properties and the material structure and composition. From the load-deflexion curves the fracture toughness is calculated and the strain measurements at various levels above the notch show the actual position of the crack tip, which is also monitored by a small microscope, its position being recorded together with all other measurements at selected loading steps or at constant intervals. Reading of all devices is steered by a computer program. During a test of the specimen as in Fig. 6 as many as 90 loading steps are realised and all date are stored in a microcomputer. Results of calculations and diagrams are obtained directly from these data analysed on the monitor and eventually printed. In Fig. 7 examples of such curves are given.

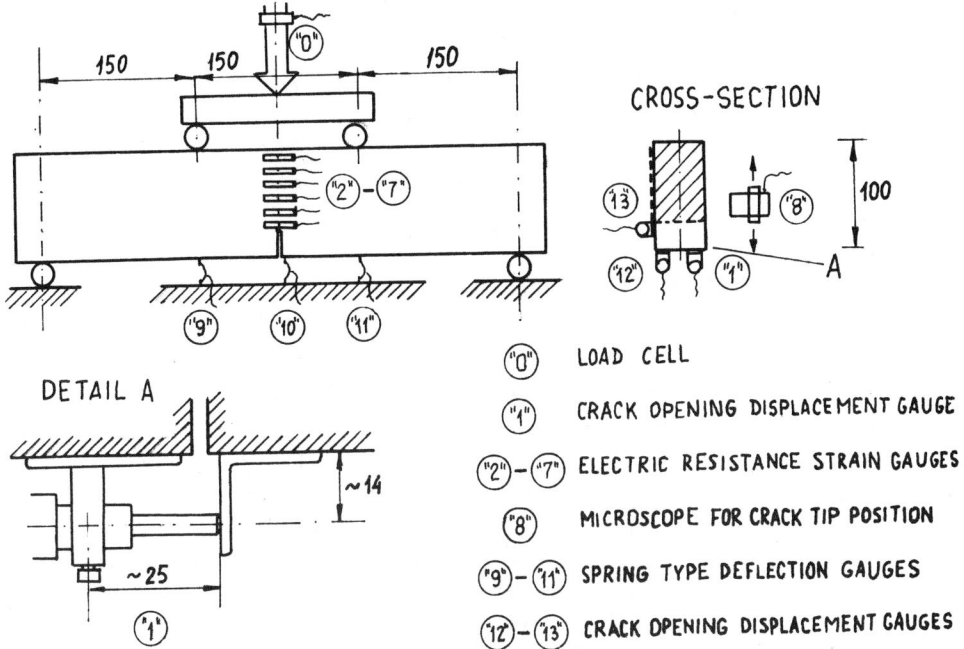

Fig. 6 Notched fibre concrete specimen subjected to flexion

[+] The tests are realised with the participation of Dr. P. Stroeven, Eng. D. Dalhuisen and Mr. L. Donker from the Stevin Laboratory and Doc. J. Kasperkiewicz and the author from the IFTR.

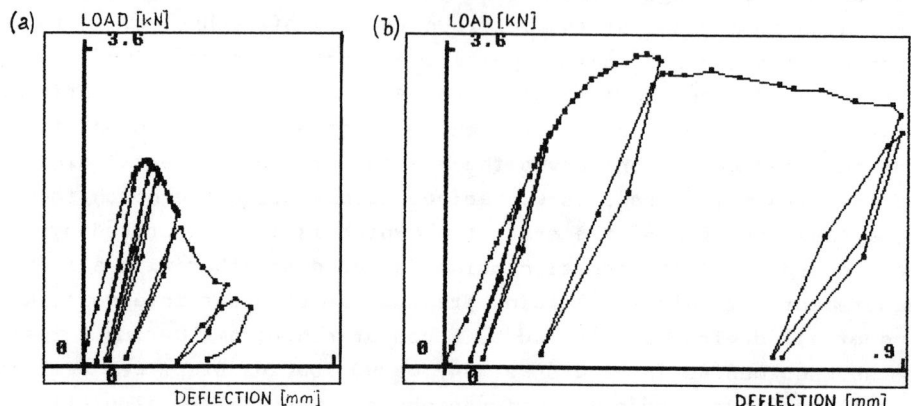

Fig. 7 Load-deflection curves for plain concrete (a) and fibre reinforced concrete (b) specimens

5. Conclusion

Present achievements in measuring techniques, microcomputers
and other connected devices allow to carry on complex test programs
which would be virtually impossible even 10 years ago. New possibilities are therefore open for material testing and should be carefully analysed when a research program is set up not to miss a possibility to gather more data on tested specimens.

6. References

1 Chhuy S., Cannard G., Robert J.L., Acker P.: Experimental investigations into the damage of cement concrete with natural aggregates,
   in: "Brittle Matrix Composites 1", A.M. Brandt and I.H.Marshall eds.
   Proc. Euromech 204, Jabłonna 12-15 Nov. 1985, pp. 341-354, Elsevier
   Appl.Sc.Publ. 1986.
2 Jaroniek M., Niezgodziński T.: Studies of fracture and the crack
   propagation in concrete and polymer concrete, as [1] , pp. 355-370.
3 Potrzebowski J.: Investigation of steel fibre debonding processes
   in cement paste, in: "Bond in Concrete", P.Bartos ed., Proc. Int.
   Conf. Paisley 14-16 June 1982, Appl.Sc.Publ. 1982. pp.51-59.
4 Weiss V., Czarnecki L.: Relationships between crack formation and
   energy changes in concrete, as [1] , pp. 311-321.
5 Kasperkiewicz J.: Fracture and crack propagation energy in plain
   concrete. Heron, 31, 2, pp. 5-14, 1986

FAILURE AND FRACTURE CRITERIA IN COMPOSITES

Pericles S. Theocaris
Dept. of Engng. Sciences
Athens Nat. Techn. University
P.O.Box 77230, GR 175-10, Greece

The paraboloidal failure criterion for isotropic materials was
extended to transtropic materials by maintaining the direction
of its axis of symmetry and the form of its surface. It was
shown that the paraboloid of revolution failure surface
becomes an elliptic paraboloid surface, it has its axis of
symmetry parallelly translated to the hydrostatic axis and its
sections normal to this axis become ellipses of ellipticity
and orientation depending on the amount of anisotropy of the
material. Examples are shown from graphite-epoxy composites.

Keywords: paraboloidal failure criterion, transtropic
          materials, composites

1. Introduction

Failure criteria are based on Hill's well known theory for
anisotropic metals [1] which is based on Mises' first attempt to
formulate failure in anisotropic solids [2]. While Hill's criterion
does not take into account the strength differential effect apparent in
all solids, Hoffman's criterion presents a further improvement by
incorporating this effect in Hill's criterion. This was achieved by
adding the linear terms in the quadratic expression of Hill's
criterion [3].

The tensor polynomial criterion introduced by Tsai and Wu [4]
constituted a flexible and mathematically elegant version of a
criterion, formulated by means of the Cartesian components of the
stress tensor, is represented by hypersurfaces in the six-dimensional
stress space, impossible to be readily visualized geometrically in a
physical stress space. Only plane sections of this hypersurface were
therefore studied representing quadric surfaces in the $(\sigma_x, \sigma_y, \sigma_{xy})$
parametric space. However, even these subspaces do not yield a direct
interrelation with the directions of the externally imposed loading and

IMEKO TC15 First Intern. Conference, Plzen Czechoslovakia May 25-28, 1987

the material strength directions a drawback causing the necessity of meticulous and delicate experiments for its definition.

Theocaris [5] has presented recently a paraboloid of revolution failure criterion for isotropic bodies, which, later on, was extended to an elliptic paraboloid, convenient for anisotropic materials [6,7]. The physical basis and properties of this elliptic paraboloid criterion is presented in this paper suitable for transtropic materials, as they are the fiber-reinforced composites.

2. The Elliptic Paraboloid Failure Surface

Consider a transtropic body with σ_{T1} and σ_{C1} the longitudinal (strong) failure strengths in tension (T) and compression (C) and σ_{T2}, σ_{C2} its respective transverse strengths on the isotropic plane, which is normal to the longitudinal (fiber) axis. When the principal stress axes coincide with the material principal strength directions the failure surface should pass through the points $A_1(\sigma_{T1},0,0), A_2(0,\sigma_{T2},0)$, $A_3(0,0,\sigma_{T2})$ and $B_1(-\sigma_{C1},0,0), B_2(0,-\sigma_{C2},0)$ and $B_3(0,0,-\sigma_{C2})$. The equation of a quadratic surface passing through these points and having its axis of symmetry parallel to the hydrostatic axis ($\sigma_1=\sigma_2=\sigma_3$) is given by [6]:

$$\frac{1}{\sigma_{T1}\sigma_{C1}}\sigma_1^2 + \frac{1}{\sigma_{T2}\sigma_{C2}}(\sigma_2^2+\sigma_3^2) - \frac{1}{\sigma_{T1}\sigma_{C1}}\sigma_1\sigma_2 + \left(\frac{1}{\sigma_{T1}\sigma_{C1}} - \frac{2}{\sigma_{T2}\sigma_{C2}}\right)\sigma_2\sigma_3 -$$

$$- \frac{1}{\sigma_{T1}\sigma_{C1}}\sigma_3\sigma_1 + \left(\frac{1}{\sigma_{T1}} - \frac{1}{\sigma_{C1}}\right)\sigma_1 + \left(\frac{1}{\sigma_{T2}} - \frac{1}{\sigma_{C2}}\right)(\sigma_2+\sigma_3) = 0 \tag{1}$$

Since relation (1) is valid only for coincidence of the principal stress directions with the material principal strength directions, for the case of an arbitrary orientation of these two systems another elliptic paraboloid is associated with these directions. This paraboloid should have its axis of symmetry parallel to the hydrostatic axis and it is expressed by the same relation (1) in which the appropriate failure strengths in these directions were calculated by using the appropriate transformation and Hoffman's criterion.

Therefore for any orientation of the principal directions of external loading the association of the convenient elliptic paraboloid defined with the appropriate strength properties of the material along the principal stress directions yields a direct clear and comprehensive view of safe loading paths on the structure and the correct evaluation of tis respective load bearing capacities in these directions.

For plane-stress states of transtropic materials along a principal stress plane, say the (σ_1,σ_2)-plane ($\sigma_3=0$), the failure locus is derived as the intersection of the ellipsoid (1) by the plane $\sigma_3=0$, it yields an ellipse whose equation is given by:

$$\frac{\sigma_1^2}{\sigma_{T1}\sigma_{C1}} + \frac{\sigma_2^2}{\sigma_{T2}\sigma_{C2}} - \frac{\sigma_1\sigma_2}{\sigma_{T1}\sigma_{C1}} + \left(\frac{1}{\sigma_{T1}} - \frac{1}{\sigma_{C1}}\right)\sigma_1 + \left(\frac{1}{\sigma_{T2}} - \frac{1}{\sigma_{C2}}\right)\sigma_2 - 1 = 0 \quad (2)$$

For a complete understanding of the topography of the failure surface and especially for the three dimensional states of stress, besides the principal intersections of the elliptic paraboloid with the planes $\sigma_1=0, \sigma_2=0$ or $\sigma_3=0$, whose expressions are analogous to relation (2), other interesting sections are worthwhile studying. Thus, the intersections of the paraboloid either with diagonal planes, expressed by $\sigma_1=\sigma_2, \sigma_2=\sigma_3$ or $\sigma_3=\sigma_1$, or with planes parallel to the deviatoric plane $(\sigma_1+\sigma_2+\sigma_3)=0$ and expressed by $(\sigma_1+\sigma_2+\sigma_3)=k$, where k is a constant, are necessary.

The intersection of the paraboloid by the diagonal plane $\sigma_2=\sigma_3$, which contains the strong principal σ_1-axis, is a parabola whose axis of symmetry is generally parallel to the projection of the hydrostatic axis on this plane. For this particular diagonal plane, containing the σ_1-strong axis, this plane does not contain both the hydrostatic axis and the axis of symmetry of the parabola. In other words, this parabola splits the paraboloid into two unequal and non-symmetric halves.

This parabola, expressed in the plane $(\sigma_1, \bar{\sigma})$, where $\bar{\sigma}=\sqrt{2}\sigma_2=\sqrt{2}\sigma_3$, is given by:

$$\frac{\sigma_1^2}{\sigma_{T1}\sigma_{C1}} + \frac{\bar{\sigma}^2}{2\sigma_{T2}\sigma_{C2}} - \frac{\sqrt{2}\sigma_1\bar{\sigma}}{\sigma_{T1}\sigma_{C1}} + \left(\frac{1}{\sigma_{T1}} - \frac{1}{\sigma_{C1}}\right)\sigma_1 + \sqrt{2}\left(\frac{1}{\sigma_{T2}} - \frac{1}{\sigma_{C2}}\right)\bar{\sigma} - 1 = 0 \quad (3)$$

Introducing the characteristic quantities of the anisotropy of the transtropic material we define the strength differential parameters along the strong direction R_L and the weak plane R_T, as well as the single parameter of anisotropy of the transtropic material R_{LT}. These quantities are expressed by:

$$R_L = \frac{\sigma_{C1}}{\sigma_{T1}}, \quad R_T = \frac{\sigma_{C2}}{\sigma_{T1}} \text{ and } R_{LT} = \frac{\sigma_{T1}}{\sigma_{T2}} \quad (4)$$

With these definitions the distance d, normalized to the strong tensile failure stress, σ_{T1}, between the hydrostatic axis and the axis of symmetry of the elliptic paraboloid for the transtropic material is expressed by:

$$\frac{d}{\sigma_{T1}} = \frac{\sqrt{6}R_L}{9}\left\{\left(1 - \frac{1}{R_L}\right) - R_{LT}\left(1 - \frac{1}{R_T}\right)\right\} \quad (5)$$

which, of course, for isotropic materials, where $R_L=R_T=R$ and $R_{LT}=1$ becomes equal to zero. Indeed, the axis of symmetry of the paraboloid for isotropic materials coincides with the hydrostatic axis.

Finally, the intersections of the symmetric elliptic paraboloid by planes $(\sigma_1+\sigma_2+\sigma_3)=k$ to the hydrostatic axis at distance p from the origin of coordinates are ellipses, which, when projected on the deviatoric plane $(\sigma_1+\sigma_2+\sigma_3)=0$ are expressed by:

$$\sqrt{3}\left(\frac{1}{\sigma_{T1}\sigma_{C1}} + \frac{1}{\sigma_{T2}\sigma_{C2}}\right)x^2 + 3\left(\frac{1}{\sigma_{T1}\sigma_{C1}} - \frac{1}{\sigma_{T2}\sigma_{C2}}\right)xy + \frac{3\sqrt{3}}{2}\left(\frac{1}{\sigma_{T1}\sigma_{C1}}\right)y^2 +$$

$$+ \frac{\sqrt{6}}{2}\left[\left(\frac{1}{\sigma_{T2}} - \frac{1}{\sigma_{C2}}\right)-\left(\frac{1}{\sigma_{T1}} - \frac{1}{\sigma_{C1}}\right)\right]x + \frac{\sqrt{2}}{2}\left[\left(\frac{1}{\sigma_{T2}} - \frac{1}{\sigma_{C2}}\right)-\left(\frac{1}{\sigma_{T1}} - \frac{1}{\sigma_{C1}}\right)\right]y +$$

$$p\left[\left(\frac{1}{\sigma_{T1}} - \frac{1}{\sigma_{C1}}\right)+2\left(\frac{1}{\sigma_{T2}} - \frac{1}{\sigma_{C2}}\right)\right]-\sqrt{3} = 0 \qquad (6)$$

Then, the distance p between the vertex of the symmetric elliptic paraboloid and the origin of the coordinate system $(\sigma_1,\sigma_2,\sigma_3)$ is expressed by:

$$\frac{p}{\sigma_{T1}} = \frac{\sqrt{3}}{\left(1 - \frac{1}{R_L}\right)+2R_{LT}\left(1 - \frac{1}{R_T}\right)} \qquad (7)$$

For an isotropic material relation (8) reduces to the well known relation [5]:

$$\frac{p}{\sigma} = \frac{R}{\sqrt{3}(R-1)} \qquad (8)$$

since $\sigma_{T1}=\sigma_{T2}=\sigma_T, \sigma_{C1}=\sigma_{C2}=\sigma_C, R_L=R$, and $R_{LT}=1$.

3. Application to a Fiber Unidirectional Composite

As a typical example for applying the elliptic paraboloid criterion we use the data known from a graphite-epoxy composite system, which is a transtropic material with the following failure strengths in its principal directions of anisotropy:

$\sigma_{T1} = 1033.50\text{MPa}, \sigma_{C1} = 689\text{MPa}, \sigma_{T2} = 41.34\text{MPa}, \sigma_{C2} = 117.13\text{MPa}$
and: $\qquad (9)$

$$R_L = 0.667, R_T = 2.833 \text{ and } R_{LT} = 25.00 .$$

Figure 1a presents the failure surface for this composite the intersections of this elliptic paraboloid by planes parallel to

the $\sigma_3=0$ plane at distances δ from this plane equal to $k=-8\sigma_{T2}$,
$-6\sigma_{T2},-4\sigma_{T2},-2_{T2},0$ and $2\sigma_{T2}$. The projection of the axis of symmetry
of the paraboloid, which passes through the centers of the elliptic
sections is inclined by an angle of 45^0 to the σ_1- or σ_2-axes. Fig.
1b presents the intersection of the elliptic paraboloid with the
plane $\sigma_3=0$.

Exactly the same patterns we derive from intersections of the
paraboloid with planes parallel to the $\sigma_2=0$ plane for the same
parametric values of δ, since the section of the elliptic paraboloid
by the $\sigma_2=0$ plane is expressed by an identical equation.

All these intersections are ellipses with their centers lying
on the axis of symmetry of the paraboloid. This axis lies parallel to
the diagonal plane $(\sigma_1,\bar{\sigma})$ at a distance

$$d = 3.025\sigma_{T1} = 3126.34 MPa$$

from the hydrostatic axis.
The ellipse derived from the intersection of the elliptic paraboloid
and the deviatoric plane $(\sigma_1+\sigma_2+\sigma_3)=0$ presents the following
characteristics:

The coordinates of its center in the deviatoric plane:

$$\sigma_{1a}/\sigma_{T1} = -\frac{\sqrt{2}R_L}{6}\left\{T_{LT}\left(1 - \frac{1}{R_T}\right)-\left(1 - \frac{1}{R_L}\right)\right\}$$

$$\bar{\sigma}_a/\sigma_{T1} = \frac{\sqrt{6}R_L}{18}\left\{T_{LT}\left(1 - \frac{1}{R_T}\right)-\left(1 - \frac{1}{R_L}\right)\right\}$$ (11)

The major axis of the ellipse subtends an angle θ_a with the
projection of the σ_1-axis on the deviatoric plane, which is
independent of the particular mechanical characteristic properties
of each transtropic material. This angle is always equal to

$$\theta_a = 30^0$$

The coordinate system $(\sigma_{1a},\bar{\sigma}_a)$ is chosen so that the σ_{1a}-axis
corresponds to the projection of the σ_1-principal stress axis on
the deviatoric plane, whereas the $\bar{\sigma}_a$-axis corresponds to the
straight line coinciding with the projections of the σ_2- and σ_3-
axes on the same plane. The positive direction of the $\bar{\sigma}_a$-axis
coincides with the positive direction of the projection of the
σ_3-axis.

Then, from the trihedron formed by the deviatoric plane, the
(σ_1,σ_{1a})-plane and the plane defined by the σ_1-axis and the major
axis of the elliptic section, it is easy to define the angle θ_a

311

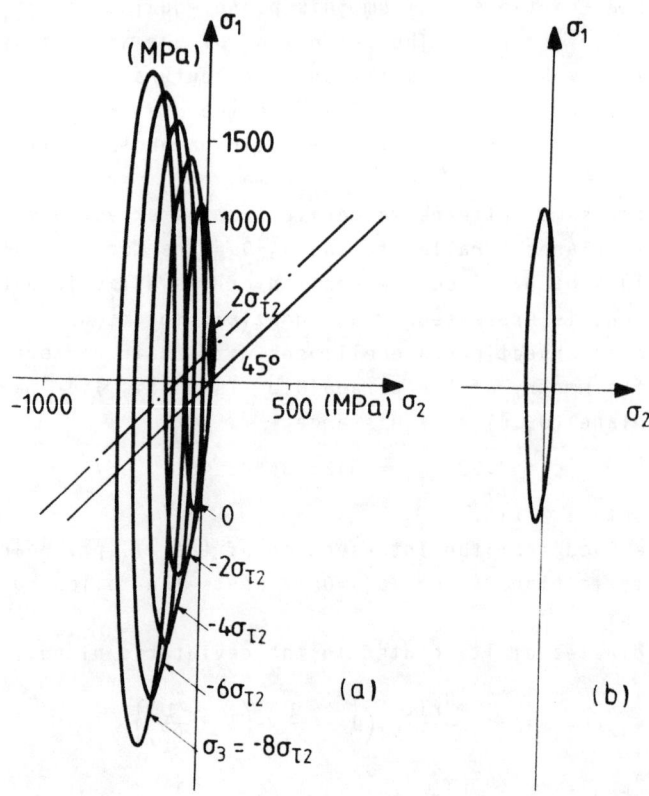

Fig. 1

(a) Intersections of the elliptic paraboloid for a graphite-
     epoxy composite by σ_3=constant planes.
(b) The failure locus in the (σ_3=0) plane.

subtended by the major axis of the ellipse and the σ_1-axis which
again is a universal constant for all transtropic materials and
equal to

$$\theta_d = 52^\circ 12' \tag{13}$$

The coordinates σ_{1a} and $\bar{\sigma}_a$ of the ellipse on the deviatoric plane
for the composite defined by the quantities (9) are as follows:

$$\sigma_{1a} = -2.618\sigma_{T1} = -2705.70 \text{MPa}$$
$$\bar{\sigma}_a = 1.517\sigma_{T1} = 1567.82 \text{MPa} \tag{14}$$

The elliptic intersection indicated in Fig.1b is expressed by:

$$\frac{1}{\sigma_{T1}\sigma_{C1}}\sigma_1^2 + \frac{1}{\sigma_{T2}\sigma_{C2}}\sigma_2^2 - \frac{1}{\sigma_{T1}\sigma_{C1}}\sigma_1\sigma_2 + \left(\frac{1}{\sigma_{T1}} - \frac{1}{\sigma_{C1}}\right)\sigma_1 + \left(\frac{1}{\sigma_{T2}} - \frac{1}{\sigma_{C2}}\right)\sigma_2 - 1 = 0$$

The coordinates of its center and the inclination of its major axis to the σ_1-axis are expressed by:

$$\frac{1}{\sigma_{T1}}(x_c,y_c) = \left\{ -R_L T_{LT} \frac{\left[\frac{1}{R_L}\left(1 - \frac{1}{R_T}\right) + \frac{2T_{LT}}{R_T}\left(1 - \frac{1}{R_L}\right)\right]}{\left(\frac{4T_{LT}^2}{R_T} - \frac{1}{R_L}\right)}, \right.$$
$$\left. -\frac{\left[\left(1 - \frac{1}{R_L}\right) + 2T_{LT}\left(1 - \frac{1}{R_T}\right)\right]}{\left(\frac{4T_{LT}^2}{R_T} - \frac{1}{R_L}\right)} \right\} \tag{15}$$

and the angle θ_c subtended by the major axis of the ellipse and the σ_1-axis is given by:

$$\theta_c = \frac{1}{2}\tan^{-1}\left[\frac{1}{(R_L T_{LR}^2/R_T - 1)}\right] \tag{16}$$

The coordinates $x_c y_c$ and the angle θ_c for the ellipse of Fig.1b are evaluated as follows:

$$x_c = 0.147\sigma_{T1} \qquad y_c = -0.036\sigma_{T1} \qquad \theta_c = \frac{1}{2}\tan^{-1}(0.00684)$$

The angle θ_c is less than 10 minutes, that is the major axis of the ellipse is approximately parallel to the σ_1-axis of the principal stress space.

From Eq.(6) it can be readily also derived that the distance between the vertex of the elliptic paraboloid and the origin of the coordinate system $(\sigma_1,\sigma_2,\sigma_3)$ is expressed by:

$$p = \frac{\sqrt{3}}{\left(\frac{1}{\sigma_{T1}} - \frac{1}{\sigma_{C1}}\right) + 2\left(\frac{1}{\sigma_{T2}} - \frac{1}{\sigma_{C2}}\right)} \tag{17}$$

It is obvious from this relation that for transtropic materials, which are more and more anisotropic, and especially as their strength differential parameters R_L and R_T take higher values, the distances of the vertices of their failure surfaces approach closer and closer the origin of the coordinates. Then, the elliptic paraboloids become more and more shallow.

On the contrary, for low anisotropies and small strength differential parameters the elliptic paraboloids have their vertices receding to infinity, their distances from the hydrostatic axis are reduced and their elliptic shapes tend to circular ones, having as a general limit the cylindrical surface defined from the Mises yield criterion valid for isotropic materials without and strength differential effect.

Fig.1a indicates this phenomenon presenting sections of a strongly anisotropic material, where the values of strength differential parameters R_L and R_T recede strongly from unity the one being below and the other above this limiting value.

It may be readily derived that as these values recede from unity and as the coefficient of anisotropy of the material R_{LT} is increasing the shape of the elliptic paraboloid becomes more and more oblong, so that its elliptic sections by the principal stress planes or by the parallel to the deviatoric planes take the shapes of "cigars".

Moreover, from the previously derived relations it may be argued that the term $\{T_{LR}(1 - 1/R_T)-(1 - 1/R_L)\}$, or similar to this term expressions, is critical to the properties of the material. It may be established that the part depending on the strong strength differential parameter $(1 - 1/R_L)$ contributes only a few percent to the characteristic dimensions of the failure surface. For instance,for the graphite-epoxy composite this term contributes only 3 percent and may be eliminated without introducing great errors.

Therefore,it may be concluded that as the parameter of anisotropy is increasing the other important characteristic quantity defining the failure surface of a composite is its transverse (weak) strength differential parameter.

Figure 2 presents the intersection of the elliptic paraboloid by the diagonal $(\sigma_1,\bar{\sigma})$-plane for a KEVLAR 49 (u_f=0.60), which at room temperature has the following mechanical properties σ_{T1}=1379MPa R_L=0.20 R_T=4.662 and R_{LT}=46.587 [8].

The distance d/σ_{T1} of the axis of symmetry of this elliptic paraboloid from the origin of coordinates is given by
$$d = 2.191\sigma_{T1} = 3021.39\text{MPa}$$
Again, for this composite, whose strength differential parameters R_L and R_T deviate strongly from unity on both sides, the paraboloid, whose section along the diagonal $(\sigma_1,\bar{\sigma})$-plane is given in Fig.2, is again a shallow one.

Figure 3 shows another aspect of the shape of the paraboloid by presenting sections of the elliptic paraboloid for the graphite-epoxy composite by planes parallel to the deviatoric plane for parametric values of $k=-\sigma_{C2}$,0 and σ_{C2}. Again the presented ellipses have the shape of cigars. It is worthwhile indicating again that this paraboloid has its vertex on the tensile-tensile-tensile octant since the ellipse for $k=\sigma_{C2}$ lies incide the other two ellipses.

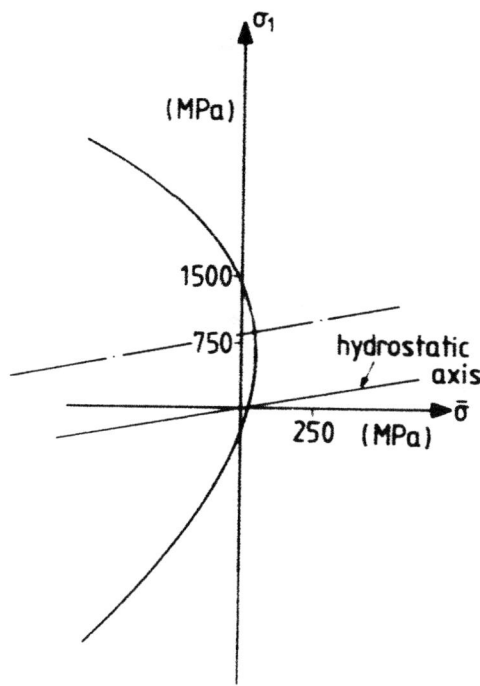

Fig. 2

Intersection of the elliptic paraboloid failure surface on the diagonal $(\sigma_2=\sigma_3)$-plane for KEVLAR 49 $(\upsilon_f=0.60)$ transtropic composite

Finally, Fig.4 presents the variation of the distances p of the vertices of the paraboloid and the distances d between their axes and the hydrostatic axis, normalized to the failure strength σ_{T1} of the material versus the parameter of anisotropy T_{LT} for a transtropic material with $R_L=0.667$ and $R_T=1.60$. It is clear from this figure, again, that while the distance d increases linearly with the parameter of anisotropy T_{LT}, the distance p reduces very rapidly as T_{LT} is increasing.

4. Conclusions

In this paper it has been shown that the failure surface for anisotropic materials, and especially for transtropic materials, is an elliptic paraboloid surface whose axis of symmetry is parallel to the hydrostatic axis.

The intersections of this surface by planes normal to the axis of symmetry are ellipses of different ellipticity and orientation. For the transtropic materials only the orientation of the major

315

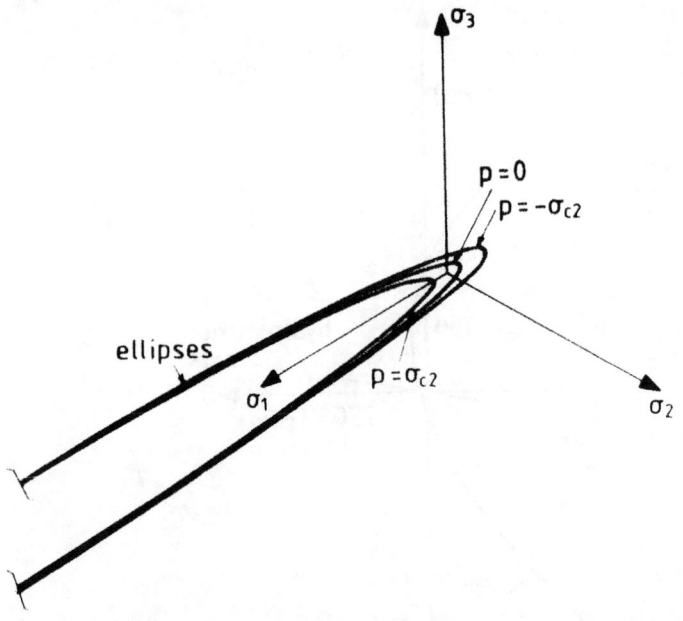

Fig. 3

Intersections of the elliptic paraboloid failure surface for the
graphite-epoxy composite by planes parallel to deviatoric plane
$(\sigma_1+\sigma_2+\sigma_3)=0$

axes of all such ellipses is constant and equal to $\theta_a=30^o$ to the
projection of the strong principal axis on the deviatoric plane.

The distance, d, between the hydrostatic and the axis of
symmetry of the paraboloids increases with increasing anisotropy.
Similarly, the ellipticity of the intersections of the paraboloid
increases and they take the form of cigars, as the strength
differential parameters recede from unity and the parameter of
anisotropy is increasing.

The critical quantities influencing the shape and position of
the elliptic paraboloid are mainly the parameter of anisotropy and
the transverse strength differential parameter.

Higher anisotropy yields shallow paraboloids, whereas weak
anisotropy results in oblong shapes along the hydrostatic direction.

References
1 R. Hill, A Theory of the Yielding and Plastic Flow of Anisotropic
    Metals", Proc. Roy. Soc. Lond., Ser.A, 193, pp.281-297 (1948).
2 R. . Mises, "Mechanik der plastischen Formänderung von

316

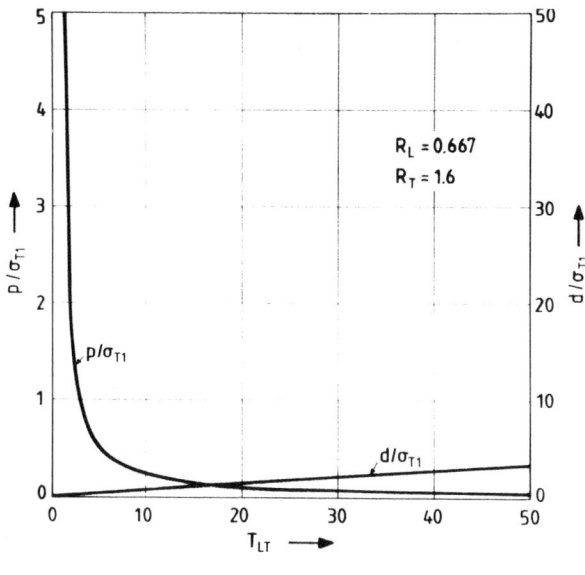

Fig. 4

The variation of the distances p and d of the vertices of the paraboloids from the deviatoric plane and the axes of the paraboloids from the hydrostatic axis respectively, normalized to the failure strength σ_{T1} versus the parameter of anisotropy T_{LT} for particular values of the strength differential parameters R_L and R_T

Kristallen", Zeit. ang. Math. and Mech., **8**, pp.161-185 (1928).

3 O. Hoffman, The Brittle Strength of Orthotropic Materials", Jnl. Comp. Mat., **1**, pp.200-206 (1967).

4 S.W. Tsai and E.M. Wu, "A General Theory of Strength for Anisotropic Materials", **5**, pp.58-80 (1971).

5 P.S. Theocaris, "Generalized Failure Criteria in the Principal Stress Space", Theoret. and Appl. Mech., Bulgarian Academy of Sciences, Vol. , No.2, pp. (1987).

6 P.S. Theocaris, "Failure Characterization of Anisotropic Materials by Means of the Elliptic Paraboloid Failure Criterion", Uspechi Mechanikii (Advances of Mechanics), 10 No.3, pp.75-109 (1987).

7 P.S. Theocaris and Th. Philippidis, "The Paraboloidal Limiting Surface of Initially Anisotropic Elastic Solids", Jnl. of Reinf. Plastics and Comp. (submitted for publication (1986)).

8 R. Narayanaswami and H.M. Adelman, "Evaluation of the Tensor Polynomial and Hoffman Strength Theories for Composite Materials", Jnl. Comp. Mat., **11**, pp.366-377 (1977)

IDENTIFICATION EXPERIMENT FOR TRACING THE CAUSE OF THE FAILURE OF A STONE CRASHER

Doc.Dr. Frigyes Thamm
Technical University Budapest Department of Engineering Mechanics
H - 1521 Budapest, Hungary

The crushing force F_{max} which led to the failure of the drum of a stone crusher was examined by a photoelastic experiment, which yielded the stress concentration factor of a mixed-mode crack which was experienced on the drum. The reduced value K/F_{max} of the s.i.f. compared with the fracture toughness of the material of the drum made a calculation of F_{max} possible, thus revealing unconvenient operatig conditions of the crusher.

Keywords: fracture mechanics, photoelasticity

The failure of the central drum of a stone crusher is shown in Fig.1. Aim of the investigation was the estimation of the value of the force which led to the failure. A photoelastic investigation was carried

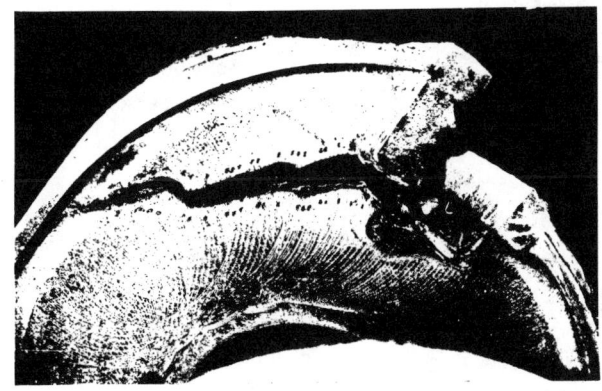

FIG.1   The cracked drum

out on a model of 1:4 scale, the dimensions of which are shown in Fig.2. At first isochromatic and isoclinic fringe patterns were taken from the model without crack /Figs. 3 and 4/ A part of the crack was approximated by the broken line OBC shown in Fig.5. From Figs. 3 and 4 the stress components σ_x, σ_y and τ_{xy} were evaluated

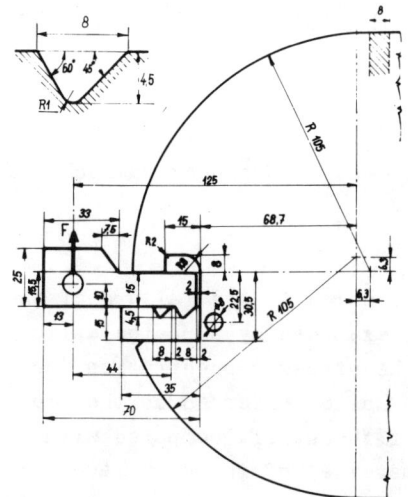

FIG.2 Dimensions of the plane photoelastic model

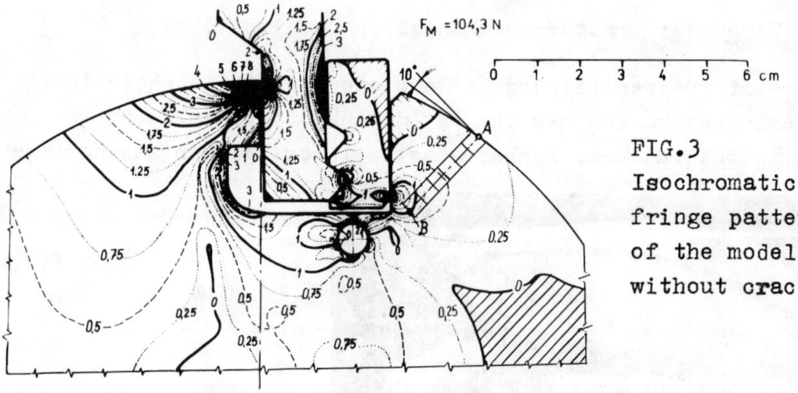

$F_M = 104,3$ N

FIG.3
Isochromatic
fringe pattern
of the model
without crack

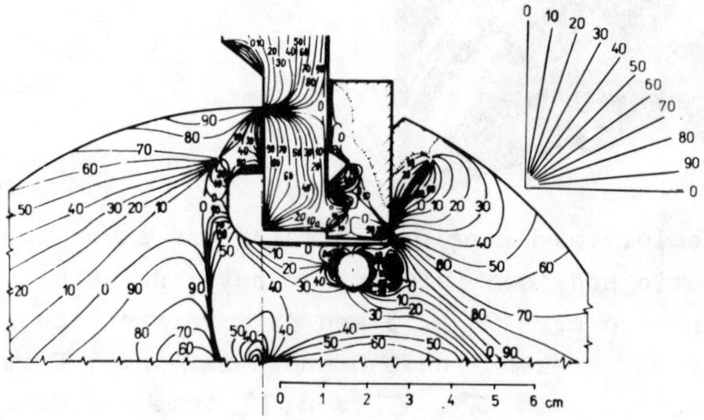

FIG.4 Isoclinic
pattern of the
model without
crack

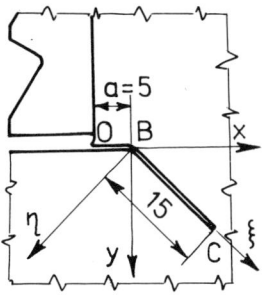

FIG.5 The investigated crack-
path OBC and the notation of
the coordinate-systems used

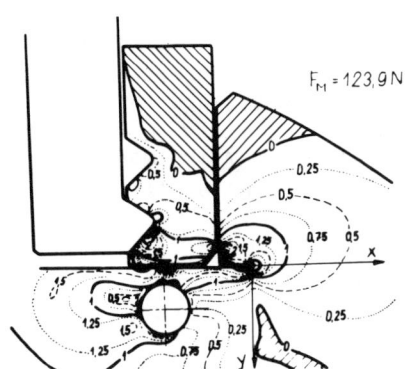

$F_M = 123,9 N$

FIG.6 Isochromatic fringe
pattern of the model with a
crack along OB

Isochromatic fringe pattern
of the model with full crack
length

FIG.7

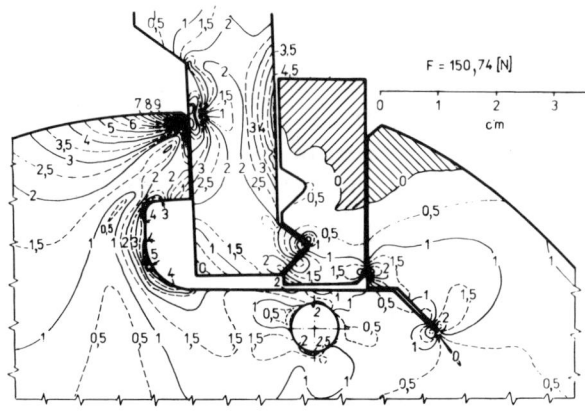

$F = 150,74$ [N]

for pont B and σ_ξ ; σ_η and $\tau_{\xi\eta}$ for pont C by the shear difference method, the path of which is shown in FIG.3. Now the crack was produced by cutting the model with a fret-saw in two steps. The fringe-patterns for crack-tips at B and C respectively are shown in FIGS. 6 and 7.

In the first case the shape of the fringe-pattern exhibited mixed cracking mode with two different stress intensity factors /s.i.f./ K_I and K_{II}. These could be obtained by the tangent method from the isochromatic fringe patterns of FIG.6 by plotting the values of the fringe orders m along the y-axis over $1/\sqrt{2\pi y}$ as shown in FIG.8, respectively plotting $m^2 \cdot 2\pi x$ along the x-axis as shown in FIG.9. The slope of the straight line drawn over the measured points in FIG.8. produced the value of K_I, while that drawn over the points in FIG.9. extrapolated for x = 0 yielded K_{II}^2. [1].

The transition of the measured s.i.f.-s from the model /subscript $_M$/ to the prototype /subscript $_P$/ could be maintained by the law of similarity using following formula

$$\frac{K_P}{K_M} = \left(\frac{a_P}{a_M}\right)^{\frac{1}{2}} \frac{l_M v_M}{l_P v_P} \frac{F_P}{F_M} = \left(\frac{l_M}{l_P}\right)^{\frac{1}{2}} \cdot \frac{v_M}{v_P} \frac{F_P}{F_M} \tag{1}$$

Denoting: F: the loading force; a: the crack length; v: the plate thickness and l: a characteristic dimension in the plane of the plate.

According to [2] the comparison of the two s.i.f.-s of a mixed mode crack with the value of K_{Ic} of the material of the prototype obtained tained by a standerdized materials test can be performed by the reduced s.i.f. K_{IR} which can be obtained in the following way.

The angle of the direction of the propagation of the crack is

$$\sin \vartheta_1 = \frac{M}{1 + 9M}\left[1 - 3\sqrt{1 + 8M^2}\right] \tag{2a}$$

with $M = K_{II}/K_I$

and finally

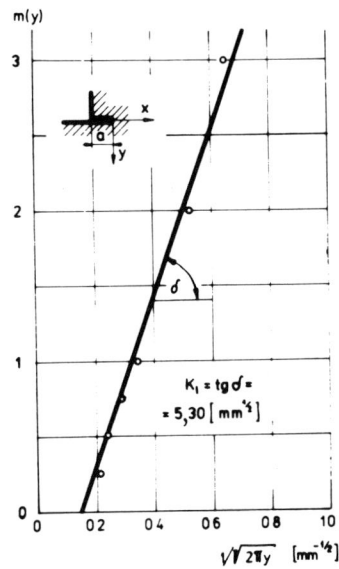

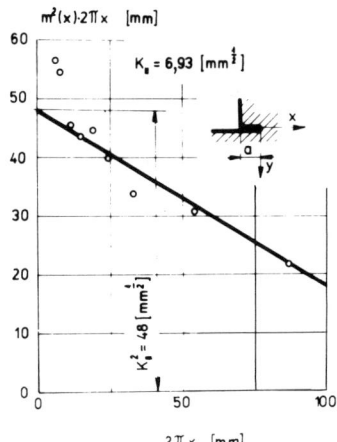

FIG.9 Extrapolation method of obtaining K_{II} from photoelastic data

FIG.8 Tangent method for obtaining K_I from photo-elastic data

$$K_{IR} = \frac{K_I}{2} \cos \frac{\vartheta_1}{2} \left[1 + \cos \vartheta_1 - 3M \sin \vartheta_1 \right] \qquad (2b)$$

The crack becomes unstable if $K_{IR} = K_{Ic}$.

For the carbon steel used for the protoype $K_{Ic} = 4350$ $Nmm^{-3/2}$ was found.

In the case of the crack-tip at point C /FIG.5 / the fringe pattern /FIG.7 / exhibited a practically pure mode I stress distribution. The s.i.f. K_I could be obtained by a plot of fringe values along the line going through the crack-tip parallel to the η -axis similar to FIG.8 and after the transition to the prototype could be directly compared with K_{Ic}.

During the hit which caused the failure, the hammer of the crusher suffered plastic deformation, the size and shape of which also permitted an estimation of the peak force, thus giving an addi-

323

tional control to the presented investigation. The values of
the peak force are collected in TABLE I. It should be noted,
that the hit occurred in the middle between two discs of the
drum and so the cracked disc suffered only half of the value
of the force. The values obtained by fracture-mechanics approach and from the investigation of the plastic deformation show a
fairly good agreement, sufficient for the aim of the investigation.

TABLE.I
CALCULATED PEAK-FORCE CAUSING THE FAILURE OF THE DRUM

		F_{max} [kN]
Calculated from the damage of the hammer based on the theory of plasticity		980
Calculated from the photo- elastic investigation	Point. B.	1007
	Point. C.	1358

References

1 THAMM,F.: Experimental determination of the stress intensity factor /in Hungarian/ GÉP XXXII/1980/ pp. 338-345

2 FISCHER,K.F.-GÜNTHER,W.: Gegenwärtiger Stand der Rissbruch-
  mechanik in Hinblick auf eine praktische Nutzung
  Maschinenbautechnik 2/1978/ pp.73-76

SOME REMARKS ON THE DETERMINATION OF STRESS INTENSITY FACTORS

Prof. Dr. Stjepan Jecić - Damir Semenski
University of Zagreb
Faculty of mechanical engineering and naval architecture
41000 ZAGREB, Đure Salaja 5, Yugoslavia

This paper deals with direct determination of the stress intensity factor from a diagram of stress concentration factors which is compared with the method of caustics. The possibility of experimental determination of the stress intensity factor in mechanically anisotropic materials is considered and discussed.

Keywords: stress intensity factors, stress concentration factors, anisotropic materials

1. Introduction

In linear elastic and mechanically isotropic materials, the stress distribution in the vicinity of the crack is given by Irwin-Williams formulae as well   known in literature. Stress intensity factors in those relations must be determined. Manogg first shows the method of caustics as a method for evaluating stress intensity factor [1]. Richard introduces factor k for calculating stress intensity factor K by means of linear characteristics of stress concentration factors [2]. The direct method without k is also given in literature [3, 4]. The extension of those methods in mechanically anisotropic materials is considered and it is discussed as one possibility.

2. Stress intensity factors given by application of stress concentration factors

Stress concentration factors are defined as a ratio of maximum stress σ_{max} and nominal stress σ_n at point:

$$\alpha = \frac{\sigma_{max}}{\sigma_n} \qquad (1)$$

The investigations leads to a linear connection between stress concentration factor and the parameter $\sqrt{a/\rho}$ (fig. 1) described in literature [3, 4].

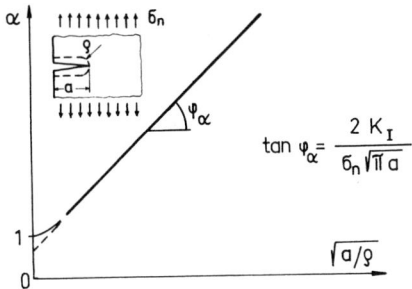

$$\tan \varphi_\alpha = \frac{2 K_I}{\sigma_n \sqrt{\pi a}}$$

Fig. 1   Connection between stress concentration factor and stress intensity factor in infinite plate with cut

The connection between the stress concentration factor and the stress intensity factor is given according to Richard [2], and in some modification [3, 4] for the cut with the small radius ϱ leads to:

$$K_I \approx 0,5 \tan \psi_\alpha \, \sigma_n \sqrt{\pi a} \qquad (2)$$

Experiments with the specimens of araldite B with a cut with radius ϱ, lead to the linear characteristic of stress concentration factor α given in the diagram (fig. 2).

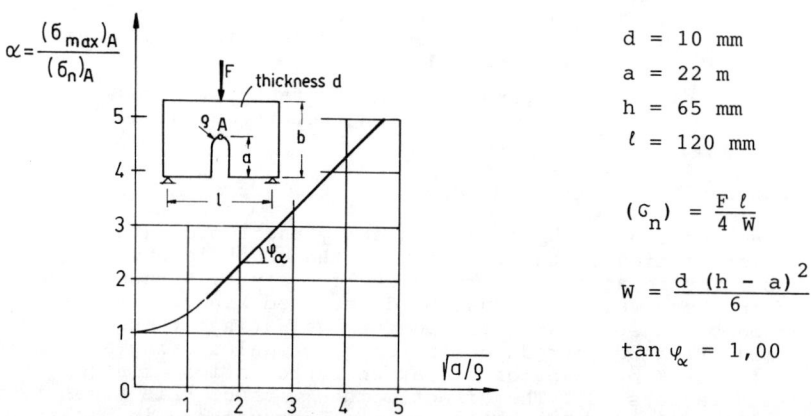

$$\alpha = \frac{(\sigma_{max})_A}{(\sigma_n)_A}$$

thickness d

$d = 10$ mm
$a = 22$ m
$h = 65$ mm
$\ell = 120$ mm

$$(\sigma_n) = \frac{F \ell}{4 \ W}$$

$$W = \frac{d \ (h - a)^2}{6}$$

$$\tan \psi_\alpha = 1,00$$

Fig. 2 Results of stress concentration factor for a specimen with a cut with radius ϱ

The stress intensity factor K_I calculated from $\tan \psi_\alpha$ is for a particular case ($F = 350$ N)

$$K_I = 0,448 \ MN/m^{3/2}$$

which follows directly from ψ_α and relation (2). For every particular case that diagram $\alpha = f \ (\sqrt{a/\varrho})$ exists, stress intensity factors K_I, K_{II} or K_{III} can be calculated by this method.

3. Stress intensity factors given by caustics

When a specimen containing a crack is illuminated with a light by a point light source, the light transmitted through the model will be deflected outwards which is caused by a reduction of the thickness of the specimen and the refractive index of the material, which is treated by Manogg [1]. This appearance results in a dark spot in a form of epycycloid on the shadow image plane which is called caustics. The stress intensity factor can be determined from the maximum diameter of caustics (e.g. [6])

$$K_I = \frac{2 \ \sqrt{2 \ \pi}}{3 \ m^{3/2} \ f^{5/2} \ c \ d \ z_o} \ D^{5/2} \qquad (3)$$

where f, and c are shadow optical constants, d is the thickness of the specimen; D is the diameter of caustics and m is a scale factor.

For optically anisotropic materials, double-caustics will be developed. The formula (3) takes the form:

$$K_I = \frac{2 \sqrt{2 \pi}}{3 \, m^{3/2} \, f_{o,i}^{5/2} \, c \, d \, z_o} \, D_{o,i}^{5/2} \qquad (4)$$

where $D_{o,i}$ is the outer (inner) diameter of caustic.

This method can be used for the specimens of transparent materials as a transmission method and for the specimens of non-transparent materials as a reflection method.

The result of linearization of stress concentration factor is compared with the result for caustics on a specimen of araldite B with a crack tip (a=22 mm)(fig. 3), and having the same dimensions. Comparing parameter

$$\frac{2 \, K_I}{\sigma_n \sqrt{\pi a}} = 0,98$$

given by caustics and tan φ_α = 1,00 given by linearization of stress concentration shows a good agreement. Expressed by stress intensity factor (for F = 350 N, K_I = 0,439 MN/m3/2 given by caustics) it shows 2% differences.

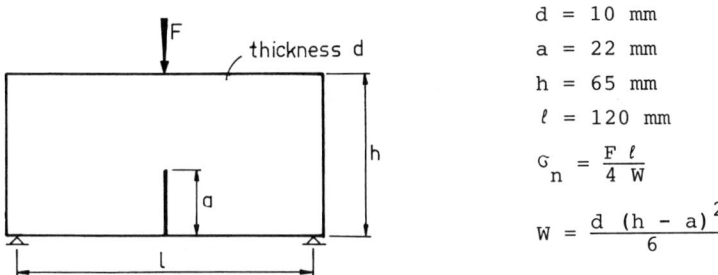

d = 10 mm
a = 22 mm
h = 65 mm
ℓ = 120 mm

$$\sigma_n = \frac{F \, \ell}{4 \, W}$$

$$W = \frac{d \, (h - a)^2}{6}$$

Fig. 3   Specimen with a crack tip

4. One posibility of extension to anisotropic materials

The stress distribution in the vicinity of the crack tip in mechanically anisotropic materials is given by the theory of elasticity [5]. For the mode I loading conditions, stress distribution is given by the following relations:

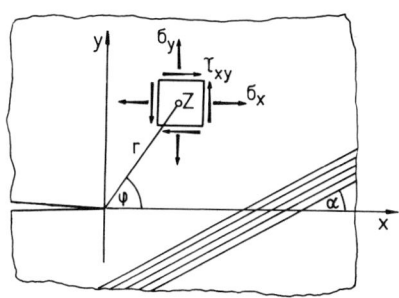

Fig. 4   Definition of stress components σ_x, σ_y, τ_{xy} in the vicinity of the crack in mechanically anisotropic materials

$$\sigma_x(r,\varphi) = \frac{K_I}{\sqrt{2r}} \, Re \, \{\frac{\mu_1 \mu_2}{\mu_1-\mu_2} \, (\frac{\mu_2}{\sqrt{\cos\varphi + \mu_2\sin\varphi}} - \frac{\mu_1}{\sqrt{\cos\varphi + \mu_1\sin\varphi}})\} + \ldots$$

$$\sigma_y(r,\varphi) = \frac{K_I}{\sqrt{2r}} \, Re \, \{\frac{1}{\mu_1-\mu_2} (\frac{\mu_1}{\sqrt{\cos\varphi + \mu_2\sin\varphi}} - \frac{\mu_2}{\sqrt{\cos\varphi + \mu_1\sin\varphi}})\} + \ldots$$

(5)

$$\tau_{xy}(r,\varphi) = \frac{K_I}{\sqrt{2r}} \, Re \, \{\frac{\mu_1 \mu_2}{\mu_1-\mu_2} \, (\frac{1}{\sqrt{\cos\varphi + \mu_1\sin\varphi}} - \frac{1}{\sqrt{\cos\varphi + \mu_2\sin\varphi}})\} + \ldots$$

with μ_1 and μ_2 as complex roots of the characteristic equation that depend on elastic properties of a material.

The problem is to evaluate the stress intensity factor in mechanically anisotropic materials. By analogy with isotropic materials, the method of stress concentration factor will perhaps lead to the stress intensity factor value.

Another possibility is the method of caustics applied to original anisotropic materials. The basic equations according to [1, 6] for the method of caustics lead to (fig. 5):

$$\vec{r}' = m \, \vec{r} + c \, d \, z_0 \, grad \, [\sigma_1 - \sigma_2 \pm \lambda(\sigma_1 - \sigma_2)] \qquad (6)$$

where λ is the constant that describes the value of the optical anisotropy. For optically isotropic materials $\lambda = 0$. Also, for reflection method $\lambda = 0$.

Using $(\sigma_x + \sigma_y = \sigma_1 + \sigma_2)$ and $\lambda = 0$ (e.g. reflection method), the formula (6) calculated for mechanically anisotropic materials, leads to the following relations:

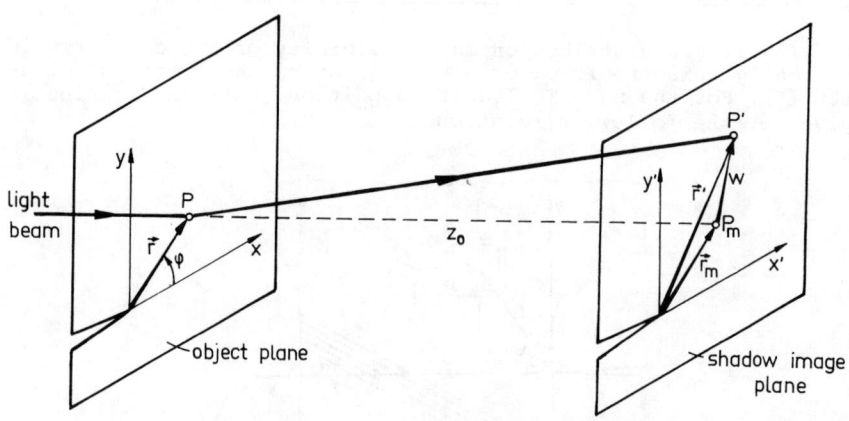

Fig. 5   Geometrical conditions of the shadow optical analysis

328

$$\vec{r}' = m\,\vec{r} + cd\,z_0\,\text{grad}\,(\sigma_x + \sigma_y)$$

$$(7)$$

$$\vec{r}' = m\,\vec{r} - cd\,z_0\,\frac{K_I}{2\sqrt{2}}\,r^{-3/2}\,Re\,\{\,[L_1(\varphi)]\,\vec{e}_r + [L_2(\varphi)]\,\vec{e}_\varphi\,\}$$

with:

$$L_1(\varphi) = \frac{1}{\mu_1 - \mu_2}\left[\frac{\mu_1\,(\mu_2 + 1)}{\sqrt{\cos\varphi + \mu_2\sin\varphi}} - \frac{\mu_2\,(\mu_1^2 + 1)}{\sqrt{\cos\varphi + \mu_1\sin\varphi}}\right]$$

$$L_2(\varphi) = \frac{1}{\mu_1 - \mu_2}\left[\frac{\mu_2\,(\mu_2^2 + 1)}{\sqrt{(\cos\varphi + \mu_1\sin\varphi)^3}}\,(\sin\varphi - \mu_1\cos\varphi) - \right. \qquad (8)$$

$$\left. - \frac{\mu_1\,(\mu_2^2 + 1)}{\sqrt{(\cos\varphi + \mu_2\sin\varphi)^3}}\,(\sin\varphi - \mu_2\cos\varphi)\right]$$

In relation (7), the function of $\vec{r}'$ represents a curve whose configuration is extremely dependent on the type of material. The problem is how to find one parameter of a curve that depends on factor K_I. The question is: which is that point or radius or diameter which is common to all materials, or does it depend on a concrete material? The measure like diameter of caustics has not been found yet.

5. References

1   Manogg, P.: Anwendung der Schattenoptik zur Untersuchung des Zerreissvorgangs von Platten. Dissertation, Freiburg, Germany, 1964.

2   Richard, H. A.: Ermittlung von Spannungsintensitätsfaktoren aus spannungsoptisch bestimmten Kerbfaktor-und Kerbspannungs- diagrammen. Forsch. Ing. - Wes. 45 (1979) Nr. 6, S. 188-199.

3   Ficker, E.-Jecić, St.: Beitrag zur Darstellung von Formzahlen. VDI-Ber. Nr. 271, S. 47-51. Düsseldorf: VDI- Verlag 1976.

4   Ficker, E.-Jecić, St.-Daffner, E.: Zur Bestimmung der Spannungen bei kleinen Krümmungsradien. Konstruktion 36 (1984) H. 1, S. 7-12.

5   Löbel, G.-Zeilinger, H.-Deska, R.: Theoretische Grundlagen für die Anwendung der Bruchmechanik auf Faserverbundwerks- toffe. Z. Werkstofftech. 15 (1984), S. 277-287.

6   Kalthoff, J. F.: Stress intensity factor determination by caustics. Proc. 1982. Spring Meeting, Hawaii

Special applications of strain analysis
in the mechanical and civil engineering
practice

EXPERIMENTAL DETERMINATION OF THE ARBITRARY CONSTANTS
IN THE SOLUTION FOR OVER-DETERMINATE SHALLOW
SPHERICAL SHELLS AT EARLY STAGE

Assoc. Prof. Dr.-Ing. Sameh S. Issa
Department of Civil Engineering, Kuwait University, P.O. Box 5969
Al-Safat, 13060 SAFAT, KUWAIT

An approach to the experimental determination of the arbitrary constants
that appear in the exact solution of axisymmetrically and partially
loaded shallow spherical shells is presented. The general solution of
these shells is over determinate and contains deflection and stress functions. An algorithm for calculating the arbitrary constants at early
stage is proposed. The deflected shape of shallow spherical shells with
large characteristic length is measured at different positions and the
arbitrary constants are determined by direct substitution in the deflection function W.

Keywords: shallow spherical shells, charactaristic length, deflection
function, over determinate analysis

1. Introduction

Shallow spherical shells subjected to axisymmetrical loads have received
considerable attention [1-6]. The characteristic length which is defined by ℓ
$= \sqrt{R^2 t^2 / [12(1-v^2)]}$, greatly affects, in case of finite shells, the course of the
exact solution. It is noted that R is the radius of the shallow spherical segment,
t is the thickness of the shell's wall and v is the Poisson's ratio of the material
used. The exact solution [1] contains deflection W, and stress F, functions that
are written in terms of Bessel-Kelvin functions whose arguement is given by

$$\lambda = a/\ell \qquad (1)$$

in which a is the normal distance from the axis of symmetry. For values of λ
greater than 6.8, the Bessel-Kelvin functions get closer either to zero or infinity. This characteristic of the Bessel-Kelvin functions along with the available
boundary conditions, lead to the elimintation of some of the invoked constants.
Thus, the problem is simplified and the exact solution presented by Reissner in [1]

is applicable for several cases. However, for large characteristic length, finite shallow spherical shells and axisymmetrical loads of circular shape acting partially, the exact solution becomes clumsy and complicated. Furthermore, when hinged boundaries are induced, an over determinate problem is present. This problem has been treated, from different aspects, by Issa in [6] and [7].

The arbitrary constants of the exact solution C1 through C5 appear in the normal displacement function W that is given by

$$W = C1 \text{ ber } \lambda + C2 \text{ bei } \lambda + C3 \text{ ker } \lambda + C4 \text{ kei } \lambda + C5 \quad . \tag{2}$$

The present work investigates the possibility of determining the arbitrary constants experimentally at very early stage, before developing the expressions for surface stresses. Normal deflections at five different, prior defined locations, are measured, then equ. (2) is used to determine the constants C1 through C5. Thus, the problem of over determination is bypassed. The results might be generalized through normalization.

For the sake of comparison, the so experimentally conducted surface stresses, in a pilot work, along with those which are obtained from a theoretical method cited in the literature [2] are illustrated in a diagram.

2. Experimental Work

Acrylic sheet of approximately 2 mm thick was used in producing shallow spherical segment that is 180 mm in base-radius and 14.8 mm in rise. Heating and vaccuming process, in the sense described by Issa in [5] was used in forming the model. Displacement transducers (DT) of the inductive type were arranged as shown in Fig. 1.

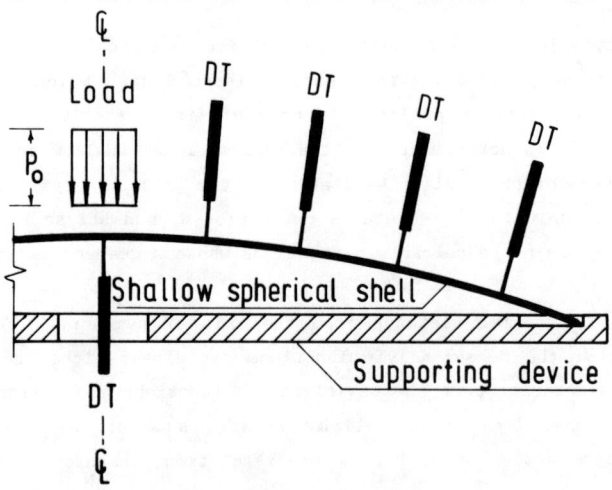

Fig. 1   Orientation of displacement transducers (DT)

At least five DT should be used. The implementation of DT removes the problems accociated with the measurements of strains on acrylic surfaces by means of electrical resistance strain gauges. Beside their dissipation of heat which affects the mechanical and physical properties of the acrylic material, strain gauges have strengthening effect on the rigidity of the tested object. This effect becomes relatively high, if strain gauges are fixed on a thin wall out of acrylic material.

Any DT whose selected location might be covered with loads, could be removed to the other side of the shell (see Fig. 1). The displacement transducers are pointed towards the center of the sphere whose segment is under consideration, in order to measure normal deflections directly.

3. Algorithm

The normal displacements are measured on a physical model that is subjected to the above mentioned loading system and is held by a supporting device which simulates hinged conditions. Consequently, the effects of the regularity, continuity and boundary conditions are implicitly included in the measured values of W. The arguements λ_i, for $i = 1,2,...5$, of the Bessel-Kelvin functions are obtained from the geometrical and mechanical properties of the model as well as from the location of each individual measuring station. Now the experimentally obtained results and the arbitrary constants are linked in a system of simultaneous equations as follows:

$$\begin{bmatrix} \text{ber } \lambda_1 & \text{bei } \lambda_1 & \text{ker } \lambda_1 & \text{kei } \lambda_1 & 1 \\ \text{ber } \lambda_2 & \text{bei } \lambda_2 & \text{ker } \lambda_2 & \text{kei } \lambda_2 & 1 \\ \text{ber } \lambda_3 & \text{bei } \lambda_3 & \text{ker } \lambda_3 & \text{kei } \lambda_3 & 1 \\ \text{ber } \lambda_4 & \text{bei } \lambda_4 & \text{ker } \lambda_4 & \text{kei } \lambda_4 & 1 \\ \text{ber } \lambda_5 & \text{bei } \lambda_5 & \text{ker } \lambda_5 & \text{kei } \lambda_5 & 1 \end{bmatrix} \begin{bmatrix} C1 \\ C2 \\ C3 \\ C4 \\ C4 \end{bmatrix} = \begin{bmatrix} W_1 \\ W_2 \\ W_3 \\ W_4 \\ W_5 \end{bmatrix} \qquad (3)$$

Equations (3) are to be solved for C1 through C5. The stress function F is expressed in terms of the same arbitrary constants,

$$F_i = (Et^2/\sqrt{12(1-\nu^2)})[C1 \text{ bei } \lambda_i - C2 \text{ ber } \lambda_i + C3 \text{ kei } \lambda_i \\ - C4 \text{ ker } \lambda_i + C6 \text{ } \ell n \text{ } \lambda_i] \text{ , for } i = 1,2,...5 \qquad (4)$$

in which E is the modulus of elasticity.

The arbitrary constant C6 does not appear in equ. (2). However, it can be obtained directly from the static condition $\sum V = 0$ as follows:

$$C_6 = -(PR/2\pi Et^2)\sqrt{12(1-\nu^2)} \qquad (5)$$

335

in which $P = P_0 \pi C^2$; P_0 is the intensity of the uniform load; C is the radius of the loaded area. Then, the appropriate formulations for the membrane and coupled forces are derived using the relations

$$N_r = \frac{1}{\ell^2} \frac{1}{\lambda} \frac{\partial F}{\partial \lambda} \qquad , \qquad N_\theta = \frac{1}{\ell^2} \frac{\partial^2 F}{\partial \lambda^2} \qquad ,$$

$$M_r = \frac{-D}{\ell^2} (\frac{\partial^2 W}{\partial \lambda^2} + \frac{\nu}{\lambda} \frac{\partial W}{\partial \lambda}) \qquad \& \qquad M_\theta = \frac{-D}{\ell^2} (\frac{1}{\lambda} \frac{\partial W}{\partial \lambda} + \nu \frac{\partial^2 W}{\partial \lambda^2})$$

(6)

in which $D = Et^3/12(1-\nu^2)$; N_r and $N\theta$ are the meridional and parallel membrane forces per unit length, respectively; M_r and $M\theta$ are the meridional and parallel couple forces per unit length, respectively. Considering the thickness of the shell, the section modulus per unit length and the internal forces of equ. (6), the membrane and couple stresses can be calculated. These stresses are superimposed to determine the maximum stresses on the upper and lower surfaces.

4. Analysis and Discussion

The expression for the characteristic length ℓ, as shown above, is a function of geometrical and mechanical properties. After the model-material and the radius of the sphere ,R, are selected, remains only the thickness as a varying parameter. The thickness of the sphere might vary within the range from the transition state, thin to thick shell, as the upper limit and maintaining stability against buckling as the lower limit. Obviously, each sphere has a unique charcteristic length while each segment of the same sphere has its own arching parameter (AP). In general, a shallow spherical shell is characterized by a very low value of its AP. The AP is defined by the altitude of the one base spherical segment divided by the diameter of its base, i.e., AP can be expressed in the form

AP = 0.5 tan (Ψ /2)   (7)

in which Ψ is the solid angle of the cone whose vertix is the center of the sphere and whose base is the base of the spherical segment.

Equation (7) shows that the arching parameter of a spherical segment is a function of the solid angle Ψ . It follows, if Ψ_s is the solid angle corresponding to a shallow spherical segment, that all segments whose solid angles are less than Ψ_s, are also shallow. Furthermore, the APs of different segments of the same spherical shell become shallower as they get closer to the apex.

The speculative assumption made by Timoshenko and Woinowsky-Krieger [8], namely the results obtained from the infinite shallow spherical shell can be used under all conditions at the boundary of a finite shell when the characteristic length is small compared with the radius of the base, is applicable only for

336

edge-value of λ ≥ 6.8. For shells with edge-values in the range 2.6 ⩽ λ ⩽ 6.8 , the over determinate approach given by Issa in [7] is workable. The hybrid analysis that is described in [6] has usefull applications. However, all the above mentioned solutions are based on deriving, first the formulations for the internal forces then determining the surface stresses.

The pilot results, obtained from the solution given herein, are compared in Fig. 2, with those achieved by Silverman et al in [2]. The solution, given in [2],

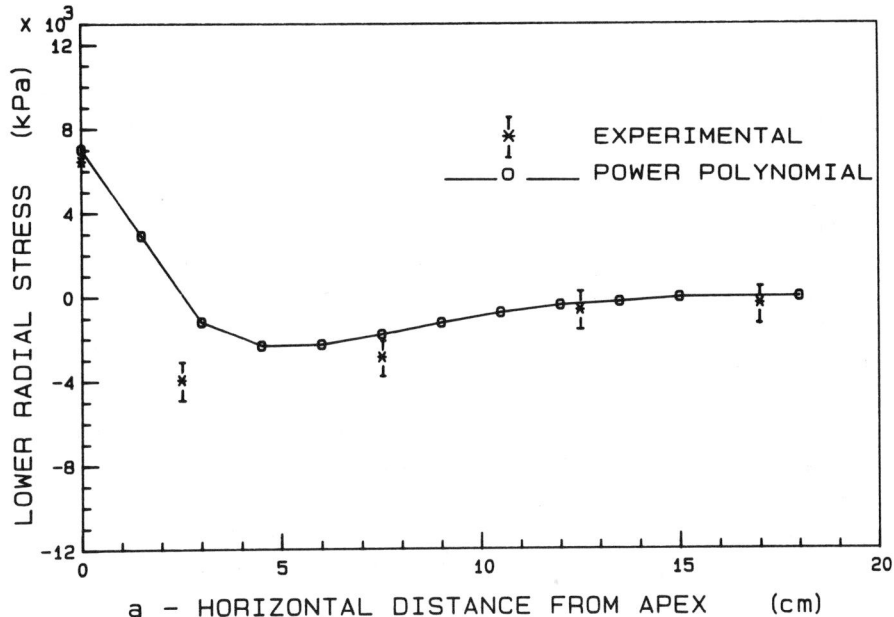

Fig. 2   Meirdional Stresses on the lower surface

is characterized by describing point loads at the apex or axisymmetrical uniformly distributed loads by a power polynomial which has a finite peak value at the apex and which diminishes to zero at an arbitrary selected point close to the apex. This solution is distinguished by bypassing the continuity conditions at the boundaries of the loaded area.

The precision of the experimental results might be improved by using greater number of DTs. This yields a more accurate description of the deformed shell which result in more realistic values of the arbitrary constants. However, the solution of eqs. (3) becomes over determinate where they can be solved according to the algorithm that is explained by Issa in [7].

5. Conclusions

The determination of the arbitrary constants which are required for the fulfillment of the exact solution of shallow spherical shells, in the present sense, can be accomplished by means of direct experimental measurements of normal deflections. This approach, beside improving the accuracy of the experimental accomplishments, saves the major bulk of the effort needed, otherwise, in considering the regularity, transition and boundary conditions. In addition, there is no need for handling over determinate problems. Finally the results can be normalized and used for a wide range of applications.

7. Acknowledgement

Support by the Research Unit of Kuwait University under Grant EV033 is greatfully acknowledged.

8. References

1 Reissner, E. : Stresses and Small Displacements of Shallow Spherical Shells I and II. Journal of Mathematics and Physics, Cambridge, Mass., 25, pp. 80-85 and 279-300, 1946.

2 Silverman, I.K. and Mays, J.R. : Shallow Spherical Shells Under Apex Loads. Journal of the Structural Division, ASCE, 100 (ST1), pp. 249-264, 1974.

3 Vasov, V.Z. : General Theory of Shells and its Applications in Engineering. Nasa Technical Translation, NASA TT F-99, 1964.

4 Zaitseva, O.B. and Rozenberg, L.B. : Partially Loaded Shallow Shells. Translated from Prikladnaya Mekhanika, 10(1), pp. 60-64, Jan. 1974.

5 Issa, S.S : Response of Spherical Shells Under Apex Load to Varying Proportions. Proc. 8th International Conference on Experimental Stress Analysis, Amsterdam, the Netherlands, May 12-16, 1986, pp. 31-40.

6 Issa, S.S. : Hybrid Analysis of Shallow Spherical Shells. Submitted for possible publication in the ASCE Journal of the Engineering Mechanics Division, 1986.

7 Issa, S.S. : Over Determinate Solution for shallow spherical Shells. submitted for possible publication in the Journal of Mechanics Research Communications, 1986.

8 Timoshenko, S. and Woinowsky-Krieger, S. : Theory of Plates and Shells, 2nd ed., McGraw-Hill Book Co., Inc., New York, 1959

LONG-TERM STRAIN OBSERVATION OF CONCRETE BRIDGES

Doc.Ing.Tibor J á v o r ,DrSc.
Research Institute of Civil Engineering /VÚIS/,Bratislava,ČSSR

Observation on prestressed concrete bridges and their
behaviour versus time allow to obtain some data on real behaviour of bridges in traffic or during construc-
tion.During the construction of various bridges acous-
tical vibrating wire gauges in number of 50 to over
300 were embedded and the measuring was automated.The-
re is presented also an appropriate way of results interpretation as well as of their processing and the
mathematical correlation of the measured strains.

Keywords: Strains,vibro-wire gauges,data loggers,creep

1. Introduction

The assesment of the behaviour is possible by various short
quality checks and tests in situ or by long-term observation and
analysis of the durability of concrete bridge structures.From an
economical point of view,it is the most important to asses for
how long a bridge will operate without repair and to compare this
with the economical or commercial life of the concrete structure.
In fact only based on the true knowledge of the long-term operation of these structures,is it possible to develop new design and
construction techniques so that works may be structurally safe
and economical.

2. The behaviour of concrete bridges under traffic

The behaviour of prestressed concrete bridges under traffic
is the object of following checking:
  - in the case of quality deficiencies during construction,
  - in the case of long-term changes of stresses owing to large
creep,long-term losses of prestressing and settlements of founa-
tions,
  - in the case of danger from overloading or fatigue of the
structure.

On the basis of observation it is possible to make a deeper analysis of the results of these investigations compare mutually the relative deformations-strains,and deflections measured and transform the measured deformations into strain so that
they may be a suitable basis for checking the static function of
the bridge and for the design of further bridge constructions
similar to the bridges in long-term observation.

The methods for measuring and checking of statical function
of bridges are divided according to the kind of values measured
as follows:
   - methods for measuring of deflections and deviations,
   - methods for measuring of strains,
   - methods for determinations of various material constants of
the structure examine.
All kinds of measuring referred to were carried out with the aim
to abanded traditional ways and to establish a high automation
of values measured being read by means of their registration and
of measurement mechanisation.

3. Automation of measurements and data processing using vibro-wire gauges during the construction and long-term

The experimental analysis of prestressed concrete bridges
is made by the Czechoslovak embedded vibro-wire gauges.For long-
term temperature measurements we used embedded vibro-wire ther-
mometers.Regarding the requirements of maximum automation of our
measurements we applied the impulse wire gauges with only two-wi-
re connection to the measurement equipments.In this case the gauge wire is set in damped oscillation ny an 0.4 ms impulse dura-
tion.After a delay of T = 10 ms there are measured 100 oscillation periods.The transformation of values to mechanical stresses
with relevant correlation effects of temperature changes is made
by a computer programme for transformation of strain to stresses
with regard to the creep,shrinkage and the Young-modul of elasticity,too.

For measuring purposes we use a measuring bus with installed
data logger,type HP 3050 with relevant digital voltmeter,frequen-
cy reader,programmed switch unit,turner and XY-recorder.The system is oriented to the internationnally codified IEC-BUS and controlled by controller HP 9826 S. The measuring equipment enables

measurements of gauge wire oscillation frequency by a velocity of 3 channels in one second.The measured values are stored in a magnetic casette unit securing a long-term storage and measurement evaluation.The measurement results are registrated in real time,in table form or graphically,by line recorder.

4.Observation of segmental box-girder bridges during construction

As example of combined,manual and mostly automatic system of observation can be reported the experimental analysis of the prestressed concrete segmental box-girder bridge near the village Podturen in Slovakia during the construction.The viaduct,is shown in Fig.1, has 17 fields with span 70 m and piers of maximum 32.9 m height.

Fig.1 The segmental prestressed concrete box-girder bridge during the erection and observation by 330 embedded vibro-wire gauges

The deflection measurements of this bridge were made by geodetic methods and inductive transducers,the inclinations by inclinomethers Maihak and Huggenberger levels,some short measurements by resistance strain gauges and the local stresses by photostress method Vishay. For the observation of strains during the construction as well as long-term 330 vibrating-wire gauges and vibro-wire thermomethers were embedded.The course of strains during the erction of the first segmental box-girder element at the corners measured by vibro-wire gauges is given in Fig.2.

During the erection and the long-term observation the creep and the shrinkage was measured on concrete samples situated in box-girder hollows and compared with samples in laboratory conditions where the changes of temperature and humidity are excluded.The long-term strain measurements of the bridge were continued after finishing the construction by automated equipments.The readings are recorded directly and the results are processed by a Orion Data-logger + IBM PC/XT Controller,by means of which the corresponding transformation of measured strains into the stresses is made also.These values are then plotted in tables and graphically registered.

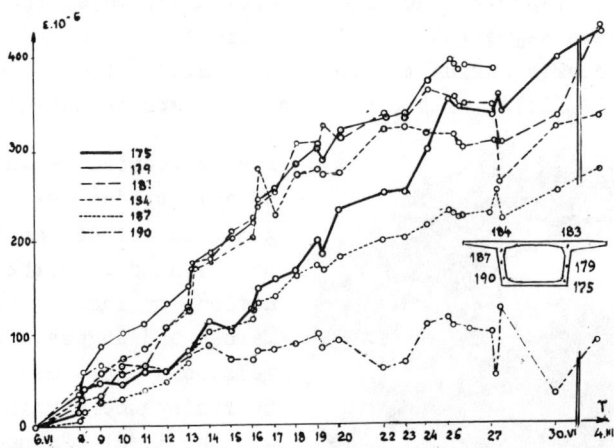

Fig.2 Course of strain at the corners of the first segmental box-girder element during the erection of the bridge

5.Long-term observation of prestressed concrete bridges

12 prestressed concrete bridges various systems have been made observation for 10 to 26 years.During the construction of the bridges in number of 50 to over 300 vibro-wire gauges were embedded.The deformations determined in this way are transformed into stresses and compared to theoretical assuptions. Fig.3 showing the course of strains at the crosssection near the piers of a prestressed concrete casted in place framework bridge of 63.40 m span with ties at the end.The deformations show,that the creep of the concrete is stabilised after 7 years and the strains oscillated following the temperature changes. The influence of temperature changes between summer and winter on the deflection was ± 1 cm.The mathematical correlation shows

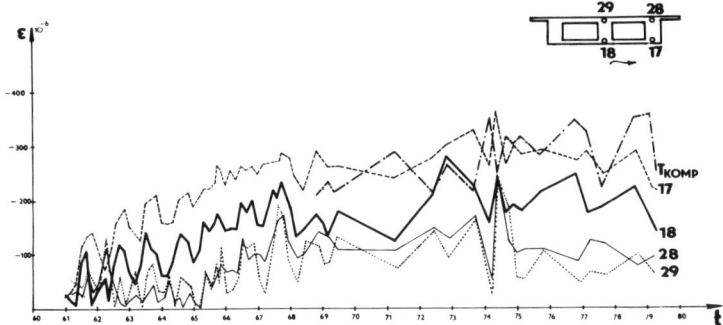

Fig.3 Strains of the cross section near the pier of the prestressed concrete framework bridge during 20 years

that the parabolical curve $t = a.v^2 + b.v$ or the logarithmical function $v = A.\ln B.t$ can replace the measured course of strains, where $t=$ time, $v=$ strain or deflection. Similar results we received by various other bridges also. Fig.4 showing us the course of strain of the double box-girder cast in place cantilever bridge of 80 m span compared with the simple function $y = A + B./\frac{1}{T}/+C.\log T$, where $A= 442;$ $B= -416;$ $C= -120.$

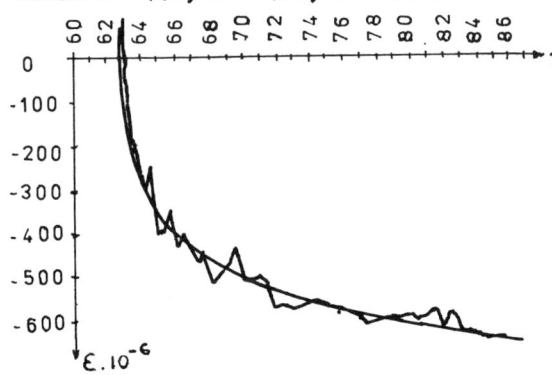

Fig.4 Course of strain of the cross section near the pier of the cantilever box-girder bridge during 26 years with the correlated simple function

For the mathematical correlation wr are used 20 functions related generally to the function $Y=A.F_1/t/+B.F_2/t/+C.F_3/t/$ Generally the best result we received using the functions like $Y= A + B.\sqrt{T} + C.T$ or $Y= A + B. \frac{T}{T + Q}$ whot is similar to the Ross creep function. A is the strain for time $T=0$, B is the increased strain, Q is the time during the structure received the half of the full long-term deformation /strain or deflection/. The mathematical correlation functions give the good results to predict the strains or deflections for the next future of the bridges.

343

STRESS AND STRAIN INVESTIGATION IN THE UMBRELLA ARCH METHOD
ON A TUNNEL

Dipl. Ing. Ferydun Nazari, CSc.
Faculty of Civil Engineering, Chair of geotechnics, TU Brno
Barvičova 85, 602 00 Brno, Czechoslovakia

The Strahov-tunnel, the largest road tunnel in Czechoslovakia, is an important part of the whole basic urban road
system in Prague. The tunnel structure proper is 2 005 m
long; it consists of three two-lane tunnels (1 536 m driven by the coring method, max. rock cover 85 m). At the
driven northern portal - the tunnel rock cover is eighteen
meters high. With respect to the placing of underground
engineering networks, communications and in order to decrease the deformations and the overall risk - it is necessary
to take specific measures, i. e. the construction of a protecting umbrella. As a convenient solution method of this
problem the experimental physical modelling offers itself
with the automatic electrical method.

Keywords: physical model, automatic electrical method,
          umbrella arch on tunnels, investigation of defor-
          mations and stresses

1. Introduction

In the northern transition section of the Strahov-tunnel - the
solidification of the rock medium is designed by means of an umbrella
of horizontal pushed-through piles. The section is driven in the vi-
nice-layers. It is the matter of 8 piles with the diameter of 1 meter,
which are pushed through in the length of 50 m, immediately behind the
tunnel portal. The horizontal piles are placed in such a way, that in
co-operation with the rock - they form an umbrella, which solidifies
the rock over the excavation and contributes to the partial formation
of the arch. In horizontal direction, the horizontal piles act at tunnel driving at first as a console, further as fixed ends beam and at
last again as a console. The surface deformation calculation of the
solidified, rock medium is complicated by the space character of the

345

problem and by the time factor of tunnel driving. That is why physical modelling Geo-Brno-86 was executed on the Chair of Geotechnics on the TU in Brno. The models were built in the scale 1:50 (model A and B). The model A represents the transition section driven under the umbrella proteetion of horizontal piles and model B - the transition section driven without the protection of horizontal piles. As both models were built parallelly by uniform working procedure, it is possible to compare - at the model experiment the influence of driving on the rock medium.

2. Equipment and equivalent materials for modelling

The size and equipment of the stand are determined by the scale, in which the reality is to be modelled by the complexity of auxiliary equipment necessary for the modelling of the problem. For the model Geo-Brno-86 a stand with the plan dimensions 200 x 100 cm was used, see fig. 1. At the design of the model production technology of equivalent materials according to the theory of similarity and dimensional analysis - it is necessary to start out of the rock medium mechanical properties of the modelled Strahov-tunnel section and of the model scale size. The basic component of the equivalent tunnel medium material is - the "ballotine" (glass beads). For the model Geo-Brno-86 - "ballotine" No. 14 was used, which has 95 % of grains in the range 0,1 to 0,24 mm. For the manufacture of lining supports organic glass was used, imported from the FRG. The organic glass is a polymethylmetacrylate, prepared by block polymerization. The supports have been milled according to the structure project. Altogether - 4 supports are placed in the models. Further - 8 Novodur horizontal piles were manufactured. Also the tunnel lining consisted of novodur.

3. Assembling and connexion technology of the measuring complex

The procedure of work on the model construction was determined precisely from the laboratory tested equivalent materials. The individual layers of equivalent materials are placed into the modelling frame. Each layer with the strips of 1,8 cm consists of several bands. When filling the stand, the condition is especially important, the equivalent material possesses along the whole model section the same properties. In order to fulfill these requirements - control samples were taken at placement of individual layers. Further, the radiometric set NZK - 202 was used four times, to specify the relative density. A pigment was used in order to differentiate the individual layers (altogether 61). The measuring complex was connec-

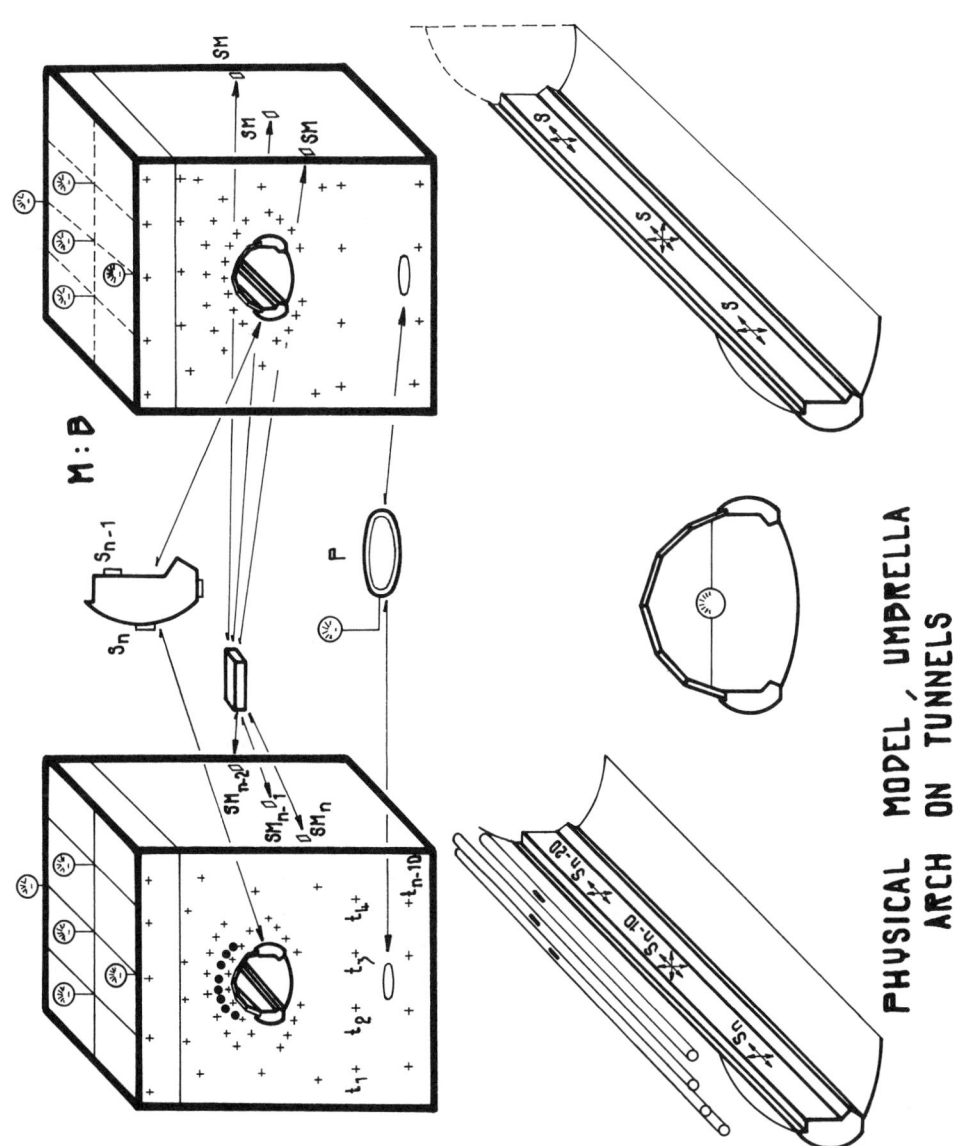

PHYSICAL MODEL, UMBRELLA ARCH ON TUNNELS

FIG. 1

347

ted by means of a digital tensometric equipment DDM 4 from the firm
Hottinger. The digital tensometric bridge DDM 4 is a measuring apparatus, which obtains static quantity measurements and converts them
into a registerable digital form. The equipment DDM 4 is equipped
with a printer, which prints automatically the complex of measured
values of the individual tensometric pickups, see fig. 2. In the
model Geo-Brno-84 as seen from the figure, the tensometers MS for
stress measurings are connected. The pickups are first of all-cali-
brated. Altogether 650 soldered joints were necessary at the assem-
bley of MS. In this manner - the 86 tensometers MS were placed into
the model. Their dislocation and connexion to the apparatus DDM 4
is to be seen in fig. 2. At support strain measuring by means of support tensometers - use was made of the resistance tensometer A 120
and the adhesive X-60. The dislocation is apparent from fig. 1. At
relative deformation measurings of horizontal piles by means of resistance tensometers - the procedure is the same as with deformation measurings of the supports. The stress and strain measurings on
the model are completed by protogrammetric measurings.

4. Execution of the experiment

The driving technology in the model had the following stages:
The first stage consisted of shield jacking and extraction of the
arch area. The abeady built - in supports of the calotte arch served as shield guides. The next stage consisted in tunnel driving in
four phases with overall length of 50 m. In the individual stages -
side adits were constructed. Further - the lining of the upper arch
was executed and the shield was gradually extruted. In the last stage - the material of the core was evacuated and the lower arch was
placed.

5. Conclusion

It was the objective of modelling to determine objectively the
state of stress and strain of the lining system with the neighbou-
ring rock massif and its utilization for a safe and economic design
and execution of the tunnel. I consider as valuable findings in comparison with the model without the protecting umbrella (model B):
The course of settlement in dependence of driving progress (the settlement value decreases by 41 %, at limit state of loading by 182 %),
and after support rotation, displacements on the front wall and the
solution of mechanical behaviour of the system of constructions (lining and umbrella) and rock, see fig. 3.

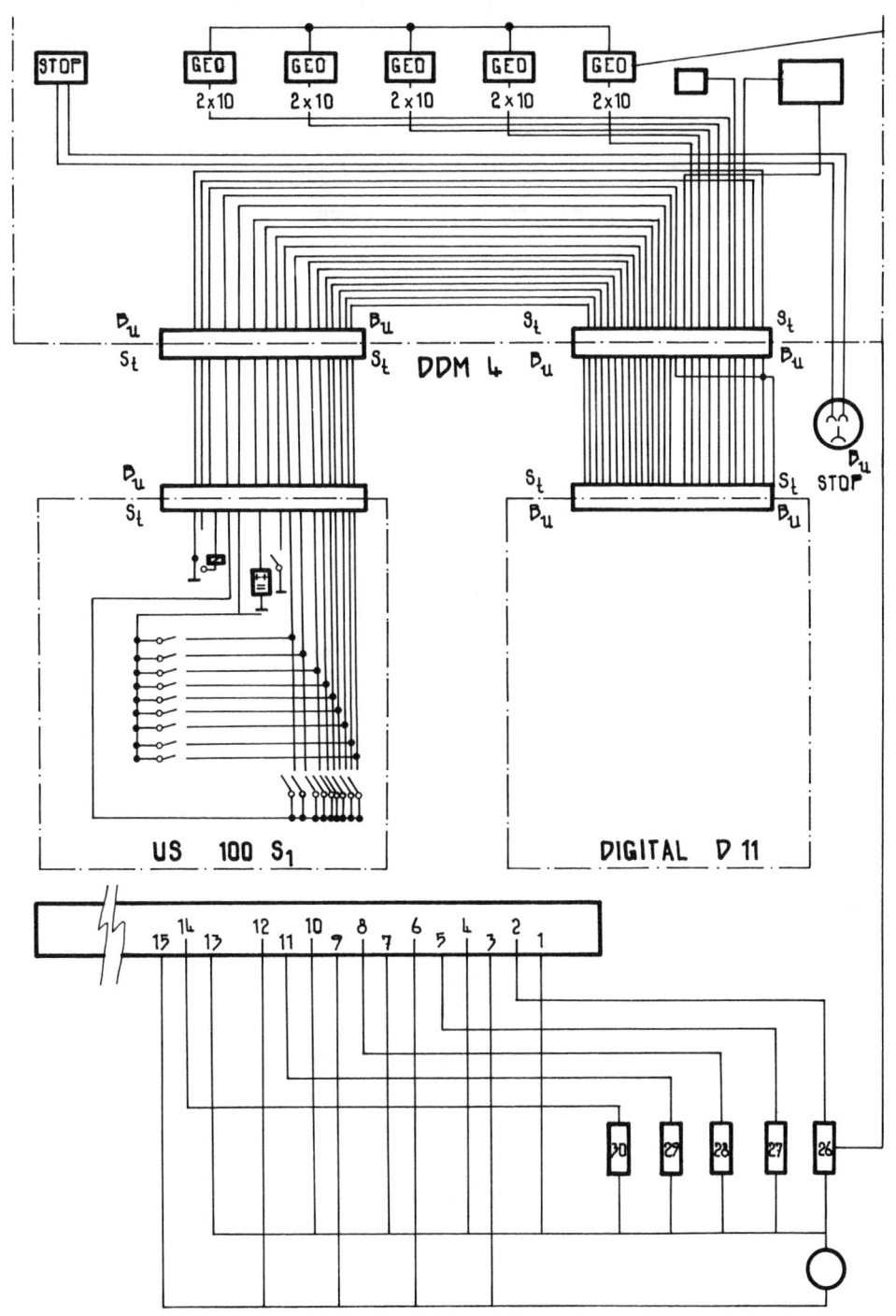

AUTOMATIC ELECTRICAL METHOD

FIG. 2

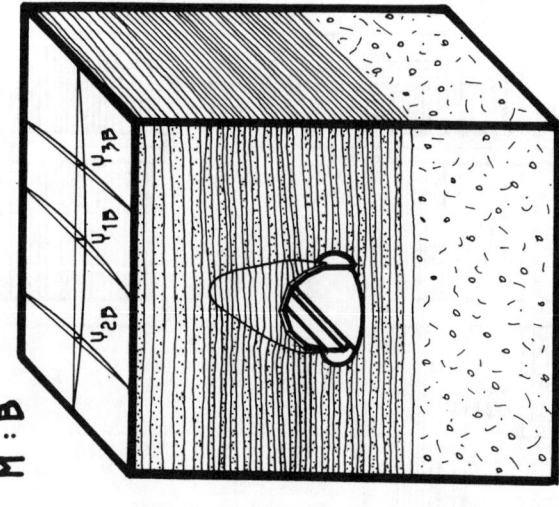

M : A

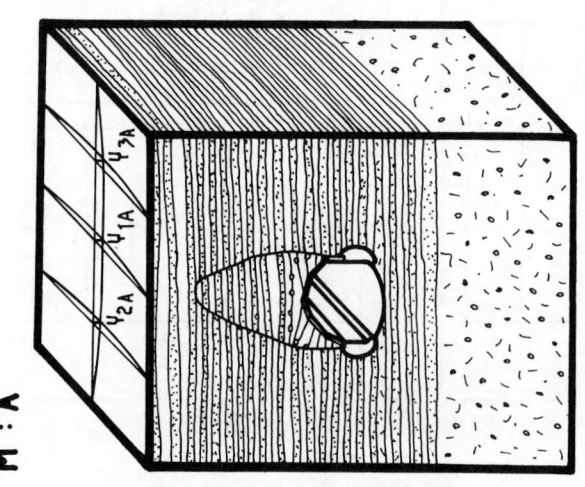

M : B

INVESTIGATION OF DEFORMATIONS AND STRESSES

FIG. 3

METHODS OF MEASUREMENT OF NORMAL STRESS IN SOIL EXCITED BY BLASTING AND THEIR COMPARISON

Dipl.Ing.Jiří Olmer, CSc.
Building Research Institute
Praha 6, Šolínova 7, Czechoslovakia

Two methods of measurement of normal stress on the surface of models embedded in soil are compared. The normal stress was excited by means of an explosion on the surface of soil. Methods using straingage cells and piezoelectric cells were used. The paper deals with these two methods and their errors. The maximal values measured are compared and analysed.

Keywords: Stress measurement, stress cells, errors

1. Introduction

The measurement was provided as a part of research of dynamical properties of simplified models of structures. Models were embedded in soil. On the surface of the soil a stress-wave was excited. The reason of these experiments was the comparison of experimental and theoretical results. This article deals only with special problems of the measurement concerned with this research.

2. Experimental procedure

The scheme of the experiment is shown in Fig. 1. A model from steel was situated in the soil. On the surface of the model two types of pressure-cells were installed. Six straingage cells were placed in section A-A and six piezo-electric cells were placed in section B-B. On the surface of the soil a stress-wave was excited. The stress-wave was

generated by an explosion of a mix of acetylene and air. Every
cell was connected across the input-terminals with suitable
amplifier. The output-signal of amplifiers was registrated by

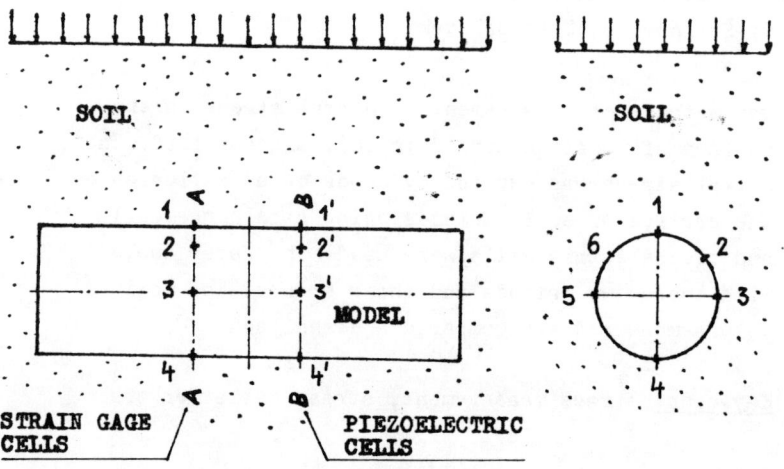

Fig. 1
Schematic drawing of the experiment

means of an electromagnetic oscillograph and by means of
a technical magnetophon. Two different types of circuits were used:

a/ the system with carrier frequency amplifiers for
straingage cells; frequency range 0 - 850 Hz /-3 dB/
b/ the system with load amplifiers for piezoelectric
cells; frequency range 5 Hz - 10 kHz /-3 dB/.

Also the cells had different properties, namely different size, rigidity, sensitivity and frequency-range. They
were calibrated by means of a normal cell in a shock-wave
tube.

Properties of the cells were following:

	strain gage cell	piezoelectric cell
measuring range	2 MPa	5 MPa
sensitivity	1 mV/V/MPa	100 mC/MPa
frequency range /in connection with amplifier/	0-850 Hz /-3dB/	5 Hz-10 kHz/-3dB/
deformation of active area correspondig to full range	0,1 mm	0,01 mm
diameter of active area	30 mm	6 mm
diameter of the cell	50 mm	11 mm

3. Analysis of results

Typical records of time-history of normal stress are in Fig. 2. They were measured by means of both types of cells.

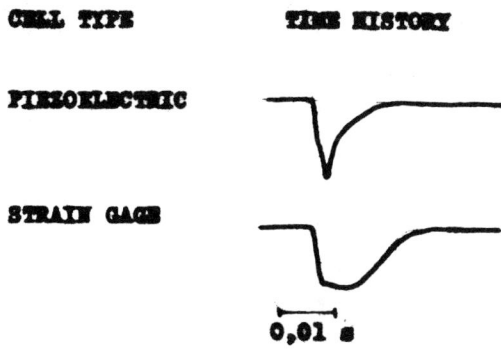

Fig. 2
Typical oscillograms of normal stresses

The records are from corresponding places on the model. Maximal values measured by means of piezoelectric cells were 30% greater than maximal values measured by means of strain gage cells. The difference 30 % is the average value from 24 repeated measurements. The reason for this difference was different size and rigidity of the active area of the cells. The next reason is the fact, that the dynamic calibration of both types of cells was carried out in the air and not in the soil.

The magnetic records were digitalised and analysed on a computer. The amplitude frequency spectrum of record of piezoelectric cell is in Fig. 3. It is obvious, that the fundamental harmonic components of high frequencies don't exceed the upper frequency limit of the measuring system with carrier-frequency amplifiers. The spectrum is probably correct in this

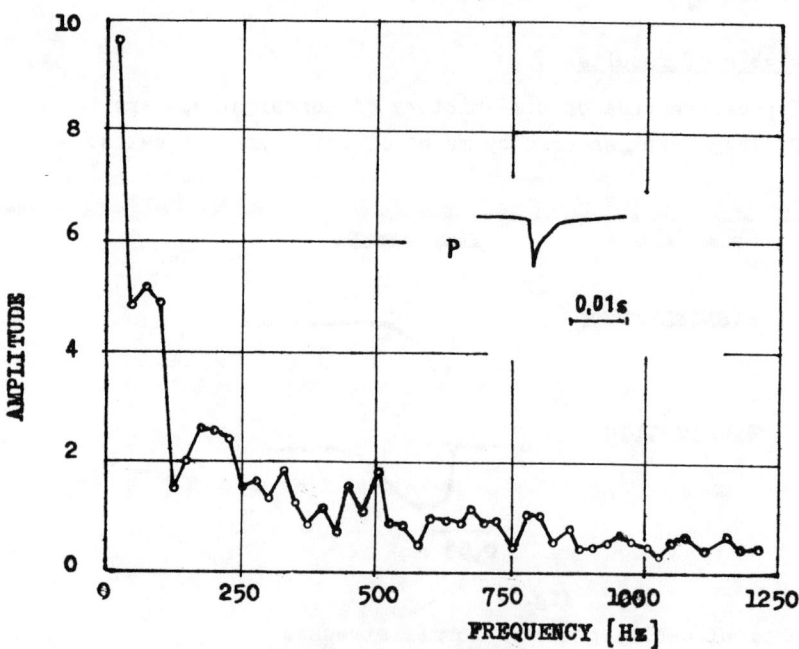

Fig. 3
Amplitude-frequency spectrum of oscillogram
of piezoelectric cell

part, because it belongs to the record of the piezoelectric cell with upper frequency limit 10 kHz. In the domain of harmonic components of low frequencies it may be significant influence of different low frequency limits of both measuring systems. This is also possible explanation of different time-histories of the records.

4. Conclusions

The article shows the problems of measurement of dynamical stress in soil, expecially if two different methods are used. The main problems are the interaction between the cell and soil and the methods of celibration of cells in laboratory.

DYNAMIC MEASUREMENTS ON THE EARTHMOVING MACHINES
THE FOUNDATION OF PROTOTYPE TESTING

Prof. Blahoslav Pacas, PhD.,DSc.
Department of Machines for Civil Engineering and Transport
Technical University in Brno
Třída Obránců míru 65, 602 00 Brno, ČSSR

The estimation of strain in the structure elements of the working mechanism and those of the carrier structure has to be related to the real working load. Through the special measurements the course of stress, pressure, acceleration and force is detected. The further experimental data processing and the corresponding conclusions form the background of the accelerated prototype testing method giving the information about the durability and reliability of the machine structure.

Keywords: random process, intensive load pattern, realization

1. Introduction

The earthmoving machines represent a group of machines determined for the heavy duty working conditions at low cost of the maintenance and with high efficiency of use provided, the carrier structure is life safe and reliable. The main technology in manufacturing these machines is the welding technology with all the joints welded in a high standard.

However, as the case is, the care of the correct technology being whatever, the welding may always be regarded as a potential source of voids, microcracks on the surface as well as inside the material of the structure elements.

The heavy working conditions form the excessive load its prevailing effect is hidden in the dynamic character of work and thus of the load. It has been proved as a random load of a wide frequency spectrum with the stress amplifying phenomena at the natural frequencies. Then any dammage on the surface or in the material of the structure element leads to a high stress concentration in such a place.

2. The dynamic character of load

To enlighten this problem the special measurements of stress in exposed spots on the carrier structure and on the elements of the working mechanism have been done [1]. Not only the stress change but also the configuration of the mechanism and the magnitude of pressure in the hydraulic control system form a picture of the load. As the working mechanism, especially that of wheeled loaders, changes during the work the mutual positions of its links it is necessary to speak about the domains of natural frequencies being typical for each working cycle of the machine. The discussion of results gained from the measurements on loaders in field has proved the existence of two basic working cycles differing in the intensity of load and in the configuration of the working mechanism [2].

a) The loading cycle:- higher speed of the machine when pointing the shovel into the heap of primary disintegrated material - greater dynamic effect (vibrations),

b) the quarrying cycle:- lower speed of the machine with the shovel prepared to penetrate into the earth and disintegrate primarily the material (greater excessive load).

To compare both the working cycles between each other there were organized the measurements in field on the real machines. The first group of machines being investigated has been taken from among the loaders due to the pure dynamic character of work. In the Fig. 1 there is shown the arrangement of pick-ups. The letter T stands for strain

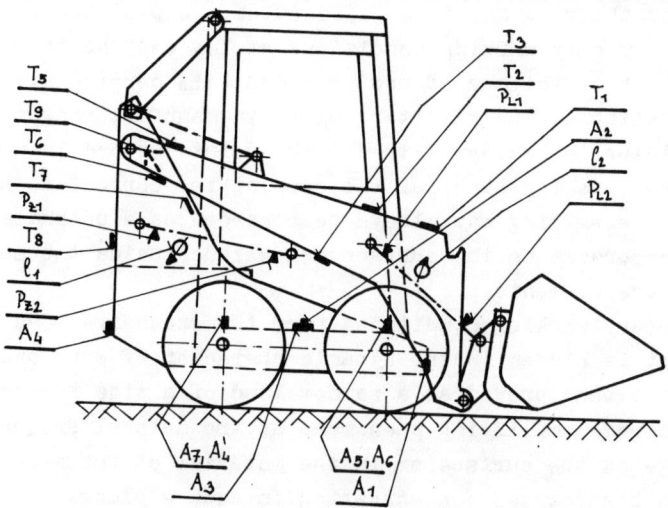

Fig.1 Pick-up arrangement

gauges, P for pressure measuring cells, A for accelerometers, l for displacement measurements. The positions where the strain gauges have been placed had been determined in advance, by means of model measurements. The models have been of artificial nature in scale 3 : 1 and the verification of the extreme stress concentration in the chosen spots has been based on the mutual confrontation of results gained by means of the reflexive photoelasticimetry method, by means of the specle method and by means of the strain gauge method. To provide the reliable accuracy of measurements there have been strictly followed the instructions given by the manufacturers of all used devices and material (as Hottinger Baldwin Messtechnik HBM , Vishay Measurement Group, Philips Devices, etc.).

The scheme of the measuring loop is shovn in Fig.2 where
$<$ means the measuring amplifier KWS 62077   HMB
$\diagdown$ means the low-pass filter TP 3554   HBM
MTR   means the measuring tape recorder Analog 714 Philips
LR   means the line recorder Hellige Recomend, Fa. Bell and Howell.

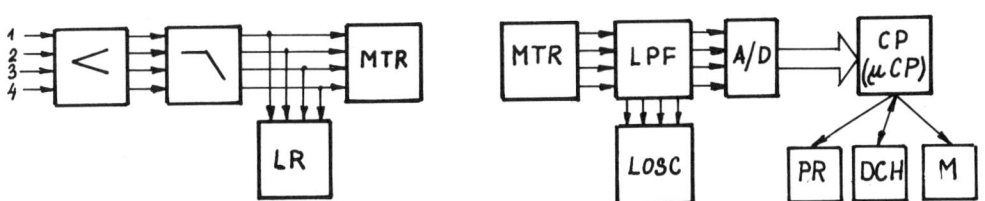

Fig.2 Measuring loop   Fig.3 Processing loop

There have been organized about 8 different measurements covering both the loading and quarrying cycles, each of them being repeated from 10 to 15 times (so called realizations) to provide the further data processing because of the random not stationary process resulting from the measurements. The mechanical properties of the loaded or quarried material are random, and the working cycle itself is formed by the succession of determined movements of the working mechanism being under the control of the crew. Each realization has represented the complete working cycle.

The experimental data processing has been done in two different ways as follows:
1. To get the statistical characteristics of the measured quantities, namely of stress course and that of the pressure in linear hydromotors of the working mechanism by means of the correlation theory.

2. To get the data suitable for the time duration (service life determination) of the structure elements using any of suitable damage cumulation theorems.

For both the mentioned ways of analysis the scheme of data processing loop is similar. The difference may be found only in the devices used in connection either with the computer (of medium size) or with the microcomputer see Fig.3. There are taken here:

MTR   the measuring tape recorder Philips Analog 714, EAM 500 Tesla
LPF   the low-pass filter e.g. 01 014 RFT
A/D   the analog-digital transducer   own production
CP, μCP the microcomputer SORD-M5, computer EMR Advance
PR   the printer or the digigraph
M   the monitor
DS   the data sette
LOSC the oscilloscope for low speed phenomena OPD 602 Tesla
DCH   the data check before the numerical processing.

For the first case the corelation method of calculation has been used. This method requires the stationary random process or at least that one stationary by parts. This the following way of analysis:

a) the complete working cycle has been detailed into so called working phases representing the specified activities within the determined time periods. The most important phases representing the extreme load (the impact of the shovel into the material and cutting off or loading the material) have been taken as the decisive loads transmitted upon the carrier structure. This part of the working cycle has been refered to as an "intensive load pattern" and has been further applied to each from among the 15 realizations [2].

b) The "intensive load pattern", formed approximately by 1500 up to 2000 samples has been tested to the stationarity and afterwards statistically analyzed. For each of the stationary "intensive load patterns" there has been found the power spectral density function in form

$$G_{xx}(f) = 2\Delta t \left[R_o + 2\sum_{k=1}^{m-1} R_k \cos \frac{\pi r k}{m} + R_m (-1)^m \right] ,$$

where for r is $r = (f/f_{max}) \cdot m$
f_{max} maximum frequency component of the function f(t)

$m \approx 0,1N$

N   number of samples.

There were taken those power spectra functions from among all the realizations their dissipation of function values has not exceeded the 95% confidency interval as a guarantee of the stationarity of the random process to form the required time series of selected realizations.

c) Such an artificially created record of the stress course, for example, has undergone the necessary "cosmetics" and has further been analyzed as a random process stationary "by parts" [3]. The probability density function of such a time series representing the course of stress in the spot given by the position T2 see Fig.1 has been collected into the Tab.1 as a macroblock of stress. The alternating character has been limited to the frequency of 5 Hz yet important as it comes out from the power spectra density function.

d) The most dangereous stress in the spot No.2 is formed by the last three blocks numbered as the blocks No.7,8,9. These blocks of stress may be then taken as the loading blocks for the prototype testing load represented by σ_a, σ_h, n_k, where

     σ_a is the stress amplitude,

     σ_h is the upper limit of stress,

     n_k is the number of stress changes.

     The second method of experimental data processing is very effective for the estimation of the fatigue life by means of the theorems of damage cumulation. There has been processed the succession of realizations stored in the data set   in each   measurement . The pick-up   signal has been sampled on-line and the samples have been stored in the memory of the microcomputer SORD-M5 with the sampling frequency 50 Hz. After filtering the data have been processed and sorted to get the extreme values of stress. These local extremes have been cumulated into classes representing the class stress intervals giving the rate of extremes in each class.

     Such a density of extremes has been further arranged into the probability table of extremes of the stress course for the evaluation of fatigue effects in the measured spot. Nowadays the "rainflow" method and others are used for the on-line processing of the measured quantities.

Block No.	σ_a MPa	σ_h MPa	n_k rate
1	5,0	-14,0	6
2	8,0	0,0	74
3	8,0	18,0	34
4	4,0	26,0	252
5	13,0	52,0	286
6	5,0	62,0	254
7	12,0	86,0	262
8	23,0	132,0	70
9	25,0	182,0	6

Tab.1 Macroblocks of stress

3. Summary

The described way how to get the necessary data and the way of processing them form a basis for creating a suitable method for the testing of prototypes. Restricting our task to the carrier structure or to the working mechanism only, their durability and life time may be determined. The processed experimental data form then a good background for the shortened accelerated testing.

The study of the power spectral density function of the loads leads further to a good estimation of suitable parameters of load necessary for the programmed long life testing of prototypes. Thus the analysis of external loads in details represents the only source for the systematical approach to either the programmed or to the accelerated testing of steel carrier structures of the earthmoving machines [4].

4. References

1 Pacas B.:"Measurement on the Earthmoving Machines in Field", research report N 4/1984, Dep. of Machines for Civil Engineering and Transport, Technical University Brno, Faculty of Mechanical Engineering /in Czech/.

2 Pacas B.:"Generalization of Measurement Results on the Earthmoving Machines in Field. Spectra of Load", res. rep.N 6/1984, Dpt.of Mach. for Civ.Eng.,TU Brno /in Czech/.

3 Bendat J.S.,Piersol A.G.:"Engineering Applications of Calculation and Spectral Analysis", John Wiley and Sons,New York 1980.

4 Pacas B., Škopán M.:"Summary of Results and Conclusions from Experimental Measurements of Working Cycles", res. rep. N 3/1985, Dpt. of Mach. for Civ. Eng., TU Brno /in Czech/

362

MODAL ANALYSIS IN MECHANICAL ENGINEERING

Prof.Dr.sc.techn. Adolf Lingener
Technical University Otto von Guericke
Department of Mechanical Engineering
3010 Magdeburg GDR
PSF 124

The paper deals with application of system identification by means of the method of experimental modal analysis to mechanical systems. The method is based on input-output measurements and data processing in frequencey domain. Measured frequency responses are approximated analytically and modal parameters are estimated. An application to a bed plate of a precision instrument is presented. The results are compared with FEM-calculations.

Keywords: modal analysis, identification, parameter estimation

1. Principle of modal analysis

Modal analysis a is method of system identification in frequency domain. It assumes the structure under investigation to be a discrete linear nonproportionally viscously damped system with n degrees of freedom and symmetric system matrices /1/ /2/.
The dynamic behaviour of such kind of system may be described completely by its frequency response matrix. Modal analysis deals with an experimental determination of the modal quantities of a mechanical model. The model usually consists of discrete masses, springs and dampers with a suitable number of degrees of freedom.
From the theory of linear systems results the following formula, describing the elements of frequency response matrix by means of the modal quantities:

$$H_{kl}(f) = \sum_{r=1}^{n}\left[\frac{x_{rk}x_{rl}/c_r}{2\ (1-f/f_{dr})+j\delta_r/2\ f_{dr}} + \frac{x_{kr}x_{rl}/c_r}{2\ (1+f/f_{dr})-j\delta_r/2\ f_{dr}}\right] \tag{1}$$

Here are $f_{dr} = \delta_{dr}/2\pi$ - r-th resonance frequency
m_r - modal mass of the r-th mode

δ_r - modal damping of the r-th mode
c_r - modal stiffness of the r-th mode
x_r - complex elements of the r-th mode

In each case, in the vicinity of the r-th resonance dominates the modal frequency response belonging to f_{dr}, which may be approximated by a circle.

Directly in resonance, $f = f_{dr}$, applies

$$H_{kl}(f_{dr}) = -j \frac{\pi f_{dr}}{c_r \delta_r} x_{rk} x_{rl} + R_{kl}(f_{dr}) = D_{kl}(f_{dr}) + R_{kl}(f_{dr}) \qquad (2)$$

Here D_{kl} is a complex quantity. Its absolute value is the diameter of the approximating circle. For fixed k and fixed r apart from a constant $D_{kl}(f_{dr})$ represents the component x_{rl} of the modal vector x_r. R_{kl} describes the influence of the other modes, not being in resonance. When checking real mechanical systems in each case a limited frequency range f_A, f_B is investigated only. For this reason, eq. (1) is dissected

$$H_{kl}(f) = -\frac{1}{M_{kl}(2\pi f)^2} + \sum_{r=1}^{m} \left[\frac{u_{klr} + j v_{klr}}{\delta_r + j2 \ (f - f_{dr})} + \frac{u_{klr} - j v_{klr}}{\delta_r + j2 \ (f + f_{dr})} \right] + S_{kl} \qquad (3)$$

In this equation means

M_{kl} - the residual mass, describing the influence of resonances in the frequency range $[0, f_A]$ on the range $[f_A, f_B]$

S_{kl} - the residual stiffnes describing the influence of resonances in the frequency range $[f_B, \infty]$ on the range $[f_A, f_B]$

m - the number of degrees of freedom, dominating in $[f_A, f_B]$, m n.

$$u_{klr} = \text{Re} \ (-j \frac{f_{dr}}{c_r \delta_r} x_{rk} x_{rl}); \quad v_{klr} = \text{Jm} \ (-j \frac{f_{dr}}{c_r \delta_r} x_{rk} x_{rl})$$

By this basic formula of modal analysis (3) it is possible to describe approximately the system behaviour in the frequency range $[f_A, f_B]$ with a small number of degrees of freedom. This is a decisive advantage of the modal view. The basic idea of computer modal consists of an approximation (curve fitting) of the analytical term (3) to the measured data as good as possible. This curve fitting is carried out by means of least square methods.

The nonlinear equations, resulting from the demand $\sum |\Delta H(f)|^2 \Rightarrow \text{Min}.$ are solved iteratively. With the correction values, obtained from these equations the starting estimates are improved. After obtaining a predetermined limit of the sum of squared deviations the calculation is stopped /3/.

2. Excitation of structures and measurements

It is not neccessary to measure the symmetric frequency response matrix completely. It is sufficient, to know one row or one column as from n measured frequency responses $H_{ij} = H_i/F_j$, $i = 1,2,...n$, j = const (one column) results $H_{ik} = H_{ij} H_{kj}/H_{jj}$ $i, k = 1,...,n$ with F_i, X_j the amplitude spectra of exciting forces f_j and system responses x_i respectively (fig. 1)

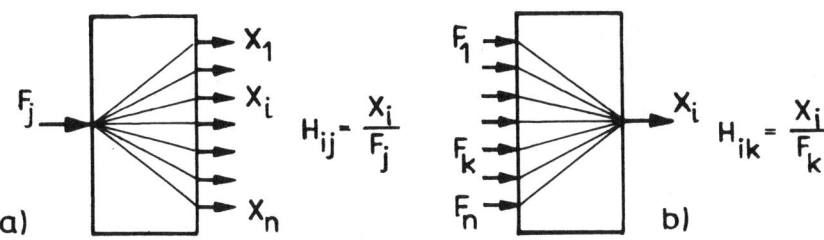

a)   b)

Fig. 1   Possibilities of excitation a): one excitation n points of measurement b): n exciting forces, one point of measurement

The kind of measurement (row or column) depends on the kind of structure. If the structure is lightly excitable for instance by a hammer method b) will be perfered.

The spectra of exciting signals in each case must comprise that frequency interval in which information about the system behaviour is wanted. Usual excitations are impacts, random and step excitations but sine and swept sine excitations are used also /4/.

Vibration measurements are carried out in the most cases with piezo-electric transducers which are easily applicable to the structures. The responses of very light structures are measured with non-contacting sensors.

3. Application
3.1. Modal analysis of a bed plate of a precision instrument

The subject under investigation is shown in fig. 2. The aim was to determine eigenfrequencies and eigenmodes in a frequency range 10-1000 Hz on a prototype. The results were meant to provide vibration-free operation. Those modes at which a deformation of the plate itself occured, were of essential interest, as the rigid body motions, depending on the supporting conditions, were not regarded of interest. The investigations were carried out theoretically by calculating

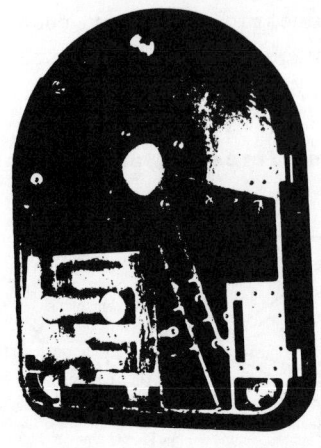

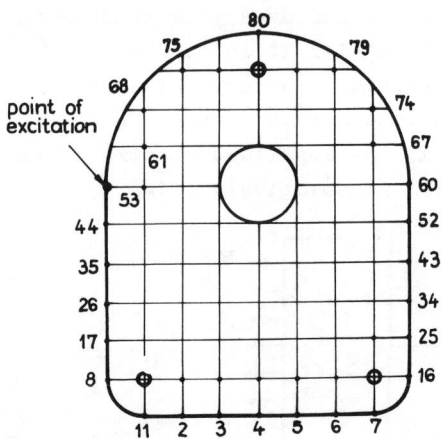

Fig. 2   Photo of the
bed plate

Fig. 3   Measurement points on
the plate (80 points)

eigenfrequencies and modes by means of FEM /5/ (without damping) and
experimentally by means of a pseudorandom excitation and subsequent
modal analysis. The plate was excited by an electrodynamic exciter
in the frequency range 10-1024 Hz. 80 points according to fig. 3
were defined as points of measurement. During the experimental investigations the plate was supported on a soft three - point rubber

Table   Modal parameters of the plate under investigation
from measuring point 70 after 7 iterations

	f_i /Hz/	$\vartheta_i(\delta_i=12,57/s)$		
calculated frequen- cies ⇓	f_1= 63 f_2=246 f_3=394 f_4=590	ϑ_1=0,032 ϑ_2=0,0081 ϑ_3=0,0051 ϑ_4=0,0034	⇐ starting estimates	
	f_5=666 f_6=748	ϑ_5=0,0030 ϑ_6=0,0028	u_i /mm/Ns/	v_i /mm/Ns/
f_1=61,9 f_2=231,9 f_3=407,7 f_4=547,3 f_5=685,9 f_6=815,8	f_1= 62,32 f_2=242,12 f_3=394,20 f_4=589,95 f_5=667,35 f_6=746,56	ϑ_1=0,0793 ϑ_2=0,0299 ϑ_3=0,0055 ϑ_4=0,0044 ϑ_5=0,0075 ϑ_6=0,0026	u_1=-,51E-4 u_2=-,42E-2 u_3=-,21E-2 u_4=-,60E-4 u_5=-,93E-4 u_6=-,17E-4	v_1=-,79E-4 v_2=-,89E-2 v_3=+,74E-2 v_4=+,17E-2 v_5=+,21E-2 v_6=+,86E-3

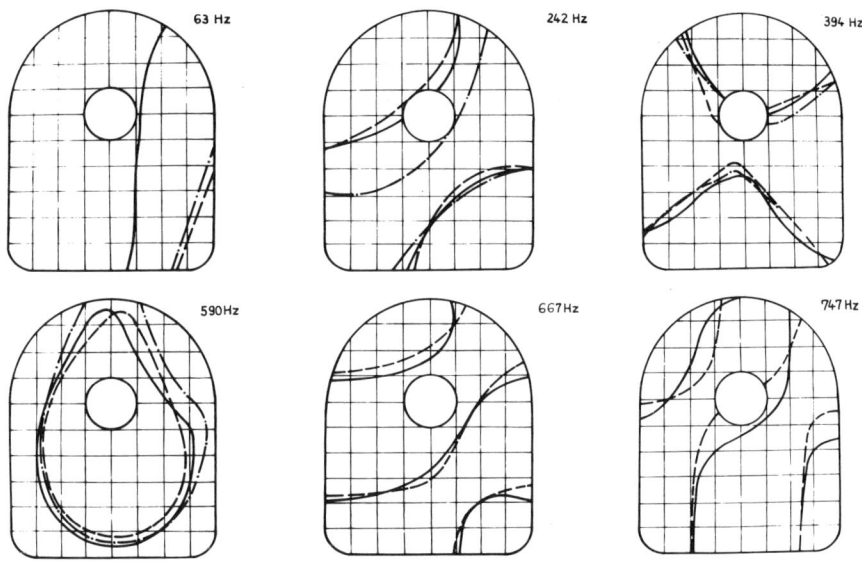

Fig. 4 Nodal lines of different modes ——— modal analysis
-·-·—harmonic excitation
— — —FEM (COSAR)

spring supporting as well as on a
foam rubber mat.
As the rigid body frequencies were
not of interest and because of the
bad coherence at low frequencies the
frequency range was limited to 50 Hz
by high - pass filtering.

3.2. Results
The frequency responses contained 6 utilizable eigenfrequencies
$f_1, \ldots, f_6$. From these $f_1 = 63$ Hz is a
rigid-body frequency. A comparison
calculation had proved, that the 6
measured eigenfrequencies comprised
all possible eigenfrequencies in the
range of interest. The starting
estimate of damping was uniformly
defined by a half-width of the
resonance peaks of $f_H = \delta/2\pi = 2$ Hz.

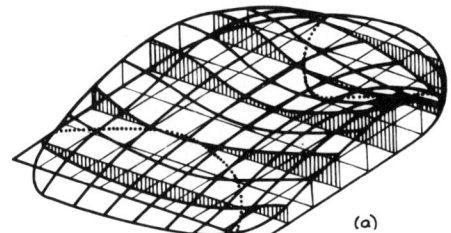

(a)

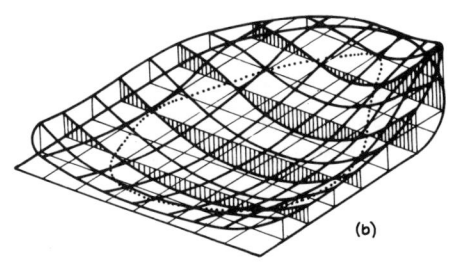

(b)

Fig. 5 Measured modes of
394 Hz (a) and 590 Hz (b)

367

From this value result different $\vartheta_i = \delta/\omega_{oi}$. The starting estimates and results for one measuring point and results of modal analysis are summarized in the table.

The nodal lines of the eigenmodes, determined by modal analysis for all 80 measuring points, by harmonic excitation (4 modes) and by FEM calculation (without damping) are compared in fig. 4.

Fig. 5 examplifies two eigenmodes with a spatial presentation of the amplitudes.

4. Conclusions

Modal analysis is an effective and proved identification method in the frequency domain. From a relatively small number of measurements on a broadband-excited structure this method yields by computer aided evaluation of the measured data (phase separation technique) all neccessary information about eigenfrequencies, eigenmodes, modal stiffness and damping. The results of modal analysis furthermore may be used to improve the calculation model and at last the dynamic behaviour of vibrating structures.

5. References

1 Ewins, D.J. Modal Testing: Theory and Practice. Research Studies Press Ltd., John Wiley a. Sons 1984
2 Natke, H.G.: Einführung in die Theorie und Praxis der Zeitreihen- und Modalanalyse. Vieweg Verlag Braunschweig; Wiesbaden 1983
3 Vasel, T.: Die Modalanalyse mechanischer Schwingungssysteme mit dem Programmsystem ASAM. Wiss. Zeitschrift Technische Hochschule Magdeburg 30 H. 7. p. 102-105 (1986)
4 Lingener, A.: Analysis of Mechanical Systems Excited by Random Vibrations. Proceedings of the IUTAM-Symposium Random Vibrations and Reliability Frankfurt/Oder 1982 Akademie-Verlag Berlin, p. 173-194 (1983)
5 Gabbert, U., Berger, H., Zehn, M., Fels, D.: Universelles FEM-Programmsystem COSAR - Übersicht über den nachnutzbaren Leistungsumfang. Maschinenbautechnik 34 H. 8 p. 352-356 (1985)

EXPERIMENTAL-NUMERICAL METHOD STRESS ANALYSIS
OF RANDOM LOADS IN METALLIC CONSTRUCTIONS

Kopecky Miroslav, Assoc.Prof.,M.Sc.Eng.,Ph.D.
Department of Mechanics and Machine Parts
Technical University of Transport and Communication
010 88 Zilina, C.S.S.R.

Designing the metallic construction, e.g. the building-
cranes with jib, must be resolved the problem its calculating estimate of the fatigue longevity. Acquisition
of the loading spectrum in operating state by theoretical
methods need not be complete enough to describe adequately the relatity always. Experimental long-time tests with
the special measuring instrument make acquisition of the
loading spectrum in operating state possible. The author
outlines the principle of instrument and the method analyse of results by probability theory and mathematical
statistics.

Keywords: fatigue-life, special measuring instrument

1. Introduction

Modern technology has advanced development of many new
materials and products which in turn have created the need for new
and advanced test methods. The method which make the subject of this
paper has been conceived in order to over come the lack of rapid
data recording and processing equipment and of fatigue testing
device meant for subassemblies of random loadings.

The method can be used with recording and computing at hand,
namely:
- transducers for various physical quantities,
- magnetic tape or paper recorder,
- programs for data statistical processing,
- access to a digital computer.

The aim of this method is that of finding out the loading
equivalent sinusoid-constant amplitude and frequency for a stationary random loading of any subassembly of metallic construction.

The results of its application to load-carrying component of building-cranes are shown.

2. Problem brief

The principal features of basic assembly of the special measuring instrument shown on Fig. 1. The special measuring instrument is in the substance the mechanical gauge cennected with the indicator. It works together with photo-cell as if starter of recording equipment. The instrument can be installed in the critical points of the metallic construction. To make use of special measuring instrument in operating state is continualy and directly. Long-time tests can be runed independently from climatic conditions. Very valuable results of experimental tests of the metallic constructions make for recording of signals of random loads under long-time operating state to render possible the special measuring instrument.

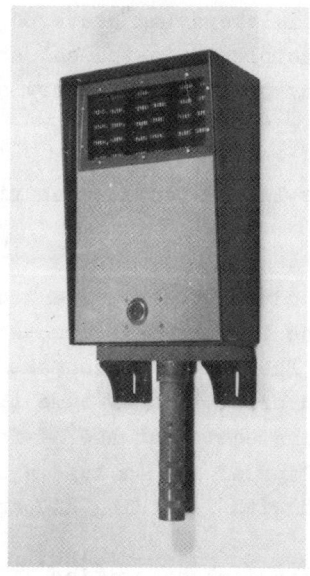

Fig. 1

View of the special measuring instrument

The usual way to determine the equivalent loading of a stationary random evolution is displayed in Fig. 2, where the stationary

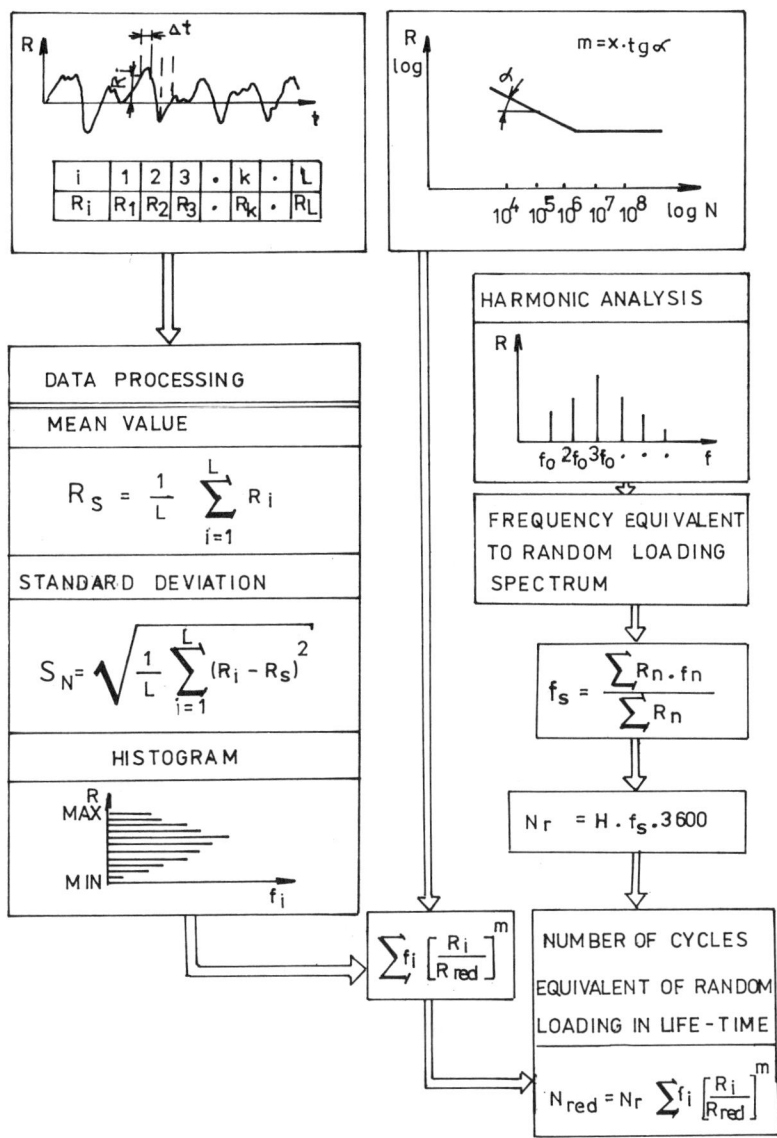

Fig.2 SCHEME OF METHOD

random signal is discretised at equal time intervals, $\triangle t$, and the Wöhler curve experimentally raised for the element or subassembly studied, for the input data. One should point out that by determining the time interval $\triangle t$ we determine the maximum frequency which we can defined as follows:

$$f_{max} = \frac{1}{2.\ t} \qquad (\ 1\ )$$

The calculation program attached to the method will furnisch the intermediate data: mean value, standard deviation, levels analysis-histogram, relatíve frequencies to the histogram and harmonic analysis of the signal taken into consideration. By these quantities one can balance the stationary random signal with an harmonic evolution of the constant amplitude and frequency:

$$N_{red} = \text{const.} , \qquad f_s = \text{const.} \qquad (\ 2\ )$$

One should specify that the equivalent frequency f_s at which the random quantity is reduced is the some for all the components R_i

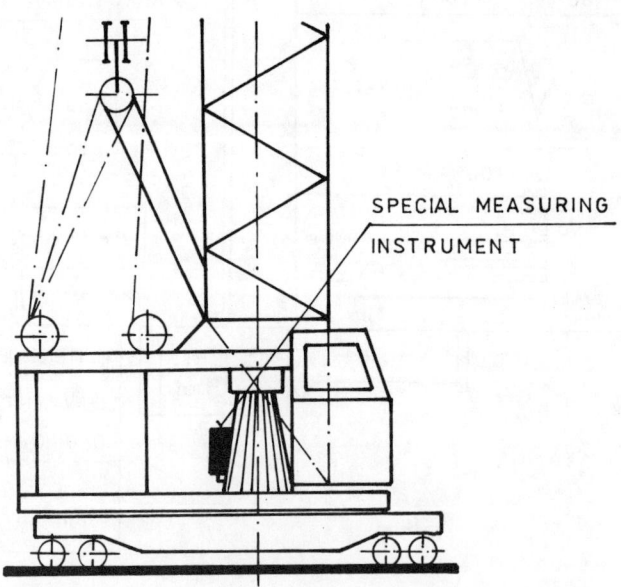

Fig. 3 SCHEME OF PARTS OF BUILDING - CRANE MB 80

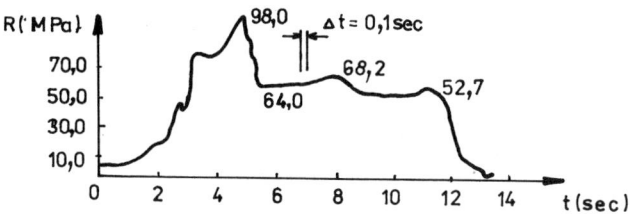

RESULTS OF METHOD

HISTOGRAM		WÖHLER CURVE	HARMONIC ANALYSIS	
f_i [%]	R_i [MPa]		f [Hz]	R_n [MPa]
14,1	1,0		0,07	22,14
8,1	7,0		0,14	12,97
1,5	12,0		0,22	3,87
2,2	17,0		0,29	5,19
1,5	23,0		0,37	1,79
1,5	28,0		0,44	1,44
0,7	33,0		0,52	1,46
2,2	39,0		0,59	1,70
11,1	44,0		0,67	0,26
13,3	49,0		0,74	0,81
.	.		.	.
4,4	92,0		4,47	0,22

$$\sum f_i\left[\frac{R}{73,2}\right]^{5,2} = 0,4265$$

$$f_s = 0,38 \text{ Hz}$$

$$H = 1870 \text{ hours}$$

$$N_r = 2,55816 \cdot 10^6 \text{ cycles}$$

$$N_{red} = 1,09105 \cdot 10^6 \text{ cycles}$$

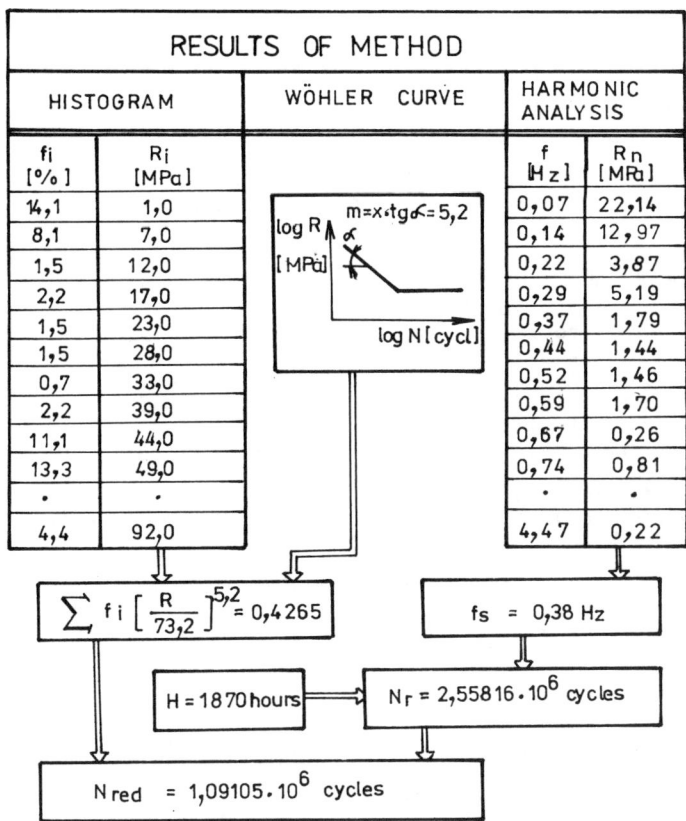

Fig. 4   RESULTS OF METHOD

in the histogram. The output value of this program is N_{red} , representing the constant level number of cycles R_{red} equivalent to the random loading.

3. Results

The application of this method which this paper is restricted is to load-carrying part of building-crane, as shown in Fig. 3. Experimental results are shown in Fig. 4.

4. Conclusions

This program is characterised by the fact that it needs Wöhler curve as input data, which usually implies laborious stand testing necessary to reproduce it. Out of these reasons it is proposed the method [2] , which is meant to determine the equivalent sinusoid of random stationary loading and of Gaussian distribution.

5. List of symbols

f - frequency in frequency analysis
f_i - relative frequency of a random loading on "i" level
f_o - fundamental frequency coresponding to sample width T
$\triangle f_n$ - frequency range in spectral analysis
f_s - frequency equivalent to random loading spectrum
H - service life
L - number of points resulting from random evolution discretisation
m - Wöhler curve index
$Nred$ - number of cycles equivalent of random loading in life-time
R - value of a random loading
R_{red} - loading equivalent to a random loading
Rs - mean value
R_i - loading value on "i" level
S_N - standard deviation

6. References

1 Bendat,J.S.,-Piersol,A.G.: Random Data-Analysis and Measurement Procedures, 1971, J.W.+S., New York.
2 Kopecky,M.: Results of Research on the Reliability Some Load-Carrying Machine Components for Transportation, In Proc. of First Conf. on Mechanics, 1987, Prague.
3 Petrescu,N.: Methods for Data Picking up and Processing with a view to Determine a Fatigue Testing Simulation Regime of Several Wehicle Element, 1977, In Proc. of Conf. UVMV Prague

AUTOMATED INVESTIGATION OF DYNAMIC PARAMETERS OF BRIDGES

Dr. Rudolf Kyska, M.Sc. - Dr. Andrej Sokolik, M.Sc.[x]
Research Institute of Civil Engineerig, Dpt. of Applied Physics
Lamacska 8, 815 37 Bratislava, CSSR
[x]University of Transport and Communications, Dpt. of Structures and
Bridges, Moyzesova 20, 010 88 Zilina, CSSR

A measuring and computing system for acquisition of dynamic parameters of a bridge construction is described. The
measuring system records analog signals of bridge responses to a normal traffic load and discrete data of the traffic. Computerized processing of the recorded data and statistical evaluation result in representative dynamic parameters of bridges. The paper presents results obtained by
the measuring and computing system during long term observation /LTO/ of influence of traffic load on two highway
bridges.

Keywords: long term observation, measurement of dynamic pa-
          rameters, highway bridges, traffic load

1. Introduction

An increase in traffic intensity on highways together with an
effort to apply the knowledge of theory of reliability to the design
of bridge constructions need to clarify many partial problems connected with the solution of the system "bridge - loading - enviroment"
from the point of service reliability of a construction. Nowadays
here is a main task to obtain a true picture of the load magnitude
and its effects on the structure. For this reason it is clear that
bridges which are exposed to considerable dead and moving load as
well as to a secondary load deserve a maximum attention.

A computer simulation of bridge responses to passing trucks, taking into consideration all characteristics of different trucks as
well as those of the bridge, is very pretentios. Based on the today's
level of technical development we could say that an easier way how
to obtain dynamic parameters and influence of traffic load on the
bridge is the investigation of responses of the structure "in situ".

2. The measuring and computing system

The system shown schematically in Fig. 1 has been developed to provide measuring time-variations of the bridge deflection, encoding the passing trucks into 8 + 2 categories and sensing their axles together with the directions of the drive. Both, analog signals /deflections/ and discrete data /traffic parameters/ are stored by the instrumentation tape recorder. The deflection sensing device consisting of a wire stretched by a spring and an inductive displacement transducer or a stretched resistance wire between the measurement point and the terrain are used to measure the bridge deflections [1].

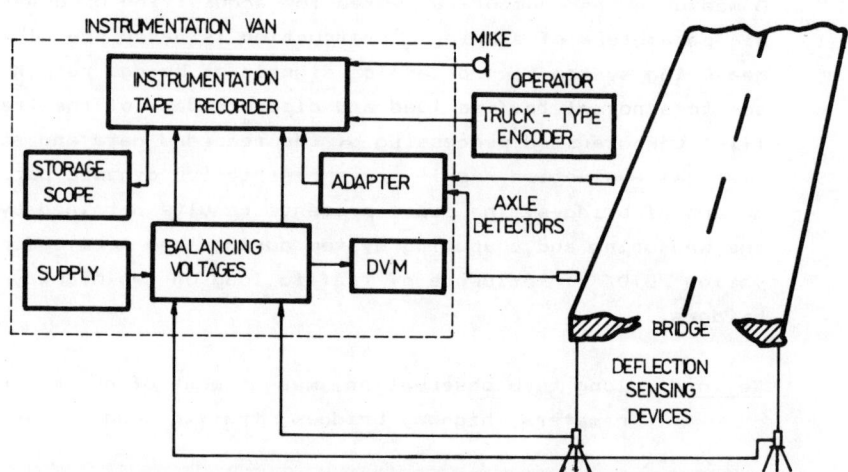

Fig. 1 The measuring system for observation of bridge deflections and of some traffic parameters

Prior to the computation of the dynamic parameters the recorded deflections are digitized and preprocessed together with the discrete data on the vehicles. The results are digital records containing just a relevant part of the digitized deflection signal and a suplementary informative block with the data concernig the categories and the directions of the drive of the trucks which have evoked previos responses. From these records of the digitized time-variations of the bridge deflection the following dynamic parameters are calculated by a Fortran IV program as they are defined by the Czechoslovak standard 73 6209 [2]: the dynamic coeficient δ_{obs}, the natural frequency f of the loaded bridge, the natural frequency $\bar{f}$ of free decayed vibrations of the bridge and the logarithmic decrement of damping ϑ . The definition and the way of manual determination of the dynamic parameters

376

according to [2] can be seen from Fig. 2.

The digital filtering is used to obtain a mean deflection curve and to select a pure dynamic portion of the signal. The computation results in the dynamic parameters for each relevant response of the bridge to the passing trucks together with the discrete data on the trucks.

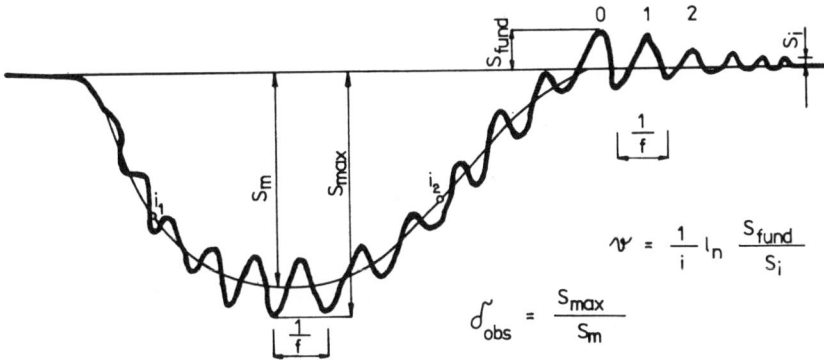

$$\vartheta = \frac{1}{i} \ln \frac{S_{fund}}{S_i}$$

$$\delta_{obs} = \frac{S_{max}}{S_m}$$

Fig. 2   Definition of the dynamic parameters according to [2]

3. The LTO of influence of the traffic load on the bridge

The long term observation /LTO/ have been carried out on two one span bridges on the first class highway. Both bridges are of the same construction system. The structure is made of standard post-tensioned concrete /50MPa/ I girders. The first one is of 26m span and its construction height is 1.25m. The second one is of 21.26m span and its

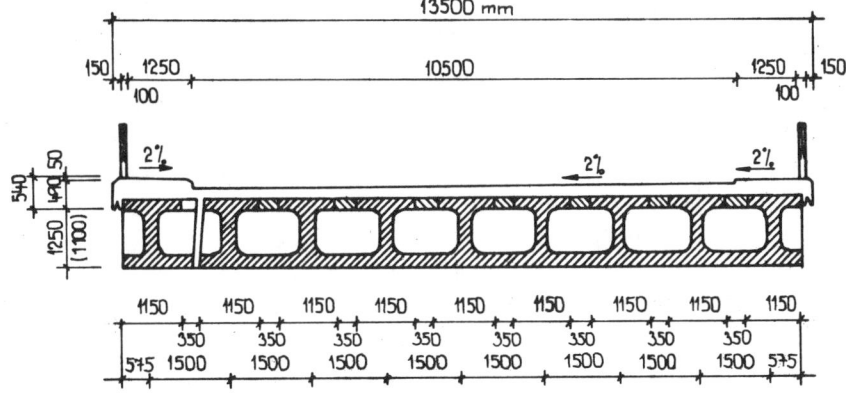

Fig. 3   Crossection of the bridges under LTO

height is 1,1m. The width of the bridges is the same. The substructures are created by massive abutement with parallel wings. The expansion joint is placed above one abutement only.

The position of the bridges on the highway network is characterized by the transport intensity. We have recorded about 2000 passing trucks on the first bridge and about 2400 on the second one. The trucks have been encoded into 8 categories /see table 1/ according to the number of axles, type of a truck, axle distances and bearing capacity, plus 2 extraordinary categories. The operator observes the traffic and operates the recorder by the truck-type encoder. He selects the pushbutton corresponding with the category and the drive direction of the passing truck or with an extraordinary situation on the bridge and makes necessary verbal comments.

After the processing of the whole group of the digitized deflection signals, histograms of the deflection distribution for each category and the drive direction of the trucks as well as for the whole group have been created. Then the statistical data of the histograms have been computed and the aproximation by the following theoretical probability distribution have been tried: Weibull's, Gumbel's, Frechet's, Raleigh's, exponential, normal and logarithmic-normal, Gama and Chi-model. We have carried out the statistical testing based on the assumption that at least one of the theoretical probability distribution is realistic for the distribution obtained from the measurement and that at least one of the introduced theoretical distribution would correspond to this distribution. The parameters of competent theoretical distribution for each group of deflection have been calculated together with the significance test of each distribution [3].

Analysing the results achieved on the first bridge /table 1/ we have found out that the Gumbel's model, at the 5% significance level, is the most suitable in 72% and the one of Weibull in 28% of all the histograms. On the second bridge we have found out that the Weibull's model can be taken as an optimal theoretical one in 30% of all the histograms. For 30% of them the logarithmic-normal model is the most suitable as well as the normal one for other 30% of the histograms. For the rest /10%/ the Frechet's model is optimal.

The analysis of the correlation between maximum deflections and the dynamic coefficients respectively the dynamic increment has shown that in all cases on both bridges the negative respectively positive coefficient of correlation /for linear as well as for logarithmic and exponential dependance/ is very high. The results have confirmed that

	SCHEME	VEHICLE TYPES	1st BRIDGE	2nd BRIDGE
			THEORETICAL MODEL	
1		VOLVO T 813 + SEMI- T 138 TRAIL.	WEIBULL'S	NORMAL
2		VOLVO T 813 + TRAILER T 138	GUMBEL'S	WEIBULL'S
3		T 111 T 813 T 138 T 148	GUMBEL'S	LOG – NORMAL
4		Š 706 + SEMI- Š 100 TRAIL.	GUMBEL'S	WEIBULL'S
5		Š 706 + TRAILER Š 100	GUMBEL'S	NORMAL
6		Š 706 Š 100	GUMBEL'S	LOG – NORMAL
7		BUSES Š 706 – RTO SL 11, Š C734 IKARUS	GUMBEL'S	LOG– NORMAL
8		V 35 – ROMAN S 5T, AVIA ROBUR, IFA	WEIBULL'S	FRECHET'S
9	UNIVERSAL GROUP	ALL VEHICLES DON'T INCLUDED IN GROUPS 1-8	GUMBEL'S	NORMAL
0	UNIQUE GROUP	PASSING AND MEETINGS 2 VEHICLES	GUMBEL'S	WEIBULL'S
	ALL PASSING TRUCKS		GUMBEL'S	WEIBULL'S

Tab. 1 The schemes of the 8 most occuring trucks plus 2 extraordinary categories and the corresponding optimal theoretical probability distribution models of the deflactions for each category as well as for the whole group of the passing trucks

with an increasing traffic load the dynamic coefficient decreases but the dynamic increment increases.

In table 2 there are the natural frequencies of the loaded and the unloaded bridges. It can be seen that the mean natural frequency of the second loaded bridge is higher aproximately by 1Hz than of the first one. This difference confirms the reality that with a growth of rigidity of construction its frequency grows up.

FREQUNCY (Hz)	1st BRIDGE			2st BRIDGE		
	MIN	MEAN	MAX	MIN	MEAN	MAX
LOADED BRIDGE	1,000	4,700	12,500	3,900	5,667	8,820
UNLOADED BRIDGE	-	-	-	5,560	5,759	8,200

Tab. 2   The frequencies of vibrations of the bridges

In conclusion we can say that for further theoretical investigation of life expectancy as well as durability of bridge constructions we could consider the Weibull's or the Gumbel's theoretical model of probability distribution of the deflection.

4. References

1   Kyska,R. - Pollak,T.: Measurement of dynamic properties of bridges. IMEKO X. Preprint, Vol. 3. Prague 1985.
2   CSN 73 6209   Zatezovaci zkousky mostu.
3   Sokolik,A. a kol.: Dynamicke charakteristiky mostov z betonu. Zaverecna sprava VU P12-526/267/E04. VSDS. Zilina 1983

GROWTH ENERGY OF RING-SHAPED FATIGUE CRACK

Sc.D.Andrzej Solecki

Mechanical - Constructional Institute of the Bielsko-Biała Branch of the Technical University in Łódź, Findera 32, 43-300 Bielsko-Biała, Poland

The way of defining of growth energy òf ring-shaped fatigue crack $/\frac{dE}{dF}/$ on the cylindrical specimens has been presented in this paper. Test material used is constructional steel 30 HM after toughening. A special testing stand has been employed. It enabled the automatic recording of the crack growth under constant deflection of the specimen and alternating load.

Keywords: fatigue crack, ring-shaped crack energy, constructional steel 30 HM

1. Introduction

The process of the fatigue crack growth in constructional kinds of steel is characterized by the rate of its growth $/\frac{dl}{dN}/$. The rate is in turn dependent on several external factors connected with the character of cycling loading and on structural factors resulting from the heat treatment of steel and metallurgical process [1,2].
It has bean found, that the speed of fatigue crack growth $\frac{dl}{dN}$ can be made dependant upon stress intensity factor K and described with the formula [3]:

$$/1/ \qquad \frac{dl}{dN} = C \, /\Delta K/^m$$

where C and m are material constants.
The speed of crack has also been made dependant on "γ" energy which is needed for creating a crack surface unit assuming that it is a material constant [4]. It has been circumscribed with the formula:

$$/2/ \qquad \frac{dl}{dN} = \frac{K_{I\,max}^2}{G_e^2} \, \psi \, /\frac{K_{I\,max}^2}{E \cdot \gamma}/, \, N, \, \frac{G_e}{E}, \, \nu \, /$$

where Ψ is undimensional function.

The value of energy used for making a unitary surface of the fatigue crack $\left(\frac{dE}{dF} = \gamma\right)$ for a cylindrical specimen with the ring-shaped notch under rotating bending and constant deflection can be calculated according to the formula:

/3/ $\frac{dE}{dF} = 0,5 \cdot \Delta Q_f \cdot d_f \cdot F^{-1}$

where: ΔQ_f - value drop of the leading force for a gi-
ven length of the fa-
tigue crack,

d_f - deflection of the specimen,

F - area of created fati-
gue crack.

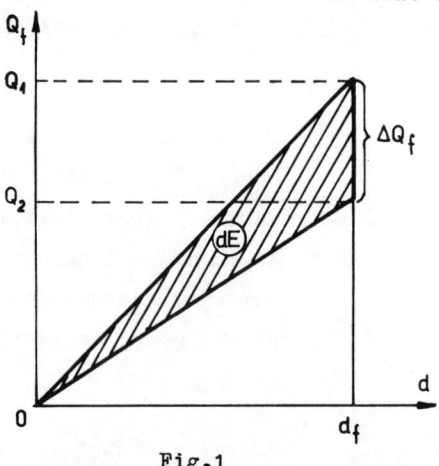

Fig.1

The change of specimen rigidity under the constant strain d_f

The purpose of the paper is mea-surment of energy $\frac{dE}{dF}$ and $K_{I\ max}$ and comparison of asses-sment eficiency $\frac{dl}{dN}$ using the same criteria.

2. Methodology of research

The test material used is constructional steel for toughening of the type 30 HM. Shape and overoll dimensions of the specimens used during the examination of the fatigue crack growth speed are shown on figure 2.

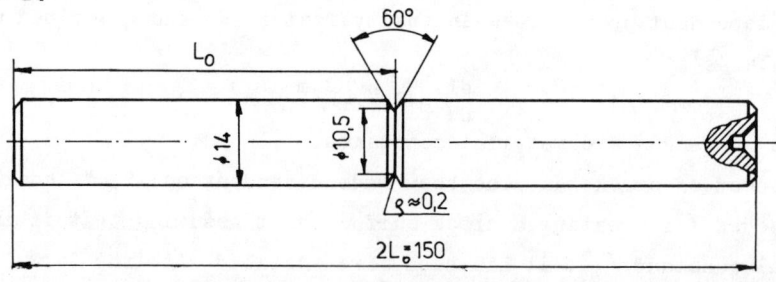

Fig. 2

Shape and dimensions of the specimens

The specimen rigidity change arises at the moment of initiation and the fatigue crack growth and is measured with the strain gauge transducer by means of the unbalanced bridge method as ΔQ_f. The measuring accuracy of the loading force amounts to $0,2 \div 0,4\%$ depending upon the range of loading. The measurement of the deflection d_f has been carried out under the fixed loading with the accuracy amounting to $0,01$ mm. The testing stand enables the research to be carried out within the range of revolutions frequencies $f = 18,3 \div 49$ Hz and is equipped with the strain gauge bridge, three - range recorder NSKV61-e, and displacement sensor. The specimens have been fixed with the constant axial force P_o, and the loading Q_f was so selected that, the fatigue stress intensity factor $K_{I\ max}$ could change within the range $6 \div 30$ MPa $\sqrt{m}$ and the deflection d_f within $0,17 \div 0,37$ mm. $K_{I\ max}$ has been calculated according to the formula /7/:

$$/4/ \quad K_{I\ max} = M \cdot D^{-\frac{5}{2}} / 1 - \varepsilon /^{\frac{1}{2}} / 28,3 - 46,9\varepsilon + 33,3\varepsilon^2 /$$

where: M - bending moment affecting the specimen,

        D - diameter of the specimen,

        ε - parameter describing the crack geometry,

$\varepsilon = \frac{d}{D}$ - d - inside diameter of the crack.

At each level of loading at least three specimens have been examined. Fig. 3 shows the diagram of the testing stand.

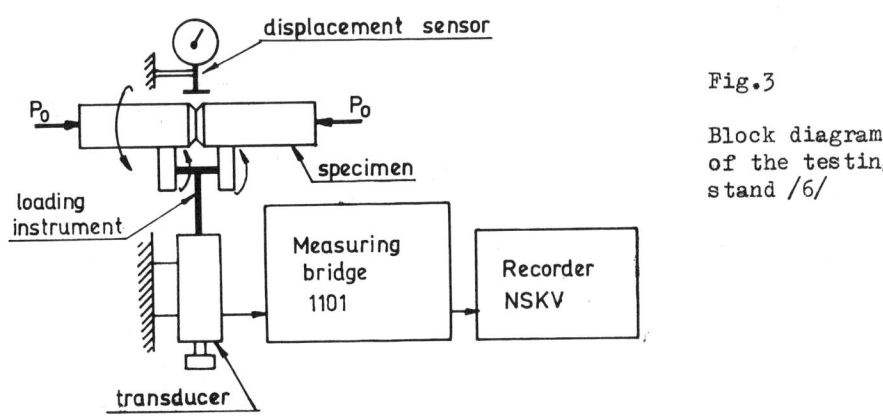

displacement sensor

Fig.3

Block diagram of the testing stand /6/

Values of $\frac{dl}{dN}$ have been defined according to prescription $/7/$, and value $\frac{dE}{dF}$ has been calculated according to the formula $/3/$.

3. Results of the research work and their discussion

The results of chemical composition analysis and of mechanical properties examination of 30 HM steel after hardening and tempering are presented in table 1.

Table 1

Chemical composition of the tested steel /%/						
C	Mn	Si	Cr	Ni	S	Mo
0,35	0,74	0,26	1,04	0,29	0,03	0,16
Mechanical Properties of the tested steel						
Hardening 1133 K/oil Tempering T		$G_{0,2}$ MPa	G_{B} MPa	A_5 %	Z %	
943 K		854	975	19	61	
823 K		1072	1167	17	59	
673 K		1468	1587	9	39	
573 K		1602	1800	8	4	
523 K		1684	1904	5	–	

According to the selected level of loading and temperature of tempering, changes in the character of recorded charts have been observed. Examples are shown on fig. 4. They changed from straight line

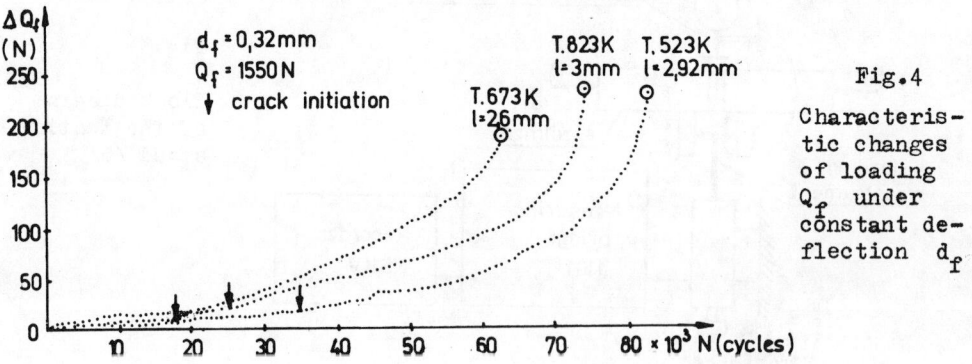

Fig.4

Characteristic changes of loading Q_f under constant deflection d_f

into parabola shape and the speed of cracks gradually increased.

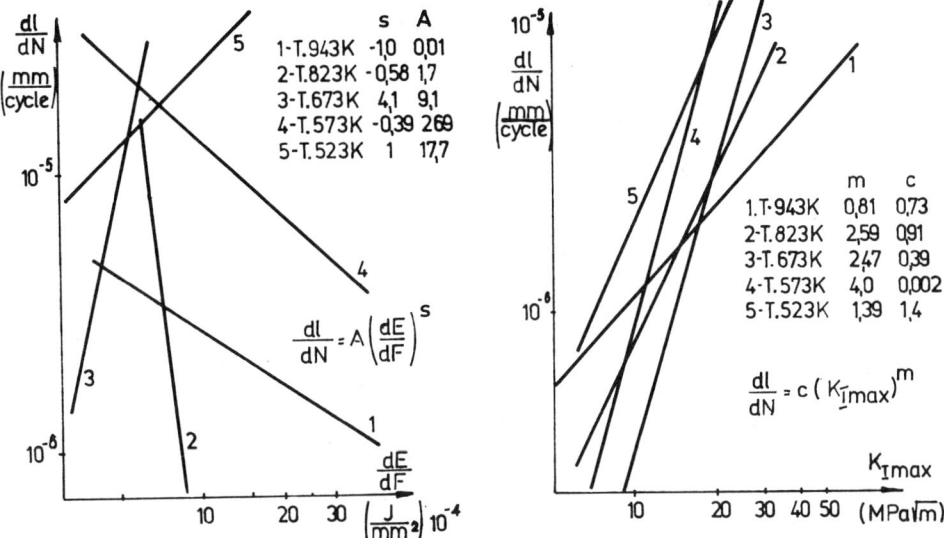

	s	A
1-T.943K	-1,0	0,01
2-T.823K	-0,58	1,7
3-T.673K	4,1	9,1
4-T.573K	-0,39	269
5-T.523K	1	17,7

$$\frac{dl}{dN} = A\left(\frac{dE}{dF}\right)^{s}$$

	m	c
1.T·943K	0,81	0,73
2-T.823K	2,59	0,91
3-T.673K	2,47	0,39
4-T.573K	4,0	0,002
5-T.523K	1,39	1,4

$$\frac{dl}{dN} = c\left(K_{Imax}\right)^{m}$$

Fig. 5

Interdependences between the fatigue crack growth speed, factor $K_{I\ max}$ a unitary energy of crack growth $\frac{dE}{dF}$

Analysis of the interdependences shown on fig. 5 allows to affirm that, the lowest speed of the crack propagation can be obtained for the variant of T. 673 K within a range up to $20 \cdot 10^{-6}$ mm/cycle, and for the variant of T. 943 K within a range of over $20 \cdot 10^{-6}$ mm/cycle. Values of $\frac{dE}{dF}$ change insignificantly for the variant 823 K and are close to stabilization. The crack propagation takes place in the quickest way for the variants T. 573 K and T. 523 K, due to low tempered martensite appearing in the microstructure.

4. Conclusion

a/ Interdependences between $\frac{dl}{dN}$ and $\frac{dE}{dF}$ for 30 HM steel can be circumscribed with the equation: $\frac{dl}{dN} = A/\frac{dE}{dF}/^{S}$ where A i s are material constants.

b/ Unitary energy of the fatigue crack growth $\frac{dE}{dF}$ can constitue

385

a heat treated constructional steel estimation criterion.

c/ The employed system of measurement and recording enables the process of the fatigue crack growth to be analyzed and the energy of its development to be defined.

5. References

1 Kocańda S.: Fatigue Destruction of Metals. Polish Scientific Publishers, Warsow 1978.

2 Frost N.E.: A Fracture Mechanics Analysis of Fatigue Crack Growth Date for Various Materials. Engineering Fracture Mechanics, 1971, Vol. 3 pp. 109-126.

3 Paris P., Erdogan F.: A Critical Analysis of Crack Propagation Laws. Journal of Basic Engineering, Trans. ASME, pp. 528-534, Dec. 1963.

4 Czerepanow G.P.: Miechanika chrupkowo rozruszienia. Nauka, Moskwa 1974.

5 Panasiuk W.W.: Mietody ocienki trieszczinostoikosti konstrukcjonnych materiałow. Naukowa Dumka, Kijów, 1977.

6 Solecki A.: Examination of Interdependences Between Capacity for Conditions Causing Decohesion for a Discrete Model. Doctor Thesis, Technical University of Łódź, 1981.

7 Fizyko-Chimiczeskaja Miechanika Materiałow 1979/3, Metodiczeskije Ukazanija, Opriedielienije charakteristik soprotiwlienija rozwitju trieszczine mietałłow pri cikliczeskom nagrużeni

VIBRATIONAL TEST OF NUCLEAR POWER PLANT FAILURE LOCK CONTROL SYSTEM FOR EARTHQUAKE-PROOFNESS

Assoc. Prof. Eng. Georgi Heruvimov
Sc. Res. Member Eng. Nikolai Georgiev
Eng. Pencho Ivanov

Lab. "Dynamics,Strength,Reliability",CNIIEM "Balkancarprogress"
9 Septemvri 126 Blvd. 1618 Sofia,Bulgaria

A real seismic spectrum envelope is reproduced to enable
heavy-duty vibration loading in laboratory conditions.The
relationship between acceleration amplitudes and frequency
is based on structure reactions spectrum at an earthquake
magnitude of 9 to MKS-64.Acceleration in various vibrational
directions has been studied.The bearing structure is optimiz-
ed to avoid working frequency resonance and unallowable loading of electrical devices.

Keywords: earthquake analysis,environmental testing,electro-
          hydraulic test rigs,digital signal processing

1. Introduction

For a nuclear power plant control system it is of extreme importance that it does not fail during an earthquake.That is why a
laboratory simulation of earthquake-type vibration loading using a
servo-hydraulic test rig is of paramount importance for electrical
devices reliable operation.

2. Seismograms and response totalized spectra

The most unfavourable spectrum of the floor spectra has been
used when studying response envelope of the nuclear reactor at the
elevation of 40.9 m with reference to the transmission of the earthquake impact.The accelerograms taken account for the vertical and
horizontal earthquake loads at a magnitude of 9 to MKS-64.

The real earthquake recordings in the form of accelerograms -
Fig.1 , Fig.2 - have been entered through a digitizer into an μECM
HP - Series 80.Three different damping decrements of the forced
vibrational process occurred as a result of the earthquake effects

have been investigated: with $\delta = 0.1$; 0.2 ; 0.4 at vertical vibrations, and with $\delta = 0.1$; 0.2 ; 0.3 at horizontal vibrations.The accelerations discrete quantities arrived at are consistent with their relevant discrete reference times.Files have been completed having both file headers of the descriptive type,and main parts of digital information consisting of two-dimensional files.

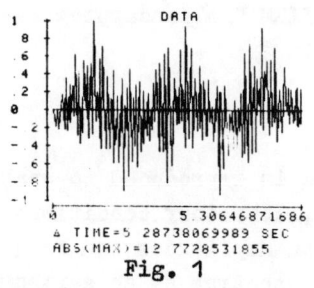

Fig. 1

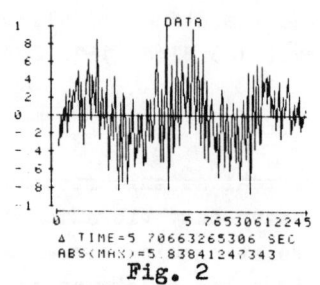

Fig. 2

Using accelerograms with identical directions of impact response totalized spectra have been constructed - Fig.3,4 - and for these,again,damping decrement has been accounted for. In this paper,response totalized spectra are presented in such a way as to ensure that different frequency components of the real earthquake spectrum are included under the envelope of the schematic totalized spectrum.

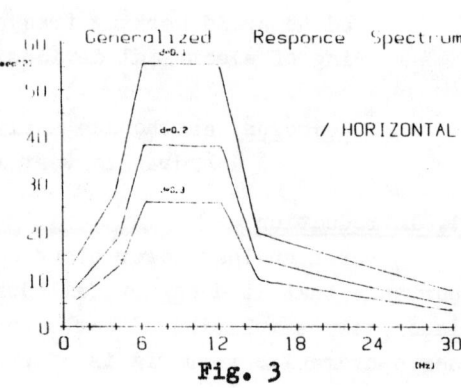

Fig. 3

3. Statistical analysis of seismograms

Analysed have been vertical as well as horizontal accelerograms with different damping decrements.Files and their discrete quantities have been processed using device driver programs to reformat

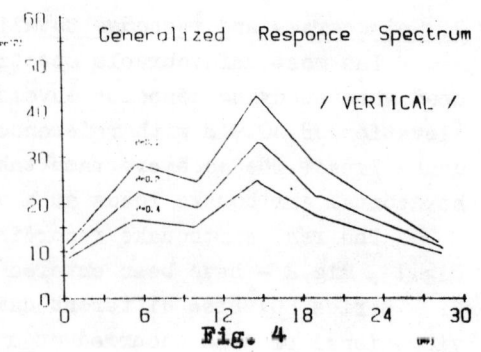

Fig. 4

388

them into more suitable types.These new formats are compatible with standard program packages [1] .

3.1. Statistics of the processes as to the amplitudes in m sec^{-2} are as follows:

A. Horizontal seismograms

With $\delta = 0.1$ at an expectation of 0.083 m sec^{-2} ; standard deviation of 2.2769 m sec^{-2} ; standard error of 18.53% ; confidence interval of 90% with bottom limit of   – 0.2237 m sec^{-2},and upper bound of 0.3696 m sec^{-2}.The extreme values of the order statistics are 5.8384 m sec^{-2} and – 4.9794 m sec^{-2} ; median of distribution . is 0.1233 m sec^{-2} as well as quantiles amounting to 25% and 75%,i.e. –1.7304 m sec^{-2} and 1.7222 m sec^{-2},respectively.The same method has been used to obtain results with $\delta = 0.2$,and with $\delta = 0.3$.

B. Vertical seismograms

With $\delta = 0.1$ at an expectation of 0.0825 m sec^{-2} ; standard deviation of 4.57 m sec^{-2} ; confidence interval of 90% with bottom limit of – 0.4472 m sec^{-2},and upper bound of 0.621 m sec^{-2}.The extreme values are 12.7729 m sec^{-2}.and –11.7922 m sec^{-2} with quantiles amounting to 25% and 75%,i.e. –2.8637 m sec^{-2} and 3.1413m sec^{-2} respectively.

3.2. It is the same method that is used to evaluate the rest of the accelerograms having as a parameter the damping decrement δ .
A statistical analysis has also been carried out for the most probable frequencies in the spectrum for:

A. Horizontal seismograms

With $\delta = 0.1$ the most probable frequency in the process is 23.84 Hz,and a confidence interval of 99% covers bottom limit of 16.88 Hz,and upper bound of 30.8 Hz.It could easily be seen that the extreme values from the order statistics – 196 Hz and 5 Hz, respectively,for the maximum as well as for the minimum frequency in the process under study,have been confirmed both by the variation factor of 137.5,and by the standard deviation.

Values with δ parameter appear to be approximately the same, and with δ increasing the lower frequency components appear more frequently.

B. Vertical seismograms

With $\delta = 0.1$ the expectation is 29.15 Hz,and 99% of the confidence interval is in the limits of 23.44 Hz and 34.85 Hz.Maximum variation frequency observed is 235.75 Hz, and 6.5486 Hz is the respective minimum.

With the vertical records of the earthquake accelerograms it appears possible that frequency components be of a higher frequency in the spectrum if compared with the horizontal accelerograms.

4. Frequency Analysis

Seismograms have also been frequency analysed, having them in a Furier order to their 30-th term at intervals of 1 Hz. It has been done by a graphic tablette via which the extreme earthquake variations have been entered. The interactive mode of entering utilises graphic subprograms which set up separate disk files while other drivers reformat the files, making them compatible as a package for curve analysing [2]. A specific feature of this analysis seems to be the arrangement of the discrete quantities which are irregularly quantized. Coefficients for this order have been obtained by digital integration of a parabola that crosses 3 consecutive points.

It is of interest to note here that the horizontal variations have their phases in a random way. With the vertical variations, however, with δ = 0.1, a phase frequency modulation could clearly be observed at about 2 Hz – Fig.5.

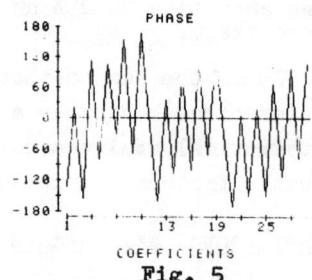

Fig. 5

The amplitude spectrum has the first coefficient of the order with a maximum value at a damping decrement δ = 0.1, both for the vertical and for the horizontal variations – Fig. 6,7.

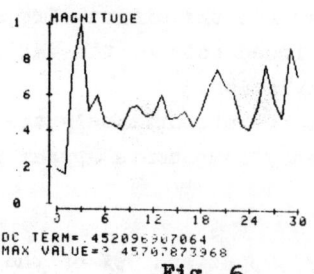

DC TERM=.452096907064
MAX VALUE=? 4576787?3968

Fig. 6

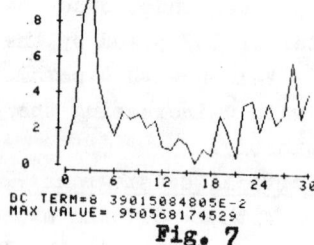

DC TERM=8 39015084805E-2
MAX VALUE=.950568174529

Fig. 7

390

With other values of δ it could be observed a comparatively uniform amplitude participation of the order coefficients in the range of 0 to 30 Hz.

Frequencies in the interval for the separate decrements and variations have been grouped, too. With the horizontal variations up to about 95% of the frequencies fall into the interval of 4 to 60Hz, and with the vertical variations the interval lies between 6 to 60 Hz. If a similar analysis is to be carried out for the acceleration amplitudes at separate levels, a bi-dimensional analysis could be performed to set up block programs for earthquake-proofness tests.

5. Results and conclusions

A. Results

Fig.8,9 show totalized spectra for vertical and for horizontal accelerograms.

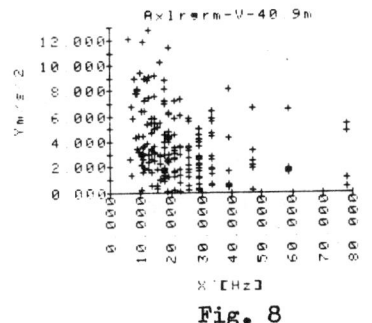

Fig. 8

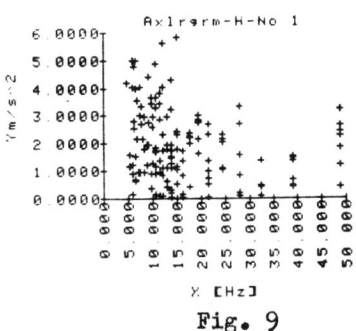

Fig. 9

As a result of the totalized spectra, and after a digital integration, functional dependences have been obtained for the two cases of actuators travel – Fig.10,11.

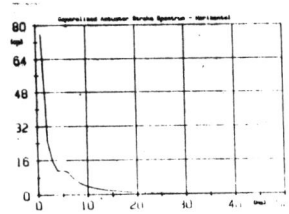

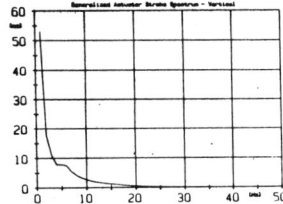

Response of the components of the emergency lock control system that are subjected to vibration loading has been analysed in terms of frequency and at characteristic points. In order to avoid resonance effects in the range of the working frequencies some modifications in the bearing structure have been suggested and carried out - e.g. modification of the mounting diagrams, increased strength of the bearing parts, reinforcement of the screwed connections as well as ensuring there are no unintentional self-unloosenings, modifications in the type of wiring used and the way it is fastened, etc. After making the above modifications, the response has been again investigated at the same points.

B. Conclusions

- Vertical spectra appear to be with a more high-frequency nature.
- Each earthquake loading analysis should be performed on the basis of certain known geophysical preliminary observations valid for a given geographic area.
- Actuators should have travel control to reach about 15 Hz, and afterwards the control is to be achieved by the force applied.

6. References

1 82807 A Statistical analysis Multipac - Hewlett Packard Co.
2 82612 A Electronics engineering Multipac - Hewlett Packard Co.

ON-LINE TESTING AND LONG-TERM INVESTIGATION OF TRENDS ESPECIALLY
DYNAMIC PARAMETERS OF MACHINE FOUNDATIONS IN POWER ENGINEERING

Ing. Václav Bohdanecký,CSc. - Doc. Ing. Pavel Novák, CSc.
Stavební ústav ČVUT Praha /Building Research Institute of TU,Prague/
Šolínova 7, 166 08 Praha 6, ČSSR

A methodology of testing and long - term investigation of
foundations of big turbo-sets in situ is presented. It is
based on the combination of theoretically and experimentally determined especially dynamic characteristics of
these structures and on the evaluation of their dynamic
behaviour over longer periods of time. The methodology
comprises determination of the actual mechanical properties of materials; identification of even concealed defects and failures; possible suggestion of modifications
or reconstruction of the foundations and/or connection
with the plant.

Keywords: dynamic testing in situ, machine foundations

1. Introduction

For the structures subjected to long-term dynamic effects of
operation a methodology of on-line testing and long-term investigation was successively developed and verified in the Building
Research Institute of the Technical University, Prague.

The methodology was primarily devised for the testing of separate foundations and multipurpose structures loaded by dynamic
effects of machine operation.
It is based on the combination and comparison of theoretically
and experimentally determined characteristics of these structures
/using generally customary methods and means/ and on the evaluation of observations of their dynamic behaviour, heat load, etc.
over longer periods of time.

It was adapted and mostly used for the testing of foundations
of big power generating machine sets [1] , [2], [3] .

2. Means - method characteristics

The contemporary possibilities of computers permit to characterize, in sufficient detail and close approximation to the actual
behaviour, both the natural and the forced vibrations of structures.

The necessary prerequisite is, naturally, an adequate design model
and such input characteristics of materials, loads, etc. as would
most truthfully correspond to actual values. At present the greatest inaccuracies and reserves are in the determination of input
characteristics.

Particularly difficult is the determination of the characteristics of building materials and operation loads with regard to
their development in time.

For the measurements and evaluation of data the methodology
applies standard apparatuses used for dynamic measurements
/displacement,acceleration,stress/ or temperature measurements.
Apparatuses made by different manufacturers can be applied, of
their parameters are satisfactory for the given purpose /for
example f_{min} = 1 Hz, displ. u_{min} = 10^{-4} mm/.

Our set comprises mostly apparatuses made by Brüel and Kjaer
/with dual channel signal analyzer type 2034/. We have in a new
foundation always about 30 sensors
- vibration pickups, thermometers, tensometer systems /new J. Ol-
mers invention/

The evaluation part, particularly in the case of field measurements, comprises a programmed calculator or computer /PC/ with
a plotter, etc. For computer evaluation of the results of measurements /e.g. recorded on magnetic tape/ a relatively large library
of standard and special programmes incl. graphic outputs is available /It is in referat from J. Král in detail described./.

The dynamic loads are produced either by standard or controlled operation of the machine set or by special vibration exciters.

Standard operation loads are used particularly for the observation of the development of instantaneous spatial vibration modes
in time, either at short-term intervals on the scale of hours, or
at longer intervals on the scale of days to months /in the form of
test repetition/.

Controlled operation loads /e.g. in the form of starting and
finishing of technological equipment/ are used particularly for
the determination of response curves in the whole network of measuring points and, consequently, for the determination of the
frequency response spectre. The use of this excitation type is
important also for the diagnostics of defects, particularly for
resonance tests of the function of visible cracks [4] . The disadvantage of operation loads lies chiefly in the fact that their

accurate description is practically never available.

These types of loads are often supplemented with loads produced by artificial vibration sources with defined magnitude and behaviour of excitation forces and moments. As a rule these are mechanical_vibration_exciters existing. We usually use the SVUSS vibration exciters with rotating or controlled force with the possibility of an infinitely variable change of revolutions. Their force effect of as many as 35 kN at 50 cps and their design are satisfactory for the testing zone of 15 - 70 cps. For the investigation of characteristics in the low frequency range it is possible to use, for example, the Vibrogir-VPB vibration exciter with a force effect of 27 kN at 6 cps.

For the testing of structures by our methodology it is possible to use also other artificial vibration sources, applied, for example, in the testing of transport structures or tall buildings, such as rocket engines, or impact hammers.

The whole conception is, naturally, based on the endeavour to achieve high automation òf measurements and evaluation in accordance with the possibilities of contemporary measuring techniques and computers [5]. In this field reserves exist particularly in_the monitoring_of_performance_vibrations_over_long_periods_of_time.

At_the_same_time_the_continuous_analysis_of_performance_vibrations_and_their_development_trends represents an important means of the presented diagnostical methodology. As a rule it represents a source of timely recognition of the overall worsening of the condition and the bearing capacity of the structure or the signalling of initiating local defects. The degree of detail of this information depends, naturally, on the density and layout of the check point network on the structures and the plant. For the foundations of a major power generating plant the principles of layout of the chaeck point network have been elaborated relatively well. In a number of works the points have been fixed and the vibration components in the directions of V, H, and A, are being regularly measured in them according to a long term programme.

An important means of assessment of the state and development of vibrations of the observed building structure or the structure--macheme system is the_working_assessment_criterion. It is a combination of the assessment of vibrations according to their absolute magnitude and the assessment of relative changes of vibrations in the course of a longer period of time.

The assessment zones signalling the possible need of checking may vary in accordance with the type of structures of plant. An exemple of auch a criterion for turbo-generator set foundations is given, for example, in [6].

The methodology principally comprizes :

a/ theoretical assessment of the building structure and its possible connection with the plant,

b/ determination of the actual mechanical properties of materials,

c/ experimental testing of particularly dynamic characteristics of foundations and their comparison with theoretical values,

d/ identification of even concealed defects and failures and their detailed testing,

e/ possible suggestion of modifications, reconstruction of the building structure and/or its connection with the plant.

The methodology enables to determine :

1/ resonance curves of both the whole structure and its individual members and, on their basis, the areas of resonance phenomena either of local or general character,

2/ deformation curves of vibration areas and, on their basis, the continuity or discontinuity of the structure,

3/ mutual interaction of cracks, connections of the individual members / e.g, longitudinal and transverse members/, connection of the building structure and the plant, particularly on the basis of resonance tests,

4/ quality of the function of expansion joints and the standard of vibration transmission from the foundations to the adjoining building structures and plant and vice versa /incl. the possibility of determination of the sources of external excitation/,

5/ mechanical characteristics of actual structures,

6/ correctness of the concept of structural analysis in the design.

On this basis it is possible to determine, for example, the character of possible failures of the building structure, the quality of the connection of the plant with the building structure, and, possibly, the character and the extent of the changes of the dynamics of the system. In the majority of cases the results enable also an analysis of the causes of the origin of failures, particularly on the basis of an analysis of long-term observations of the development of vibrations in the structure.

3. Conclusion

The above described methodology of diagnostics of building structures and systems subjected to dynamic loads affords sufficient means for an assessment of both their local and general state. The outputs of the methodology, i.e. the results of detailed theoretical and experimental testing, based on the knowledge and assessment of the behaviour of the structure in the course of a longer period of time, represent far-reaching and mutually supplementing data for a more integral discription of the dynamics of structures, among others also for the identification of their concelaed defects and failures, the reduction of their bearing capacity in time, etc.

For assessments of the residual life and further usability of structures subjected to long-term dynamic loads it is necessary to know further technical and technological data, such as long-term information on vibration levels, loading history, executed repairs, history of material characteristics, etc. For this purpose it is necessary to elaborate methods enabling good quality prognoses of the origin of defects and estimates of the residual life of such structures.

4. References

1 Bohdanecký V. - Novák P. : Dynamic measurements of foundations of a turbo-generator /in Czech/, Report SÚ ČVUT, Prague, 1982

2 Bohdanecký V. - Novák P. : Vibration of turbo-set foundations with elements of higher compliance. Proc. of the XIVth Conference "Dynamics of machines", Prague-Liblice, 1983

3 Bohdanecký V. : Diagnostik der Tischfundamente. Proc. of the Kolloguium "Dynamisch beanspruchte Industriebauten", Dresden, 1986

4 Bohdanecký V. - Novák P. : Determinations of failures of foundations subjected to dynamic loads /in Czech/. Proc. of the IIIrd Seminar "Interaction of building structures and plant of power plants and similar industrial buildings". Plzeň, 1983

5 Bohdanecký V. - Bouška P. - Holický M. - Novák P. : Statistical analysis of random vibrations of building structures subjected to dynamic loads. Prace naukove institutu budownictwa Politechniki Wroclawskiej. Wroclaw, 1986

6 Novák P. - Bohdanecký V. : Service life of foundations of
   large machine sets subjected to dynamic loads /in Czech/.
   Stavebnický časopis 11/12, 1982

AUTHOR INDEX

Aben, H., 3
Achmetzyanov, M., 11
Albertini, C., 129
Amberg, C., 155

Bohdanecky, V., 393
Boleslav, A., 173
Borbás, L., 47
Brandt, A.M., 3ol
Brémand, F., 115
Bugakov, I., 17

Dalakishvili, G.L., 85
Dally, J.W., 283
Drdácky, M., 221
Drzik, M., 135
Dveres, M.N., 215

Genkin, M., 161
Goja, Z., 233
Grosser, V., 91

Hanna, A.M.**H.**, 167
Heruvimov, G., 387
Heymann, J., 2o9

Issa, S.S., 333

Jávor, T., 339
Jecic, S., 325

Karmowski, W., 23
Kopecky, M., 369
Kyska, R., 375

Laermann, K.H., 73, 193
Lingener, A., 363
Lukas, J., 185

Macura, P., 67
Mazurkiewicz, S., 29, 33
Milewski, G., 227

Nazari, F., 345

Olmer, J., 351

Osten, W., 241

Pacas, B., 357
Pietrzyk, A., 263

Sciammarella, C.A., 97
Solecki, A., 381
Stanley, P., 141
Stefanescu, D.M., 179
Stroeven, P., 147

Thamm, F., 319
Theocaris, P.S., 3o7

Videnova, J., 53
Vrba, K., 269

Wahl, F., 251
Walczak, W., 39
Will, P., 277

Zhilkin, V.A., 1o7
Zsáry, Á., 59